OLD REGIME FRANCE AND ITS JETONS

Pointillist History and Numismatics

JAMES E. MCCLELLAN III

Numismatic Studies 41

AMERICAN NUMISMATIC SOCIETY
NEW YORK
2020

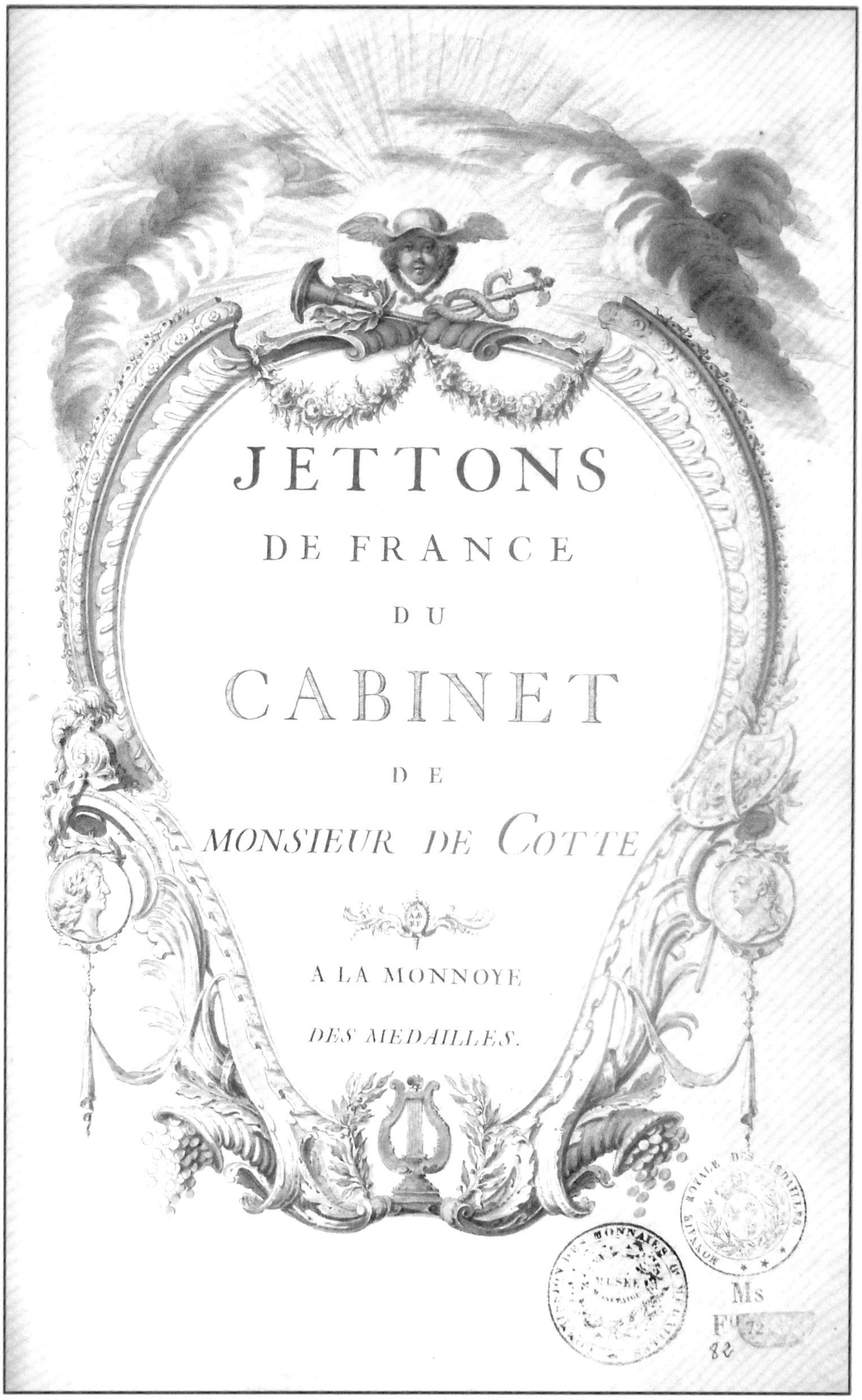

FRONTISPIECE: Title page of manuscript jeton catalog of Robert de Cotte, 1727.

TITLE PAGE: Obverse of the jeton of the Académie Royale des Sciences, Belles-Lettres et Arts of Lyon, 1749.

To see a world in a grain of sand
And a heaven in a wild flower,
Hold infinity in the palm of your hand
And eternity in an hour.

—William Blake (1803)

Pour la précision des renseignements chronologiques, géographiques et même linguistiques qu'elle apporte, la Numismatique mériterait de prendre à la tête des sciences historiques une place qu'on ne lui accorde pas volontiers. Est-ce à cause de ses rapports fréquents avec l'Art que la Numismatique fut peut-être trop négligée et l'a-t-on considérée comme appartenant plus à l'Art qu'à l'Histoire proprement dite ? C'est précisément pour ce double caractère artistique et historique que la Numismatique devrait être estimée encore plus qu'elle ne l'est.

—Adrien Blanchet (1912)

L'étude des jetons n'est intéressante que si elle donne occasion à des recherches historiques et si elle fournit le moyen d'élucider quelques points controversés.

—Maurice Prou (1888)

Et cependant l'Histoire peut et doit retirer quelque profit de la compréhension des types et devises de tant de petits monuments.

—Adrien Blanchet (1950)

Il n'entre pas dans mon plan de donner ici l'histoire du jeton, si souvent faite depuis quelques années, mais qui néanmoins est un peu comme le Dictionnaire de l'Académie *qui, fait, bien fait et refait, reste toujours à faire.*

—G. Vallier (1879)

ISBN 978-0-89722-362-1
ISSN 0517-404X

Library of Congress Control Number: 2020933251

Printed in Canada

CONTENTS

OLD REGIME FRANCE AND ITS JETONS

Pointillist History and Numismatics

Jeton of the Marine Royale, 1712

PREFACE

This study concerns a particular kind of non-monetary token known as jetons and the institutional and cultural history of Old Regime France that we can write using jetons as our entrée into this topic.

Three particular notions underlie this work.

First is an historiographical starting point grounded in the literature about Old Regime France. We confront at the outset an unexpected cleft in the historiographical landscape separating the discipline of history from the discipline of numismatics (or the study of coins and currency). In a word, numismatists and historians specializing in early modern France constitute two separate and highly developed scholarly communities that, certain exceptions aside, turn out to have had strikingly little overlap or intellectual cross-fertilization. By contrast, the disciplines of history and numismatics are quite close in the study of classical antiquity, where coinage and chronologies based on coinage provide a notable pillar on which historians and other students of the ancient world base their work. This historiographical realization comes as a surprise. It is worth noting in its own right; it calls out for explanation, and its investigation was one prompt for this research.

The second starting point concerns jetons themselves. Jetons have a long and complicated history stretching back to counters used on counting tables in medieval counting houses, and their history extends down to today, in old subway or telephone tokens, for example, or poker chips (*jetons* in French). It turns out that the use and significance of jetons exploded in France and the Low Countries in numismatically and historically extraordinary ways in the period roughly 1500–1789. Jetons are another distinctive element of early modern France and especially characteristic of France under the Bourbon monarchy (1589–1792). Numismatists have thoroughly inventoried the tens of thousands of different French jetons that appeared in this period, and they have produced a substantial literature about these tokens. Overwhelmingly, this literature is in French, and many specialized numismatic studies of jetons date from the nineteenth and early twentieth centuries and appear in relatively obscure journals and reviews. Jetons are of interest in their own right and deserve to be better understood. The present monograph adds new research and presents the first comprehensive English-language study of these distinctive tokens, and in this way one of its purposes is as a contribution to the numismatic literature.

Thirdly, and most importantly, however, this study derives its rationale and originality from the remarkable cultural and institutional history of Old Regime France that it becomes possible to write when one folds the numismatic literature into existing historical studies of early modern

France. The idea here is not to use history to understand jetons (as some numismatists have done in looking at monetary legislation, for example), nor is it to use jetons as sources for history per se (as in information gleaned from a commemorative coin, for example). Rather, the purpose here is to gather together and examine the collective historical phenomena that were jetons in order to shed light on the larger society and culture that produced them. Like Blake's grains of sand, jetons are revealing sociocultural and material artifacts, and looking through the lens of jetons, virtually the whole of Old Regime society and culture comes into focus. Jetons had many uses and formed part of an ensemble of cultural practices that are revealing in many ways, as we will see throughout. (To give but a single example here, in the Old Regime they were commonly offered as New Year's Day presents in certain segments of society.) Jetons notably encapsulate the historical development of the French state, and jetons represent a remarkable efflorescence of technology and coinage that is telling of a distinct society as well. Some jetons are strikingly beautiful and call out for aesthetic evaluation in the context of contemporary art. Jetons reflect the society that produced them, and they reveal much about contemporary French life and culture. This monograph thus finds its rationale and intends first and foremost to contribute to a deeper historical appreciation of early modern and Old Regime France made possible by examining jetons.

How I came to write this volume proves once again just how accidental and contingent the practice of historical research actually is.[1] From the writings of Roger Hahn, Daniel Roche, and Charles Gillispie and from my own work on European learned societies of the eighteenth century I long ago knew that little silver tokens, *jetons de présence*, were doled out to academicians at French royal academies for attendance.[2] One afternoon in the 1980s leaving the old Bibliothèque Nationale on the rue Richelieu in Paris and heading to the Bourse métro station, I once again passed one of the many coin stores that dot the neighborhood, and on a whim I went into one to ask if they might have any of these jetons. They did, and I bought one for about 10 dollars, an undated and unremarkable Louis XV silver jeton of the Académie Royale des Sciences. So began an unexpected odyssey that reaches its destination in the present work nearly four decades later.

I confess I began collecting jetons. On various trips to Paris for other research I'd make the rounds of the coin stores, and for a while I was active in the jeton market that popped up on eBay. Over the years I compiled a representative assortment of about 750 jetons spanning the period from the Middle Ages forward to today, although the focus of the collection was always on French jetons of the Old Regime.

Collecting does have its pleasures...the joys of finding rare things, the intellectual excitement of learning and exploring new territory, the solipsistic satisfactions of sheer possession.[3] As a professional historian, however, in principle I am against collecting, and I felt guilty being a collector and an antiquarian among fellow historians with goal-directed research imperatives who, like me, spurn dilettantism and connoisseurship, much less commercial motivations. As Maurice Prou (1861–1930), curator at the Cabinet des médailles at the Bibliothèque nationale de France, later professor and finally director at the École des Chartes in Paris, put it already in 1888 in one of

1. For this historiographical pet peeve, see McClellan (2005).
2. Hahn (1971); Roche (1978); Gillispie (1980); McClellan (1985).
3. On the psychology of collectors, see Sarmant (2003), pp. 366–68; Pastoureau (1984), p. 27; Boutin (1966). Although he doesn't use the term, Mazard (1992) in his study, "The psychoanalysis of the numismatist," makes clear from a Freudian perspective the anal character of collecting.

the epigraphs that heads this work: "The study of jetons is interesting only insofar as it gives rise to historical research and if it provides means for elucidating some controversial points."[4] The historian and numismatist Philip Grierson offered his unvarnished views on the subject in 1951, saying, "...collecting involve[s] an expenditure of time and money which can be better employed in other ways."[5]

I vaguely began to wonder if I could rescue myself professionally as an historian and make something of the collection I was slowly amassing. Fortunately for me, I zeroed in on the historiographical realization referred to above, namely, that historians of early modern France and the numismatic community expert in the same period are not congruent, and I began to wonder what would result if we brought them together. I furthermore came to see that what is known about jetons is scattered about in the highly specialized numismatic literature in French just mentioned, so to write an account in English might be of service to scholars and experts who might not otherwise run across this material. Only then did I realize that I could address my jetons in a more comprehensive fashion and thereby embrace in a general way the totality of the society and culture of Old Regime France. Suddenly, I had a legitimate and fascinating historical study to pursue.

ACKNOWLEDGMENTS

This monograph is not, then, a work of numismatics per se nor the presentation of a collection, but it is an attempt to capture the world of early modern and Old Regime France using jetons and the materiality of jetons as an entrée into that world. Because it has been so long in the making, this work has accumulated a formidable set of debts that I am pleased to acknowledge with warmth and gratitude.

The American Philosophical Society in Philadelphia is first on my list. A sabbatical fellowship from the APS back in 1999–2000 allowed me to glean covertly what I could of jetons while doing archival work for the nominal project. After he became Co-Executive Officer of the APS Dr. Richard Dunn instituted a series of annual conferences for APS sabbatical fellows. I later whined to Dunn that there were no such wonderful things when I was a fellow, so he allowed me to participate in the 2007 conference, but only if I presented new work, which I did in the form of a first report on jetons, history, and numismatics, receiving terrific feedback from a wonderful group of fellow scholars. I have also most warmly to thank Ms. Mary McDonald of the APS for the moral support she lent this work and its author, sustaining him and his efforts over many years of research and writing.

I gave an updated version of my APS Fellows presentation at my own institution, Stevens Institute of Technology, and more recently at the 2015 summer seminar of the American Numismatic Society, which allowed me to hone my thinking and presentation further. My thanks go to Matt Wittmann, then of the ANS, for the invitation and opportunity. I also owe extended thanks to Stevens Institute of Technology for its sabbatical support in 1999–2000 and yet another sabbatical in 2011–2012 that was wholly devoted to research for the present work.

Other debts accumulated as research proceeded. The Département des Monnaies, Médailles et Antiques of the Bibliothèque nationale de France (the BnF) holds the world's largest and rich-

4. "L'étude des jetons n'est intéressante que si elle donne occasion à des recherches historiques et si elle fournit le moyen d'élucider quelques points controversés"; Prou (1888), p. 348; quoted also in Roy-Chevrier (1919), p. v.
5. Grierson (1951), p. 15; see also p. 13.

est collection of jetons.[6] My connection to this wonderful establishment goes back to the early 1980s tracking down the jeton of the royal scientific society established in colonial Haiti, the Cercle des Philadelphes, which I was studying at the time. That jeton and its story figure again below. More recently, I have benefited from the scholarship and generosity of Thierry Sarmant, then of the Département des Monnaies of the BnF and now of the historical services of the French Ministry of Defense, for sharing his insights and his data; his work and the catalogs published with François Ploton-Nicollet are pillars on which the present work rests. More recently still, I am indebted to Ms. Inès Villela-Petit of same department of the BnF for showing me the full collection of jetons and medals held in its vaults and for other assistance as I explored and photographed that treasure-trove. At another point, the staff of the Département des Estampes of the Bibliothèque Nationale was instrumental in helping me locate the wonderful engraving of the parish of St. Germain l'Auxerrois presented in chapter seven below. I need to thank the personnel and new friends at the Bibliothèque nationale de France, site François Mitterand at Tolbiac in the 13th arrondissement in Paris, where I spent many productive days reading and taking notes. The library catalog of the BnF available over the web and the amazing electronic resources of the BnF's electronic library, Gallica, have equally proven a godsend in preparing the present text.

I worked and was pleasurably assisted by the personnel at the French national archives in Paris (the CARAN), and it was fortunate for me that the archival material I needed remains in the Marais. I thank Jean-Marie Darnis, now retired, for his expertise and assistance (as well as his scholarship) when I did archival research at the national mint, La Monnaie de Paris, on the Quai de Conti. Those archives have since moved to the Centre des Archives Économiques et Financières (CAEF) in Savigny-le-Temple outside of Paris, and I appreciate the assistance I received from the helpful staff when I commuted there. Back in Paris, I worked in several departments and libraries of the Institut de France, and I have especially to thank M. Ben-Zerook of the archives of the Institut for his assistance and my late and much lamented friend, Claudine Pouret, of the archives of the Académie des Sciences. I wish to thank others of the staff of the archives of the Académie des Sciences, including Mmes Christiane Demeulenaere-Douyère and Florence Greffe, former heads of the institution, and my old friend, M. Le Roy, the *magasinier* who has now retired back to India.

The coin dealers I mentioned above play a key role in contemporary French numismatics, and I have to thank many of them for the education I received at their hands. I owe Daniel Renaud of the venerable Maison Platt on the rue Richlieu my deepest debt in this regard. I have likewise to thank Pierre Crinon and his team at the O.G.N. establishment on the rue Richelieu as well, and I cannot omit mention of the pleasure I had in meeting the late Michel Prieur of the formidable firm C.G.B. (now cgb.fr) on the rue Vivienne and discussing the Grateful Dead.

I am most fortunate that the American Numismatic Society and its remarkable numismatic library are located on Varick Street in Manhattan, right across the Hudson River from where I live in Hoboken, New Jersey. That locale, the open stacks of the library, its exquisitely detailed electronic catalog, and the amiable help I received from the librarian, Ms. Elizabeth Hahn Benge, her successor David Hill, and other of the staff and numismatic professionals at the ANS proved an enormous boon to this project. Dr. Gilles Bransbourg, Executive Director of the ANS, read the manuscript at a late stage, and his keen eye and erudition made for many improvements for which I am immensely grateful. Mostly however, I am indebted to the American Numismatic Society

6. La Tour (1897), p. xxi.

and Dr. Andrew Reinhard, its Director of Publications, for boldly taking on this work for publication. Falling as it might seem between two stools—history and numismatics—the venture risked rejection by the guardians of both fortresses. Andrew saw its value and that at the least it deserved public presentation, and for that I cannot sufficiently express my gratitude.

I owe a word of thanks to Ms. Mary Ellen Valverde of the interlibrary loan department of S. C. Williams Library at Stevens Institute of Technology for her much appreciated assistance and to historical collections specialist Caroline Bradford at the National Museum of Health and Medicine in Silver Spring, Maryland, for her help in tracking down the Horatio R. Storer collection of jetons of the Paris Faculty of Medicine that ended up in the United States. An undergraduate assistant, Mr. Nicolas Chapel, lent his aid in the compilation of Table 3.1. I am very much indebted to Mr. Ridge Montes, a most impressive young man, whose command of Latin and his assistance in translating Latin inscriptions saved the day. I similarly thank Dr. Pamela J. Russell and my archaeologist brother, Dr. Murray C McClellan, both of whom also helped with the Latin and the annotated list of the figures that accompany the text. Last, but hardly least, I again express tons and tons of love and heartfelt thanks to my wife, Jackie, for her highly expert editing, for the tremendous labor she put in on this project, especially in its late stages, and for her companionship in life.

Unless otherwise indicated all translations are by the author; the original French is given in the footnotes; original orthography is preserved in the notes without otherwise being noted.

I have previously used jetons as illustrations in earlier works of mine, most notably in *The Colonial Machine: French Science and Overseas Expansion*, a scholarly tome I wrote with François Regourd.[7] There, these seemingly innocent, trivial jetons are integrated into a larger explanatory narrative, particularly to illustrate the bureaucratic offices that superintended the French colonial and overseas enterprise under the Bourbon monarchs. Here, finally, in the present monograph I am most pleased at last to focus on jetons themselves and on all that their story brings into view.

Hoboken, New Jersey
October 14, 2019

7. McClellan and Regourd (2011); see also McClellan (1992/2010), (2003), (2005).

(TOP) **FIGURE IN.1:** Jeton of the guild of architects.
(BOTTOM) **FIGURE IN.2:** Jeton of the carpenters' guild, 1773.

INTRODUCTION:
Historians and Numismatists

History is the study of the past. Numismatics is the study of currency, coins, medals, and tokens. Professional historians and professional numismatists both practice their crafts within expert and highly specialized communities of long standing. The point of departure of the present study is, as mentioned, an historiographical one and lies in recognizing a divide separating communities of historians and numismatists specializing in early modern France. Interpretive nuance is called for and this divide has been breached, but by and large for medieval, early modern, and Old Regime France through 1789 these two highly developed communities—historians and numismatists—have not taken cognizance of one another and have produced essentially separate expert literatures. The disjuncture between these two communities is not really in dispute, and as we progress, we will see what can be had by bringing history and numismatics together to bear on Old Regime society.

Historians and Jetons

Historians have examined an immense variety of topics pertaining to early modern and Old Regime France, and the literature they have produced is vast and wonderful. One need only consider all that is known about the French Enlightenment, say, or the court of Louis XIV at Versailles to be persuaded of the point. Consider, too, all that has emerged from the *Annales* school over the decades regarding the sociocultural and economic history of the Old Regime in the hands of such practitioners as François Furet, Daniel Roche, and their predecessors. As seen in works from Peter Gay to Robert Darnton and a host of others, the community of Anglo-American historians specializing in French history before the French Revolution is also substantial and sophisticated. The historical literature pertaining to Old Regime France is certainly among the historiographically most developed literatures, and arguably it is the most developed.

But jetons, the particular tokens under the microscope here, have not attracted the attention of these selfsame historians. As the eminent Michel Pastoureau noted in 1984, "The jeton is an unknown document for most historians."[1] By way of confirming this point, I asked three great historians of the Old Regime—Robert Darnton, Daniel Roche, and Charles Gillispie—

1. Pastoureau (1984), p. 34: "Le jeton est un document inconnu de la majorité des historiens." Pastoureau is professor at the École des Hautes Études in Paris where he holds a chair in the history of Western symbolism. De Callataÿ (2014) shows further that the numismatic literature is underrepresented in bibliometric reporting of the humanities in general and thus even less drawing the attention of historians.

what they know about jetons.[2] All three were aware of these little things and were familiar with *jetons de présence* handed out at meetings of the academies, and Darnton went on to remind me of d'Alembert's reputation as a *jetonnier* (or one who attended meetings especially to pick up the dime) at the Académie Française. Roche recalled the pleasure of having been given one of the jetons of the Académie de Stanislas in Nancy when he did research there. Given other, rich resources available to these remarkable scholars and to historians of Old Regime France generally, there is no special reason jetons should have caught their eye or drawn sustained attention. It is not that jetons never appear in historical studies, as we will see further, but with respect to the correspondingly rich numismatic literature to which we turn next, it is still telling for present purposes that jetons remain essentially invisible in the history literature.

FIGURE IN.3: Jeton: Académie de Stanislas, 1750

The "Science" of Numismatics

Readers will probably have an intuitive sense of the discipline of history and historical studies, but they may be less familiar with the separate field of numismatics. According to the 2001 *Dictionnaire de numismatique*, numismatics is a "science that deals with the description and history of money, medals, jetons, tokens, paper currency, and every other means of exchange."[3] In the seventeenth century the great numismatist, Charles Patin (1633–1693), already referred to the study of historical medals as a science, and numismatics remains so for most practitioners today.[4]

Self-evidently, it takes considerable learning and expertise to pursue any such work seriously. The origins of modern numismatics date back to the Renaissance and to Renaissance humanism, and the field retains the hallmark of high scholarship and erudition. Although amateurs still populate the field, numismatics is led by expert, trained professionals. Numismatics is highly developed and a flourishing domain of research and publication. The sheer scope of numismatics as a field and the depth and level of detail mined by numismatists is stunning. Consider, for example, the bibliographic survey, *Numismatic Literature*, published by the American Numis-

2. Longtime professor of history at Princeton, Robert Darnton is presently Carl H. Pforzheimer University Professor and Harvard University Librarian Emeritus; Daniel Roche is emeritus holder of the *Chaire des lumières* at the Collège de France; the late Charles Gillispie was Dayton-Stockton Professor of History of Science emeritus at Princeton.

3. "Science qui traite de la description et de l'histoire des monnaies, médailles, jetons, méreaux, papier-monnaie et tout autre moyen d'échange," Amandry (2001), pp. 412–13; see also Darnis (1996/1999), vol. 2, p. 406, and Babelon (1958) on this notion of numismatics as a science. Grierson (1951), pp. 3–6, 12, classifies numismatics as a subdomain of archaeology.

4. Patin (1695), p. 129. Joseph-Étienne de Fontenay makes like claims for numismatics as a science in Fontenay (1850), p. 181; see also his (1854), p. 14, and early remarks about numismatics as a science by Blanchet (1890), vol. 1, pp. v–vi; (1899), pp. v, vii. In his survey of "the present state of the science of numismatics," de Witte (1893) gives a world tour of institutions and periodicals devoted to numismatics. Sarmant (2003) characterizes Ernest Babelon's monumental *Traité des monnaies grecques et romaines* and especially vol. 1 (*Théorie et doctrine*, 1901) as the best history of numismatic science. See also Clain-Stefanelli (1965).

matic Society from 1947 through 2007. This series enumerates roughly forty areas of specialist work in numismatics, including ancient Greek, Roman, Sassanian, Byzantine, Visigothic, Jewish, Medieval/Modern Europe, Africa, Islamic/India, Hawaii, Armenia, Far East, U.S., Canada, Latin America, the Philippines, Australia/New Zealand, and it includes special categories for tokens and medals.

Numismatics has long possessed its own organizations and specialized publications. The earliest of these, the Royal Numismatic Society, was founded in London in 1836, the *Société royale de numismatique de Belgique* came into being in 1841, the American Numismatic Society in New York dates from 1858, and the *Société française de Numismatique* saw the light of day in 1865.[5] Brussels hosted the first international numismatic congress in 1891, the second took place in Paris in 1901, and such conclaves are now held every six years under the aegis of the International Numismatic Council that was founded in 1927. The still extant French *Revue numismatique* dates from 1836; its continuously published counterpart, the *Revue belge de numismatique*, dates from 1842. The oldest German language numismatic journal is the Austrian *Numismatische Zeitschrift*, published since 1869. Another noteworthy German series, *Numismatisches Literature-Blatt*, appeared from 1880 to 1939 under the editorship of Dr. Max von Bahrfeldt, and today there are thousands of amateur, professional, and commercial organizations and publications devoted to coins and coinage.

The Numismatic Literature on Jetons

As will be detailed momentarily, a rich literature has grown up around jetons, yet jetons (and tokens in general) occupy a peripheral and generally looked down upon place in the overall universe of numismatics. It is telling, for example, that the entry concerning jetons in the 628-page *Dictionnaire de numismatique* of 2001 contains but two small paragraphs about jetons, and although not inaccurate in describing the accounting functions and later remunerative and ceremonial functions played by jetons, the entry hardly captures the historical or cultural richness of jetons or the place they deserve in numismatics.[6] This disdain goes a long way back. In the middle of the nineteenth century, a local historian in Burgundy, Joseph-Étienne de Fontenay, sought to establish the study of jetons as "a new branch of numismatics."[7] In his 1897 catalog of the holdings of the Cabinet des Médailles of the Bibliothèque Nationale, the curator, Henri de La Tour, called for giving jetons greater importance in numismatic studies.[8] So too did the president of the Société Archéologique of Montpellier, Émile Bonnet, two years later in his 1899 study of the jetons of the Estates of Languedoc, saying that jetons "well merited being retrieved from the condescending oblivion to which they have been consigned for only too long."[9] Yet, in the 1980s Michel Pastoureau could still label jetons as marginal to numismatics as a whole, and he condemned "the contempt that numismatists have always had for these 'small monuments' [*petits monuments*]," as jetons have always been known.[10]

5. On the Société Française de Numismatique, see especially Jambu (2015), passim.
6. Amandry (2001), p. 297; on this point see also Jambu (2015), p. 863.
7. "...une nouvelle branche de numismatique," Fontenay (1850), p. 5.
8. La Tour (1897), pp. xxvii–xxxi.
9. "...ils méritaient bien d'être tirés du dédaigneux oubli dans lequel ils n'ont été que trop longtemps laissés," Bonnet (1899) in Depeyrot (2007), p. 131.
10. "...le mépris qu'ont toujours eu les numismates pour ces « petits monuments »...," Pastoureau (1984), p. 30. On the greater attention given to historical medals over jetons, see Mainguy (1999), pp. 57–61.

Nonetheless, the specialized literature surrounding jetons is vast.[11] Largely inclusive through 1979, the bibliography compiled by Victor Jaumonet lists 76 numismatic works pertaining specifically to jetons.[12] Of this and related literature, almost all of it is in French, mostly in short articles, and mostly buried in specialized numismatic periodicals and regional and local histories. Numismatists have inventoried and studied jetons largely from the point of view of numismatics itself, not history, and the aims and purposes of this literature have overwhelmingly been to serve the numismatic interests of authors, readers, scholars, and collectors.[13] The emphasis in these studies has been on erudition, and jetons have attracted the close attention of learned antiquarians, genealogists, and experts interested in heraldry as well, the latter because royal, regional, civic, and family crests often appear on jetons.[14] Collectors and hobbyists receive more than their due in this literature, too.

In works published in 1982 and 1984 the historian Michel Pastoureau saw the situation clearly: "Jetons and church tokens are documents that are hardly known. Their study constitutes a marginal branch of numismatics and has accordingly privileged collectors and amateur archeologists more often than historians....the interest of these two types of coinage is, however, manifest."[15] He went on to say:

> The history of jetons remains to be written. I mean by that that we do not have available any scientific work on the evolution and transformation in western Europe from the thirteenth through the nineteenth century concerning the minting, use, purposes, or circulation of this small token. As always in numismatics, we are submerged by catalogs and monographs and possess very few synthetic studies. As always, the collector has had priority over the historian.[16]

Pastoureau further noted that "...the study of jetons is a yet unformed discipline. Everything is empirical and everything remains to be built," and he went on to remark that since Francis Pierrepont Barnard's classic 1916 work, *The Casting-Counter and the Counting-Board, A Chapter in the History of Numismatics and Early Arithmetic*: "...jetons have not been the object of any work of synthesis. Even more, if collectors have multiplied, researchers interested in jetons have become more rare to the point where today there is, so to say, no European specialist in this type of coin."[17]

11. In addition to the works discussed next, for introductions to jetons, see first and foremost Sarmant and Ploton-Nicollet (2010/2014) passim and especially vol. 1, pp. 11–51, and then La Tour (1897), "Introduction," pp. i–xlvi; and Barnard (1916); Lerouge and Lerouge (1977) and (1979). See also Elliott-Smith (1952), Müller (1984) in German, Mackay (1994), and Rostron (1993) for brief, not always accurate, overviews of jetons. These later works plainly have numismatic audiences, not historians, in mind.

12. Jaumonet (1985/1986).

13. Mainguy (1999), p. 7, makes this point.

14. On heraldry and numismatics, see the formidable 1986 work by Pierre Corre and earlier J. Florange (1902, 1907); also Dauban (1859), pp. 3–4.

15. "Jetons et méreaux sont des documents peu connus. Leur étude constitute une branche marginale de la numismatique et, de ce fait, a souvent tenté davantage les collectionneurs et les archéologues amateurs que les historiens...L'intérêt de ces deux types de pièces est pourtant manifeste," Pastoureau (1984), p. 7.

16. "L'histoire du jeton reste à écrire. J'entends par là que nous ne disposons d'aucun travail scientifique sur l'évolution et les transformations de la frappe, de l'emploi, de la destination et de la circulation de cette petite pièce à compter en Europe occidentale du XIIIe au XIXe siècle. Comme toujours en numismatique, nous sommes submergés de catalogues et de monographies et ne possédons que très peu d'études de synthèse. Comme toujours, le collectionneur a eu priorité sur l'historien," Pastoureau in Rouyer and Hucher (1982), preface, p i.

17. "...l'étude des jetons est une discipline qui n'est pas encore constituée. Tout y est empirique, tout reste à construire... le jeton n'a fait l'objet d'aucun travail de synthèse. En outre, si les collectionneurs se sont multipliés, les chercheurs s'intéressant aux jetons se sont raféfiés ; au point qu'à l'heure actuelle, il n'existe pour ainsi dire aucun spécialiste

More recent work by such scholars as Josèphe Jacquiot, Thierry Sarmant, Michael Mitchiner, Sylvie de Turckheim-Pey, Anne Mainguy and others belie Pastoureau's harsh judgment of nearly 40 years ago, but such works do not, as yet, significantly redirect the inertial mass and momentum of the literature—historical or numismatic—dealing with jetons.

Early Collectors

Collecting and the study of coins and medals trace their roots back to antiquity. The collection and analysis of ancient coinage represents one of the foundations on which humanists of the Renaissance based their recovery of the ancient world.[18] Jetons spread on the counting tables of Europe in the High Middle Ages, but they did not attract the attention of collectors or the learned until the seventeenth century, however, well after the apparatus of high scholarship had established itself on the continent. (The explanation for this lag is that medieval jetons—little pieces of tin or bronze on casting tables—were seen as unworthy of attention.) Jetons were looked down upon by curators and aficionados of medals.[19] By the middle of the seventeenth century, however, the situation changed. As the social function of jetons altered and as the aesthetics of these objects became more appealing, collectors began to seek out jetons and assembled prized collections. Jetons have now been collected and studied for over 300 years.[20]

With a collection of roughly 100 jetons, the oldest-known amateur and jeton collector seems to have been Pierre de l'Estoile (1546–1611), the chronicler of the reigns of Henri III and Henri IV.[21] In Jacques de Bie's *La France métallique* (1636) and then in the Jesuit Claude-François Ménestrier's *Histoire du roy Louis le Grand par les medailles, emblêmes, devises, jettons, inscriptions, armoiries, et autres monumens publics* (1689) we find experts collecting and inventorying mostly historical medals, but also jetons.[22] Already by 1665, according to the abovementioned physician and numismatist Charles Patin, contemporary collectors hoarded jetons in pretty purses and sought out the most beautiful specimens.[23] And so soon, scholars, collectors, churchmen, officers, doctors, lawyers, jurists, antiquarians, government officials and others—the *robe* and not the *épée*, undertook to build up collections of jetons.[24] A private collection of 1670 numbered 1,800 jetons, and by 1692 the genealogist, antiquarian, and collector, François Roger de Gaignières (1642–1715), had assembled his collection of 3,000 different jetons.[25] Jetons ordered and paid for by the king for various administrative units of the French state were required to be deposited at the royal Cabinet des Médailles (a subunit of the royal library), and by 1670 the Cabinet des Médailles possessed a collection of over 2,000 jetons.[26] In 1739 we find an estate sale of 3,500 jetons, "the majority being

européen de ce type de pièces," Pastoureau (1984), p. 30. Pastoureau goes on to point out that Bernard was an archaeologist, not a numismatist.

18. Mazard (1992); Amandry (2001), entry "numismatiques."
19. Sarmant (1994), pp. 178–79; see remarks to this effect by La Tour (1897), pp. i–iii; see also Mainguy (1999) and below, chapter 1.
20. On this point see Blanchet (1930), p. 238.
21. Boutin (1966), p. 241. On early collectors, see also Valin (2016), vol. 1, pp. 16–18; Wellington (2015a), pp. 29–34; d'Affry de la Monnoye (1878), pp. ii–iv; Barnard (1916), pp. 84–85.
22. Ménestrier (1689); see also Wellington (2015a), pp. 127–44; Mainguy (1999), pp. 173–77; La Tour (1897), p. vi.
23. Patin (1695), chapter xiv, p. 129; see also d'Affry de la Monnoye (1878), p. ii; Schodt (1883), p. 3. Chapter 14 of Patin's postumous *Histoire des médailles* (1695) was devoted to "gettons."
24. Sarmant (2003), p. 141. Notable private collectors in the seventeenth century included the antiquary and humanist Peiresc and Gaston d'Orléans, brother of Louis XIII.
25. On Roger de Gaignière and his collection, see Valin (2016), vol. 1, pp. 16, 344–52; Sarmant (2003), p. 69; Schodt (1873), p. 3. See also d'Affry de la Monnoye (1878), pp. ii–iv, for these and other private collectors of the seventeenth and eighteenth centuries.
26. d'Affry de la Monnoye (1878), p. ii; Blanchet (1930), pp. 210–11. See, by comparison, CAEF, Fol. Ms. 152, "Monnaie de Paris, Médaillier du Roi, 1756–1767," an important catalog that excludes jetons.

mint strikings for government offices, faculties, Paris merchants, city mayors, and families."[27] In 1759, to pick a further example from our period, the Académie Royale des Inscriptions and Belles-Lettres refused a donation of 5,000 jetons, although in 1765 it turned around and accepted one of 4,000 jetons collected by one of its members.[28]

Jetons struck in the Low Countries in the fifteenth, sixteenth, and seventeenth centuries occupy a special place in world of jetons. (They circulated in huge numbers as propaganda pieces in the Dutch revolt against Spanish sovereignty, described below.) In 1687 the abbé and church canon, Pierre Bizot (1630–1696), published in Paris his *Histoire métallique de la République de Hollande* (his illustrated collection of 167 Dutch medals), and then in 1723–1731 the Dutch jurist, historian, and numismatist, Gerard van Loon (1683–1758), published in the Hague his monumental four-volume *Beschrijving der Nederlandsche Historiepenningen* (Description of Dutch historical tokens) that cataloged Dutch jetons from 1555–1716.[29] A five-volume French translation, published in the Hague, *Histoire métallique des dix-sept provinces des Pays-Bas,* appeared in 1732–1737, and these 2,500 folio pages of material made their mark on the world of French numismatists and collectors. In 1732–1735 Frans van Mieris (1689–1763) followed up on van Loon's work and published three volumes in Dutch on rulers of the Low Countries as depicted in Dutch jetons and medals prior to 1555.[30]

These early French and Dutch collectors and numismatic specialists were primarily interested in jetons and medals as sources for historical accounts or what was known as building up an *histoire métallique* or histories retold through medals and tokens. The published results are noteworthy historiographically, but they do not engage the history of medals and jetons themselves and are little more than elaborate illustrated catalogs.

Systematic Study and Collection in the Nineteenth Century

Collecting and studying are two different, if related, activities, of course, and in the nineteenth century French numismatists began to systematically organize and categorize the thousands of jetons from the previous five hundred years that had gradually come to their attention.[31] These *érudits* were amateurs in the sense that they were not institutionally affiliated or employed as professional numismatists, although the level of their learning and expertise was high. Joseph-Étienne de Fontenay (1811–1859), the local historian in Burgundy mentioned above, provides an early example. Fontenay began publishing in the 1840s, with his *Nouvelle étude des jetons* appearing in 1850 and his *Manuel de l'amateur de jetons* in 1854. Fontenay knows a lot, yet his attention is focused on individual jetons for the most part, and as the title of his 1854 work indicates, his orientation is towards amateur collectors.

Similarly, with his resonant and for us appropriate name, Alfred d'Affry de la Monnoye (1811–1869), was a tax collector for the city of Paris, and he formed his collection of jetons and art objects privately; his book on the jetons of the city council of Paris, published posthumously in 1878,

27. "...la plupart à fleur de coin, des administrations, facultés, prévôts des marchands, maires de villes, universités, familles," per Blanchet (1930), p. 238; on eighteenth-century collectors, see also Blanchet (1939).

28. Sarmant (2003), pp. 315–16; (1994), pp. 132–33; AI, A65, January 11, 1765.

29. Chautard (1881), p. 335; van Beek (1986), pp. 201–05.

30. This is van Mieris, the younger, the grandson of the seventeenth-century Dutch painter, van Mieris the elder; see comment by Jaumonet (1985), p. 13, that van Mieris the younger "constitue avec Van Loon une documentation exceptionnelle sur les jetons et médailles des Pays-Bays de 1405 à 1716." On these early collectors, see also La Tour (1897), pp. vi–vii; de Witte (1893), pp. 5–6.

31. On this historiography, see also Valin (2016), vol. 1, pp. 19–23; Pastoureau (1984), p. 28.

presents a solid historical and thematic introduction to jetons, but his, again, is an illustrated catalog of over 400 pages and his audience is the antiquarian and the heraldist.[32]

The career and work of Jules Rouyer (1820–1898) best typifies the state of jeton scholarship in the nineteenth century. In his day job Rouyer headed the post office in provincial Nancy (Henri de La Tour calls him a "modest postal employee"[33]), yet on the side Rouyer assembled the largest jeton collection to date, and in the process he became the preeminent contemporary authority on jetons, publishing extensively on them from the 1850s through the 1890s. He coauthored with Eugène Hucher, a public servant working in Le Mans, who later resigned to dedicate his life to local archeology, history, and numismatics, *Histoire du jeton au Moyen-Âge,* which appeared in 1858 and remains a touchstone in the numismatic literature. Rouyer bequeathed his collection to the Département des Médailles of the Bibliothèque Nationale, and the associate curator, Henri de La Tour, published Rouyer's collection in two volumes (1899, 1910) documenting 5,025 jetons. Rouyer displayed a special interest in the local history of Lorraine and his work has a whiff of antiquarianism about it; La Tour calls Rouyer a "learned collector" and labels the published catalog "a useful resource for the increasing number of researchers, scholars, the curious, and collectors who are passionate about this so typically French series of jeton."[34] Yet to Rouyer's credit, according to La Tour, Rouyer "loved these *petits monuments* not only for their rarity or beauty, but especially for what they taught him about mores, customs, literature, the arts and history."[35]

Henri de La Tour himself (1855–1913) is noteworthy in this connection, at least in part because of his professional position in the Département des Monnaies at the Bibliothèque Nationale. His publication of the Rouyer catalog is a landmark in the numismatic literature, as is his own 1897 edition that inventories and lightly annotates 2,334 royal jetons held in the Bibliothèque Nationale. Noteworthy is the fact that La Tour categorizes his own work as simply "a numismatic catalog."[36] That jetons might be thought of in larger cultural or historiographical contexts was not part of the picture at the turn of the twentieth century.

Félix Feuardent and the Twentieth Century

Developing an adequate system for classifying and organizing jetons was a major issue facing nineteenth-century numismatists as they sought to get their arms around the wealth of material they encountered and inherited.[37] For better or worse, this problem was trumped by the monumental three-volume illustrated catalog of jetons and assorted tokens published in 1904, 1907, and 1915 by the harmoniously named Félix-Bienaimé Feuardent (1819–1907).

Félix Feuardent was a provincial librarian from Cherbourg who moved to Paris in 1846 and married into the Rollin family.[38] The patriarch Charles-Louis Rollin (1777–1853) began his career as a money-changer in the galleries of the Palais Royal and published his first coin sales list in

32. See d'Affry de la Monnoye (1878), passim and especially. his "Introduction à l'étude des jetons," pp. i–xxviii. The work of Jules Guiffrey (1840–1918) is not to be overlooked in this connection. An historian of French art and member of the Académie des Beaux-Arts, Guiffrey headed the Gobelins dye works; given his administrative position, the manufacture of jetons caught his attention, and he published several works of note for the present study; see bibliography and Fenaille (1923)
33. "...modeste employé des Postes," La Tour (1899), p. iv; see also pp. i–iii, viii.
34. "...un utile instrument de travail pour les chercheurs, les érudits, les curieux et les collectionneurs de plus en plus nombreux qui se passionnent pour cette série si française du jeton," La Tour (1899), p. xvii.
35. La Tour (1899), p. xii: "Il aimait ces petits monuments non seulement pour leur rareté ou leur beauté, mais aussi et surtout pour ce qu'ils lui enseignaient...sur les moeurs, les coutumes, la littérature, les arts, l'histoire."
36. "...un catalogue de numismatique," La Tour (1897), p. xliii; on this work, see Pastoureau (1984), p. 28 n.
37. On this point, see La Tour (1899), pp. xii–xiii.
38. These details are drawn from various web sources and Sarmant (2003), pp. 353–54.

1811. His son, Claude-Camille Rollin (1813–1883) continued the family business, moving it in the 1840s to the rue Vivienne opposite the Bibliothèque Royale. Feuardent married Claude-Camille's daughter, Louise-Antoinette, and became a partner in the enterprise which claimed to be the world's oldest proper coin dealership. As scholars, dealers, publishers, and auctioneers Rollin & Feuardent became preeminent in fin-de-siècle Paris, and in 1867 the firm opened a branch in London. Claude-Camille Rollin died in 1883 at which point Feuardent assumed control with at least one of his brothers, operating under the name Feuardent Frères. The business continued in Paris until it ran out of steam in 1953.[39]

Feuardent's *Jetons et méreaux depuis Louis IX jusqu'à la fin du Consulat de Bonaparte* (1904–1915) is a numismatic monument and remains the touchstone for any study of French and Belgian jetons. Although Feuardent did not succeed in offering the hoped-for complete catalog of all jetons that ever existed, in three volumes and over 1,500 pages Feuardent heroically enumerates close to 16,000 different jetons and *méreaux* (church tokens), and jetons today are reported by their Feuardent numbers or as variants of jetons with Feuardent numbers.[40] Feuardent provides plates illustrating many jetons and especially busts, and he gives an inventory of jeton engravers, an alphabetical list of Latin and French inscriptions, as well as a detailed table of contents, all of which make "Feuardent" an indispensible tool for any jeton collector or scholar. In 1966 Serge Boutin called Feuardent's volumes "the breviary of the collector today," and the Paris coin dealership of Maison Platt further cemented Feuardent's place in the numismatic literature in 1995 by republishing his work along with the catalogs of the estate sale in 1928 and 1929 that contained additional plates of rare jetons in the Feuardent collection.[41] It is noteworthy for present purposes that the Maison Platt edition is explicitly intended for "the specialized collector/numismatist... who will doubtless find here a new ally to advance the love/hobby of jetons (*jetonophilie*)."[42]

Feuardent's classification system has become the de facto standard. As essential as Feuardent is to any study or collection of jetons, it hardly needs to be pointed out that Feuardent's is not an historical work nor did he see himself as part of a pursuit by historians in the quest for historical understanding. The stern Michel Pastoureau calls Feuardent's opus, "a mediocre amateur work, filled with errors and imprecisions, endowed with a scientifically dubious classification system, yet a catalog with an advantage over every other inventory (and especially the all-too-thin records of public collections) of covering 15,126 jetons and being graced with an alphabetical index of inscriptions."[43]

Félix Feuardent did have his peers, and his younger contemporary, Jules Florange (1863–1937), stands alongside Feuardent as a giant in the world of jeton numismatics in the early years of the twentieth century.[44] From Lorraine and an expert on its numismatics, Jules Florange is best known for his still useful two-volume work published in 1902/1907 devoted to jetons struck by and for

39. Daniel Renaud, personal communication.

40. The last jeton in the catalog is numbered 15,126, but Feuardent, as he himself remarks (vol. 3, p. 3), had to make many additions after he had established his number system, so that number variants often appear, e.g., one finds for the royal chamber of accounts #1748, #1748a, #1748b, etc. On church tokens or *méreaux*, see below. On Feuardent's goal of creating a complete catalog, see Josset (1978), p. 340.

41. "le bréviaire du collectionneur," Boutin (1966), p. 241: Feuardent (1995); see also Blanchet (1930), p. 238.

42. "[le] collectionneur-numismate spécialisé…trouvera sans nul doute un nouvel allié pour faire progresser la jetonophilie," Feuardent (1995), vol. 1, "Note des Éditeurs."

43. "Médiocre travail d'amateur, rempli d'erreurs et d'imprécisions, doté d'un classement scientifiquement discutable, ce catalogue présente néanmoins sur tous les autres répertoires – et spécialement sur les trop minces catalogues de collections publiques – l'avantage de recencer 15126 jetons et d'être pouvu d'une table alphabétique des legendes," Pastoreau (1984), p. 28.

44. Boutin (1966), p. 241, adds Florange to "the breviary of the collector today."

particular individuals and families (as opposed to Feuardent whose work has a more institutional bent): the *Armorial du Jetonophile: Guide de l'Amateur des Jetons Armoriés*.[45] The first volume inventories 1,300 different jetons, the second records 1,900 more, and the title alone ("The book of crests for the jeton lover: guide for the amateur of heraldic jetons") makes plain Florange's orientation towards the amateur collector and enthusiast. By documenting coats of arms, family crests, heraldic emblems, and other insignia appearing on jetons, Florange's work furthermore serves as a contribution to heraldry and genealogy.[46] Not to take away from Florange's signal accomplishment, like Feuardent's, the work is not a work of history or intended for historians.

In the present context we must mention the *opera omnia* of Adrien Blanchet (1866–1957).[47] Blanchet was a professional numismatist of the highest order and a prolific scholar. For many years he headed the Département des Monnaies at the Bibliothèque Nationale in Paris, and he was a member of the Académie des Inscriptions et Belles-Lettres of the Institut de France. His major works include a two-volume *Nouveau manuel de numismatique du moyen-âge et moderne* (1890) and with Adolphe Dieudonné a four-volume *Manuel de numismatique française* (1912–1936). The 600-page third volume (by Blanchet alone) published in 1930 deals with French medals, jetons, and *méreaux*, with 300 pages given over to jetons alone. Despite its extensive character—and the point is true of Blanchet's other work on jetons as well, Blanchet is mostly descriptive and provides few general perspectives or analyses. The volume just referred to, for example, is largely a catalog addressed to scholars and collectors ("*les érudits et les collectionneurs*"). Nevertheless, much fascinating detail crops up here and there in Blanchet's work that has proven of great value in preparing the present account.[48]

Adrien Blanchet published his books in French and in francophone numismatic journals. The literature examined so far underscores the point that, with rare exceptions, what was (and is) known about jetons in numismatic circles did not (and largely does not) escape the scholarly orbit of the francophone world. One of those exceptions—and an early one—is the aforementioned Francis Pierrepont Barnard's *The Casting-Counter and the Counting-Board, A Chapter in the History of Numismatics and Early Arithmetic* (1916). As his title indicates, Barnard is interested in the history of accounting techniques, and in this way he was drawn to jetons which found their first uses as counters on medieval and early modern counting tables. Barnard's is an impressive, knowledgeable scholarly study, and he offers clarifying details on accounting and reckoning methods using jetons. That said, the body of his book is given over to inventories of jetons and counting tables.[49] By and large, Barnard does not set jetons themselves in a larger cultural context nor see that they might help define a larger cultural history.

The reproduction of images has long been a feature of the literature about jetons and medals as well as numismatics in general, and a word might be added here about the changing technologies of reproducing images. Although woodblocks may have been used early on, plainly the technology of intaglio printing or copperplate engraving predominated in representing jetons from the seventeenth century through the nineteenth. The 1689 work by Claude-François Ménestrier, for example, his *Histoire du roy Louis le Grand*, consists entirely of 61 engraved plates of images *and* of text. Diderot reproduces images of jetons and coinage in the *Encyclopédie* using this technolo-

45. Florange issued a second edition in 1921.
46. On the usefulness of jetons in genealogy, see also Chalon (1867), p. 333.
47. See bibliography; on Blanchet, see Babelon; Christ. The prolific Blanchet won the medal of the Royal Numismatic Society in 1919, and the 1934 Huntington medal from the American Numismatic Society.
48. A 12-page article Blanchet published in 1920, "Jetons et méreaux," might possibly be the best introduction to jetons available; see Blanchet (1920b) and works cited in the bibliography.
49. Barnard (1916), pp. 93–292.

gy.[50] Lithography or printing from etched stone slabs developed and ultimately predominated in the nineteenth century for the reproduction of numismatic images as well as images in general. Fontenay's *Manuel de l'amateur de jetons* of 1854 is a representative example, and surprisingly the use of lithography continued into the twentieth century, as can be seen in Florange and Strowski's 1927 *Les Assemblées du Clergé de France avant 1789 et leurs jetons commémoratifs.* The surprise comes from the fact that by the end of the nineteenth century photography and the development of photolithography, phototypography (what Henri de La Tour and Feuardent call "phototyping"), heliogravure, and other photomechanical reproduction techniques (as distinct from photography itself) became standard.[51] These developments had the important effect of making photographic images of jetons available rather than the mere reproduction of drawings or simply, as was generally the case before the turn of the twentieth century, of having to rely on written descriptions of jetons. La Tour's 1897 catalog of the jetons of the Bibliothèque Nationale, for example, includes 36 photographically reproduced plates depicting approximately 500 jetons; his two volumes of the Rouyer collection of 1889 and 1910 offer 56 such plates and images of some 800 jetons.[52] Further developments in offset printing/offset lithography since the 1960s improved the image quality of jetons reproduced in catalogs and scholarly works, and today sophisticated digital, electronic printing techniques permit exquisite reproductions, often in color, that capture extraordinary detail and nuance. The 2004 sale catalog issued with the title *Jetons & méreaux du moyen*-âge by Roelandt et al., for example, with its beautiful color reproductions of the obverse and reverse of 1,000 medieval jetons contrasts remarkably with the 17 engraved plates depicting 150 jetons in the 1858 original of the same title by Rouyer and Hucher appended to this very same catalog of 2004. The two volumes by Thierry Sarmant and François Ploton-Nicollet published in 2010 and 2014, *Jetons des institutions centrales de l'Ancien Régime*, and their color images evidence the extraordinary number and quality of images of jetons (and coinage in general) available today. Such is the panorama of technological change in reproducing images since the seventeenth century that is necessarily part of the history and historiography of jetons.

Dealers and Hobbyists

The high level of scholarship produced by professional numismatists and erudite amateurs should already be plain. In addition to great museums, learned numismatic societies, and prestigious specialized publications along with their staffs, coin dealers and coin dealerships are also major pillars defining numismatics today.[53] The expertise and experience commanded by dealers is formidable. They bring commerce and learning together, and they stand at the interface between savants and collectors.[54] Coin dealers relevant to this study dot the Bourse neighborhood in Paris in particular, some with special expertise in jetons and medals, such as the abovementioned establishments Maison Platt, O.G.N. (*Or Gestion Numismatique*), and C.G.F. (*Comptoir Général Financier*). It is worth pointing out that trade organizations exist, such as the SNENNP in France (the *Syndicat National des Experts Numismates et Numismates Professionnels*), that certify these expert dealers and coin merchants.

Coin dealers regularly issue sale catalogs dedicated wholly or in part to jetons. They form an essential part of the bibliographical literature for numismatics and are resources not to be over-

50. See notably Gillispie (1959), vol. 1, pl. 208.
51. On these and several related processes, see Fisch (1905).
52. See La Tour (1897), p. xlv; the plates in this volume were printed by "Imp. Phot. Chène et Longuet, Paris." Roy-Chevrier (1919), pp. v, xxi–xxii points to the new technology of photo reproduction; see also Dillemann (1997), p. 5.
53. Cf. Grierson (1951), p. 17.
54. See related remarks by Sarmant (2003), pp. 363, 370–74.

FIGURE IN.4: Maison Platt, Paris.

127. Jeton CU 29, les États, 1746. (Cuivre, 29,5 mm, 6 h, 9,28 g).
A/ LUD• XV• REX - CHRISTIANISS•. (Louis XV roi très chrétien). Buste cuirassé et lauré à droite de Louis XV, au-dessous signature D•V•. **R/** COMITIA BURGUNDIÆ 1746•. (États de Bourgogne). Armes de Bourgogne sur un manteau d'hermines. Tranche : lisse. **G/** Jean Duvivier.
F. 9849.
TTB .... *(223FF)* 34€
"Après la mort du cardinal de Fleury, cet ardent ami de la paix, la France eut à combattre la Savoie, la Hollande, l'Angleterre et l'Autriche. Louis XV fit sa première campagne en 1744, et l'année suivante les États lui posèrent une couronne de laurier sur la tête".

FIGURE IN.5: Jeton catalog entry.

looked in thinking about jetons. Sale catalogs contain a wealth of information and documentation and, these days, images of jetons. Those issued by Victor Gadoury of Monaco in the 1980s, for example, are beautiful numismatic landmarks, and his one-page account of jetons is remarkable in capturing their story.[55] More recent catalogs with improved reproduction techniques are exquisite as well as informative, and catalogs often come with specialized bibliographies and much other information, including historical information about jetons. Consider, for example, the self-evidently rich source represented by this entry for the 1746 jeton of the Estates of Burgundy from the 2002 catalog, *Jetons XVII* [*sic*] issued by Michel Prieur and Stéphan Sombart of the Comptoir Général Financier.[56]

55. Gadoury (1986), p. 3; his first catalog dates from 1983.
56. Prieur and Sombart (2002), p. 16.

The entry (Fig. In.5) pictures the jeton front and back; it dates, gives a physical description, and otherwise describes the jeton. It transcribes (and translates) the Latin inscriptions. It identifies the engraver and provides the Feuardent number of the jeton. Significantly for our purposes the listing adds detailed historical background to explain the laurels that rest atop the head of Louis XV in this jeton. But more to the point right now this entry grades the coin and sets a purchase price. Coin dealerships and auction houses are businesses; they establish markets for jetons, and we cannot overlook the commercial dimensions of numismatics or coins (including jetons) as investment instruments. Grading a coin is an expert craft in and of itself that numismatic scholars seemingly eschew, but the grade of a coin is key to determining its market value.[57] Prices of jetons vary of course. The value of a handsome late seventeenth- or early eighteenth-century jeton in silver on today's market would be in the neighborhood of $200 to $300; the unremarkable 1746 example in copper is listed for 34 euros in the Prieur/Sombart catalog above. A truly rare jeton coveted by collectors can reach as high as $1,500-$2,000, and something wholly exceptional, a mint jeton in gold of a rare striking with a royal provenance, say, if such still exist, could possibly fetch $5,000 to $10,000. But these rarities are outliers that do not show up in the general market for jetons.

Numismatics and jetons have migrated to the World Wide Web along with so much else associated with the Internet over the last decades. Coin dealerships and auction houses have an active presence on the Web. These sites offer precision color scans of every variety of coinage including jetons; they provide descriptions and other information and are portals for purchasing jetons.[58] In addition, informal markets for jetons have arisen around online auction and shopping websites such as eBay, a development that only amplifies awareness and the availability of jetons as well as other glories and collectables in the world of contemporary numismatics.

Beyond all that extends an immense sea of hobbyists and amateur collectors. Coin collection is a popular pastime, and as mentioned above jeton collectors have built and maintained jeton collections since the seventeenth century. Already by the mid-nineteenth century, authors offered advice and encouragement for beginning jeton collectors.[59] Such guidance continues to be given in more recent times. For example, in 1966 Serge Boutin in his short article, "The Jeton Collector," ventured the following for the uninitiated:

> In the universe of jetons each collector can find various series of personal interest. For example, among the jetons of guilds or administrative departments, each can choose those that correspond with his occupation or profession. Just as easily he can collect the jetons of his province or town and thanks to them vivify and enjoy local history.[60]

57. Corre (1986) and the estimations he provides in his appendix seem exceptional; see related Wellington (2018) for related pricing information. Note that cleaning and maintaining coins and coin collections also involves nontrivial expertise; see Welter (1976).

58. See, for example www.ogn-numismatique.fr/; www.cgb.fr/; www.maison-platt.fr/; accessed October 23, 2018.

59. As seen with Joseph-Étienne de Fontenay's *Manuel de l'amateur de jetons* of 1854; Charles Even's *Le jetonophile* of 1877 and Jules Florange's *Armorial du jetonophile: Guide de l'Amateur des Jetons Armoriés* (1902–1921) fit this tradition. See also Bauer (1894) writing in English. NB how much Daniel (1980), e.g., p. 75, is still oriented to the "*amateur*" and the "*collectionneur*."

60. "Dans cette masse, chaque collectionneur peut trouver des séries qui l'intéressent ; par exemple, dans les corporations ou les administrations, chacun choisira ceux qui lui rappelleront son métier ou sa profession ; il pourra également rassembler ceux de sa province ou de sa ville et faire revivre grâce à eux la petite histoire," Boutin (1966), p. 241; Boutin's language suggests that most collectors are likely to be male. See also Goldfiem (1976), p. 95, and Mackay (1994), p. 52, for similar exhortations regarding jetons "for beginners."

Jeton collectors have had their own organizations. In North America a group called the International Society of Jeton Collectors existed briefly around 1989.[61] In France, however, the Association des Collectionneurs de Jetons-Monnaie has been in continuous existence from 1989, and although a small organization it has issued an impressive series of jeton catalogs and otherwise well serves collectors.[62] Distinct from the more scholarly American Numismatic Society in New York, the American Numismatic Association of Colorado Springs, Colorado, is a general organization for collectors, hobbyists, and enthusiasts founded in 1891. The ancillary businesses and activities that support collecting and the hobby of collecting cannot be overlooked: coin shows, consignment houses, storage and grading services, materials for collections (albums, cases, labels, scales, calipers, etc.), insurance, reference publications, and how-to books that form another part of the world of numismatics and jeton collecting today.

Historical Studies Involving Jetons

Already in 1888, as previously mentioned, the learned Maurice Prou from the Cabinet des Médailles, future professor and director at the École des Chartes in Paris, remarked that "the study of jetons is interesting only insofar as it prompts historical study and if it provides the means of elucidating points of debate."[63] Adrien Blanchet agreed in 1892 that "history and numismatics lend each other mutual aid, and that's why I placed a short historical introduction, as brief as possible, at the head of all my chapters."[64] J. P. Barnard in 1916 saw "medallic jetons" as a rich source for history.[65] A number of efforts over the last 150 years have put advice into practice and have used jetons as sources in historical research. Madeleine Desavis' 1933 dissertation on the jetons of the medical faculty of the University of Paris is an example, and she, too, is explicit that "numismatics is more than simply a subject for distraction; it is a source for history."[66]

To date, jetons have been used primarily in writing local or regional histories. Joseph Roy-Chevrier's 1919 book on the Chalon-sur-Saône region as seen in its jetons, medals, seals and other tokens is a work of 650 pages and 27 plates.[67] Several studies like this using jetons have been about French provincial estates. A pioneering one (and one of the best) is Claude Rossignol's 1851 piece on the liberties of Burgundy as seen in the jetons of the Estates of Burgundy.[68] Certainly for the time (but not without a grain of partisanship), Rossignol succeeds in using jetons to recount the

61. Edited by Ted Banning, in 1989 this group published at least one number of *The Exchequer*, "an occasional newsletter containing both introductory and more advanced articles on the jetons of Venice, Nuremberg, France, Flanders and other places," (donum/numismatics.org/cgi-bin/kpha/opac-search.pl?q=the+exchecquer).

62. See http://www.acjm.fr/; accessed September 23, 2018.

63. Prou (1888), p. 348.

64. "L'histoire et la numismatique se prêtent un mutuel appui. C'est pourquoi on a placé en tête de presque tous les chapitres une petite introduction historique, aussi brève que possible," Blanchet (1890), vol. 1, p. viii; de Witte (1893), p. 3, made the same point at the same time, writing, "...des connaissances numismatiques [sont], hélas!, trop peu appréciées et dont l'ensemble constitue, cependant, la plus utile, la plus sérieuse, la plus vaste des sciences auxiliaires de l'histoire."

65. Barnard (1916), pp. 39–42.

66. Desavis (1933), p. 10, "...la Numismatique est plus qu'un sujet de rêverie : elle est une source de l'Histoire," see also pp. 169-171. Still, hers is more a simple catalog of the deans of the Faculté de Médecine than an analytical history. For example, the work is intended as a visitor's guide to the medical history exhibitions at the medical school and for collectors and amateurs of medical numismatics, in addition to bringing to life the history of the Paris medical faculty. Grierson (1951), p. 13, in a like vein points to historical information embedded in numismatic literature and the value of numismatics in general as a rich source of historical information and insight.

67. Roy-Chevier (1919); Joseph Roman's 1911 study focused on the jetons of the Dauphiné is an earlier example.

68. Rossignol, passim and p. 42, where he notes jetons as "digests of local history" ("*abrégés d'histoire locale*").

history of the Burgundy estates and France from 1580 through the French Revolution.[69] Émile Bonnet and Gabriel Amardel in a like vein published on jetons and the Estates of Languedoc in 1899 and 1901 respectively. Much more recent works by Georges Depeyrot (1980 and 2007) have brought these studies up to date by focusing on the Estates of Languedoc and their jetons.[70] Joseph Daniel's 1980 survey of the jetons of the Estates of Brittany fits in this category.[71]

The same can be said of towns. Born in Lyon and curator at the Musée des Beaux-Arts there, Jean Tricou (1890–1977) published extensively from the 1950s through the 1970s on Old Regime Lyon as illustrated in and by its jetons.[72] Earlier in 1907–1909, Léon Charvet published a 360-page tome on the medals and jetons of the city of Lyon.[73] Like Tricou and Charvet, Alexis de Évrard de Fayolle earlier in the twentieth century similarly depicted his town, Bordeaux, in a number of articles that used civic jetons and medals as their sources.[74] Even earlier in the nineteenth century one Charles-Aimé Dauban used local jetons for his histories of the mayors of Tours and Nantes, and contemporaneously, Joseph Napoléon Loir captured the history of the town of Mantes using "jetons historiques."[75] In much more recent times Professor Judi Loach's 1995 examination of the consular jetons in seventeenth-century Lyon is a substantial and high-quality historical investigation likewise oriented around one urban setting.

The book, *Les Assemblées du Clergé de France avant 1789 et leurs jetons commémoratifs* (The general assemblies of the clergy through 1789 and their commemorative jetons) written by Charles Florange and Stéphane Strowski and published in 1927 is another successful example of using jetons to write solid history.[76] George Berry's short piece of 1977 on Louis XIII is another, to which might be added McClellan and Regourd (2011) on the "colonial machine" of Old Regime France. Beyond that, a variety of miscellaneous, specialized studies have tapped jetons as a source for historical research and writing.[77] Jules Florange, for example, published on French mines and mining, evoking jetons to identify tools and techniques employed in mines.[78] Love and marriage, the royal passion for gambling, notaries, archery, sanitation, and historical images of horses and asses have variously been the focus of yet other works that have used jetons and medals as sources.[79]

69. Yet, Rossignol characterizes himself as a "simple antiquarian," and he gives over 50 pages to a list of the jetons he discusses in his text; see Rossignol (1851), p. 203 and chapter xxiv, pp. 254–301. Yet, already in 1899 Bonnet in Depeyrot (2007), p. 131, criticized Rossignol and his study of the jetons of the Estates of Burdundy, saying one cannot trace the history of a province (solely) through its jetons, although they can add to our understanding of its overall history.

70. Depeyrot (2007), p. 11. In his bibliography Depeyrot lists twelve sources as specialized studies of jetons and the Estates of Languedoc. See also Vian (1949), p. 36, for a related emphasis on local history of the province.

71. Daniel (1980). But Daniel, e.g., p. 75, is still oriented to the *amateur* and the *collectionneur*. On nineteenth-century studies/compilations of jetons in various regions in France, see also La Tour (1897), p. vii.

72. See bibliography. On Tricou, see Lafaurie (1976) and also Loach (1995), p. 24.

73. Charvet (1907–1909), p. 269, where he labels his "*une partie, non sans intérêt, de l'histoire de la ville de Lyon.*" Charvet's work is particularly oriented toward the aesthetics of medals and jetons and the history of art.

74. See bibliography; Évrard de Fayolle (1902), p. 89 is explicit about serving local history.

75. See Dauban (1858) and (1859); Loir (1859). Vallier (1879) on the Dauphiné Chambre des Comptes fits in this constellation, as he explicitly sets his study, p. 5, in the context of regional, provincial history.

76. Florange and Strowski document these important assemblies, but theirs is a history that uses jetons more as illustrations accompanying the text than as sources of that history.

77. On collectors specializing in local, regional or thematic jetons, see also Pastoureau (1984), p. 27.

78. J. Florange (1904), especially p. 6.

79. See Rostron (1997); Goldfiem (1976); J. Florange (1899) and (1936); Storer (1887). Veillon (2005) and Gombert (1974) on Hercules fits this category. Jacquiot (1989) provides a related example of a specialized study of jetons and medals themselves, in this case of sayings of Virgil used as legends on these pieces. For a failed effort, see Pédery-Hunt on the history of Canada as seen through jetons and like tokens. It is noteworthy that, regarding Lerouge and Lerouge (1977) on French notaries, the authors (p. 23) enumerate their somewhat less than strictly historical goals: to catalog the jetons issued by notarial bodies, to provide a practical guide and resource for numismatists and collectors, and to increase interest of notaries in these "petits monuments traditionnels."

We will take up all of these jetons and their issuing institutions in due course, but for now, while researchers have indeed used jetons to pursue historical projects, overall efforts have centered on relatively narrow topics, some still with a numismatic tinge and some verging on the antiquarian. Although similar work continues to appear—some of it quite good—much of this literature comes from a bygone era. In a word, jetons have been well studied by numismatists and have on occasion been tapped by historians in more focused projects, but historians seem not to have addressed jetons as a general historical phenomenon or asked what they in their totality might have to say about contemporary French society. The architectural historian Judi Loach grasped this possibility in her abovementioned 1995 study of seventeenth-century jetons of Lyon when she spoke about "the larger rereading of seventeenth-century France as a whole, which is surely imposed on us once we accede to the messages which were conveyed by the reverses at the time of their original conception and reception."[80]

The Current State of Play

In work extending from the 1960s to the 1990s Josèphe Jacquiot (1910–1995) effected a sea change in the historiography concerning jetons and medals. Herself a curator at the Cabinet des Médailles of the Bibliothèque nationale de France and professor at the École du Louvre, she is mainly known for her monumental four-volume presentation of the medals and jetons of Louis XIV that appeared in 1968–1970, but her life and work extend beyond that.[81] According to Anne Mainguy, before Jacquiot "a thick and almost impregnable wall seemed to separate numismatics from every other more general domain, such as history, that tried to approach it." And with Jacquiot, Mainguy continues, "Jetons acquired their letters of nobility as historical sources in and of themselves."[82] In like terms, Michel Pastoureau's own 1984 volume on medieval jetons, medals, and church tokens is another landmark in the relatively more recent historically-oriented literature. The three substantial volumes on "jetons, medalets & tokens" by the wide-ranging numismatist, Michael Mitchiner, published in the decade 1988–1998 need to be mentioned here, especially volume two which is devoted to the jetons of France and the Low Countries.[83] Written in English, Mitchiner's opus of 2,160 pages is not especially a work of synthesis; it is largely a catalog (and a richly illustrated one), and it possesses something of the character of being a modern Feuardent. Still, Mitchiner's is a scholarly work of the highest order and a gold mine of historical and other information regarding jetons, including, for example, details concerning the metallurgy and chemistry of jeton alloys.[84] More than any other source, including this one, Mitchiner's presents the richest history of jetons themselves.

In the meantime, since the present work first germinated a new generation of scholars interested in medals and jetons has arisen, and a succession of talented French numismatists have continued to turn towards historical sources and historical perspectives in pursuing their work. Numismatics practiced at its highest levels today rejects amateurism, dilettantism, antiquarianism, and an abstruse erudition too often associated with the field, and the trend has indeed been

80. Loach (1995), p. 52.
81. See bibliography and Josèphe Jacquiot, "Vie publique et intellectuelle," Société d'Histoire Locale de Montgeron, http://shlm.sharepoint.com/Pages/JosepheJacquiot.aspx (consulted in 2016). The town of Montgeron, 12 miles from Paris, houses the Musée Municipal Josèphe Jacquiot.
82. "...un mur épais et quasi infranchissable semblait séparer la numismatique de tout autre domaine plus généraliste —telle que l'histoire—qui tente de l'approcher...[Avec Jacquiot] le jeton acquiert ses lettres de noblesse en tant que source historique à part entière," Mainguy (1999), p. 8.
83. See bibliography and especially Mitchiner, vol. 2 (1991), section six, pp. 978–1356.
84. See especially Mitchiner, vol. 2 (1991), pp. 727–54.

towards reliance on more critical, professional and "scientific" standards that are inclusive of the wider world of history and its methods.[85] Anne Mainguy's 1999 dissertation on the jetons of various departments of royal government and administration from 1695 to 1758 evidences this shift toward the historical and away from numismatics more narrowly conceived. But Mainguy concentrates exclusively on jetons of official government units, her chronological scope is restricted, and she puts a heavy emphasis on evaluating jetons as art objects.[86] In this connection historians and numismatists must salute Thierry Sarmant and François Ploton-Nicollet for their exquisite two-volume catalog, *Jetons des institutions centrales de l'Ancien Régime* that appeared in 2010 and 2014. The first volume logs 825 representative jetons held in the collections of the Cabinet des Médailles of the Bibliothèque Nationale de France; the second adds an additional 1,487. Although limited to administrative jetons, this work is handsomely illustrated, as mentioned, and reproduces key archival documents. Sarmant's general historical introduction to jetons is especially to be praised.[87] More recent work still by Inès Villela-Petit, Jérôme Jambu (the two now with Département des Monnaies, Médailles et Antiques of the Bibliothèque Nationale), Marion Delcamp, Sabrina Valin, Arnaud Clairand, Ludovic Jouvet and other recent PhDs, graduate students, and up-and-coming commercial numismatists testify to the vitality of current scholarly work in numismatics in general and on jetons in particular and to a methodological embrace of history.[88] Marion Delcamp's 2015 study of female mint workers, for example, is a landmark in this regard, as is Sabrina Valin's monumental 2016 dissertation enumerating some 400 royal jetons struck from 1610 through 1661. Marion Delcamp and Jérôme Jambu's focused, but fascinating 2018 article on colonial trade and the mint workers of Bordeaux is likewise exemplary. And then, among this new generation, Robert Wellington, Senior Lecturer in the Centre for Art History and Art Theory at the Australian National University, is exceptional in writing in English, and he is to be especially lauded for his 2015 volume, *Antiquarianism and the Visual Histories of Louis XIV: Artefacts for a Future Past*, a very model for writing history using numismatics.[89]

By and large, however, numismatics remains its own, separate domain, "à part entière," as the French say. As much as the discipline has tapped larger historical sources and socio-economic and political perspectives in recent decades, its thrust for the most part remains self-referential and directed to synchronic case studies, to the formal description of particular pieces, to the preparation of detailed catalogs, and, relatedly, to specialized art historical interpretation. In this, numismatics—as rich and wonderful as it is—presents itself as an auxiliary to history, a puissant auxiliary to be sure, but one that seems largely passed over by ordinary historians. Regardless, an account of Old Regime France as told through the jetons of the era has yet to be attempted. Certainly nothing like it exists in English or premised on the historiographical divide separating history and numismatics. The archival record and secondary literature are rich and allow a close-up, diachronic study. Although the effort may be foolhardy, there seems to be a place for a book like the present one.

85. On this point see especially Jambu (2015), passim and pp. 864–65.

86. Mainguy (1999) is explicit about this fine art orientation, pp. 14, 16 and per her part III (pp. 183–301).

87. Sarmant and Ploton-Nicollet (2010, 2014), passim and especially vol. 1, pp. 11–57. Two further volumes of this catalog of the collections of the jetons of the Département des Monnaies are projected for a total of 6,080 jetons described and cataloged. Thierry Sarmant's books on the history of the Cabinet des Médailles of the Bibliothèque Nationale (1994) and on numismatists and numismatic collections in the seventeenth and eighteenth centuries (2003) are notable contributions to the historical literature under review here.

88. See bibliography.

89. See review by Jouvet (2016).

The Contrast with Antiquity

Another factor indirectly framing this study concerns the fact that numismatics is firmly integrated into the historiography of the ancient world and has long been a valuable resource for historians of antiquity.[90] Numismatic expertise can be found in university Classics departments, and numismatics has further connections to archaeology and ancient epigraphy. This circumstance contrasts remarkably with the relative segregation of the disciplines of history and numismatics as practiced by specialists concerned with early modern France, as we have just seen. The reason for this discrepancy in how history and numismatics are connected (or not) in these contrasting eras is not hard to identify: Historians and other students of antiquity have fewer sources, so in establishing chronologies, for example, coins and coinage can speak loudly; ancient hoards, to give another example, usually say a lot. On the other hand, as mentioned at the outset, historians of early modern France have had an abundance of sources to work with, wherein coinage, while not irrelevant in principle, inevitably carries less weight and has drawn less attention in historical research; modern hoards say less. The historian and numismatist Philip Grierson recognized that historical interest in numismatic sources varies inversely as the volume of documentation from other sources.[91] Thierry Sarmant said so in 2003 when he wrote, "The importance of numismatics decreases according to the period studied; essential for antiquity, useful for the Middle Ages, it is ignored by modernists, given competition with the ever greater triumph of written sources."[92] And here lies the key for understanding the division between historians and numismatists regarding France in the Old Regime. Certainly, Old Regime historians have had plenty to work with for quite a long time and with increasing sophistication, and whatever inattention there has been towards these otherwise compelling little jetons becomes more understandable. This sharp distinction between how numismatics is treated by ancient and by early modern historians is not merely an historiographical observation. Rather, the integration of numismatics into ancient history points to what can be done for early modern history using jetons in particular.[93]

Jetons and Cultural History

Already in 1873 one Alphonse de Schodt noted that "jetons plainly can be a powerful aid in the study of history and archeology," and who would disagree?[94] But the question becomes how are we to enlist that aid and use jetons to frame any history, much less a full-fledged institutional or cultural history of the Old Regime. In his short piece of 1951, *Numismatics and History*, after which the present study is entitled, Philip Grierson (1910–2006) discussed some of the ways in which numismatics might in fact serve history. Grierson was a British don—fellow of Gonville and Caius College, Cambridge and professor of numismatics at Cambridge and at Brussels. Trained in medieval history and turning to numismatics thereafter, at various points Grierson was director of the Royal Historical Society and president of the Royal Numismatic Society. His volumes on medieval

90. Murray McClellan, private communication. Every ancient historian I have spoken to confirms this point. Fontenay (1847), pp. 3–4, already mentions the much more developed study of numismatics in antiquity and the uses of numismatics for understanding ancient history; see also Grierson (1951), pp. 5–7; Meadows and Shipton (2001), p. v and passim; Mörkholm (1979–1980).

91. Grierson (1951), p. 8; Mainguy (1999), p. 9, makes something like this point.

92. "L'importance de la numismatique décroît pourtant à travers les âges éudiés : essentielle pour l'Antiquité, utile pour le Moyen Âge, elle est ignorée par les modernistes, car concurrence de plus en plus victorieusement par les sources écrites," Sarmant (2003), p. 368.

93. Suffice it to say that the literature on ancient coinage is vast; see, as a least example, the nine volumes of *The Roman Imperial Coinage* by Mattingly et al. (1927–1981); Balmuth (2001) on hordes; almost any issue of the *American Journal of Numismatics* likewise exemplifies this point.

94. "Les jetons sont évidemment d'un secours puissant pour l'étude de l'histoire et de l'archéologie," Schodt (1873), p. 4.

and Byzantine coinage are standard tomes. Grierson was thoughtful and knowledgeable—someone to pay attention to on the subject of numismatics and history. The published piece in question is the English translation of his inaugural lecture in French in 1948 as professor of numismatics and the history of coinage at the University of Brussels. Grierson is well aware of the disciplinary divide separating numismatics and history, and he privileges history over numismatics, adding, "Furthermore, it is easier for the historian to acquire the elements of numismatics than it is for the numismatist to acquire the outlook of the historian."[95] Nevertheless, Grierson argues for numismatics as overlooked in the historians' toolkit and suggests what's to be gained. Thus, per Grierson, coins are primary sources and entrées into society and culture, and they offer opportunities to fill in gaps in our knowledge. In particular, European coins are connected to and document the rise of the modern nation-state. Numismatics is relevant for the historian, especially economic and political historians, but historians of art as well.[96] Grierson chides historians for neglecting numismatics. He blames this neglect on historians for an overreliance on written records, as noted, but also on numismatists, quoting Marc Bloch to say, "Numismatics has trouble leaving cabinets of curiosities."[97]

Notably for Grierson, the historian needs to come to grips with numismatic evidence as a whole, in its totality, not just some piece of it.[98] As an historian writing on the jetons of Old Regime France, I hope I have followed Grierson's sage advice. Part of that totality of numismatic evidence concerns the actual production of jetons and the technology involved in their manufacture. The numismatic literature on jetons has generally overlooked this side of the story, and a later chapter is given over to examining the production process as an element of the larger tale of jetons and the society that produced them. Even adopting Grierson's larger perspectives, however, the question remains: How best to translate these intentions into practice in writing a history of Old Regime France and its jetons.

The answer comes from the domains of cultural anthropology and cultural history and historiographically from what we get in combining microhistories with *longue durée* history. Regarding the former, we need only wave at the transformative work of the anthropologist Clifford Geertz (1926-2006) and his notion of an emic, "thick" description of the dynamics of a culture.[99] More than that, Princeton historians Natalie Z. Davis and Robert Darnton led the way in applying Geertz's approach to cultural *history* and to evocations of Old Regime France, Davis notably with her 1983 book, *The Return of Martin Guerre* and Darnton with his *The Great Cat Massacre and Other Episodes in French Cultural History* of 1984. (Geertz, Davis, and Darton were professorial colleagues at the Institute for Advanced Study and at Princeton University respectively.) The present work hopes to follow Geertz, Davis, and Darnton in seeing jetons as mirrors of a larger society, woven into contemporary culture, and capable of teasing out resonances with the larger culture and lived experience of those whose world was littered with jetons. In an anthropological spirit, this book presents and decodes many little vignettes in hope of insight. To give another example, hundreds of gold and silver jetons of the Estates of Burgundy played a part in ceremonies surrounding the triennial presentation of the provincial *cahiers de doléances* to the king, and it is

95. Grierson (1951), p. 13.
96. Grierson (1951), pp. 8–11.
97. "...la numismatique a peine à sortir des cabinets de curiosités." The short entry "Numismatique et histoire" that appears in Amandry's *Dictionnaire de numismatique* of 2001, pp. 412–13, doesn't go beyond Grierson, i.e., that coins need to be treated as "un vestige matériel au service de l'histoire" and for their use by historians for messages decoded for understanding politics, the economy, and sociology.
98. Grierson (1951), p. 9.
99. See notably Geertz (1973). For a textbook introduction to cultural anthropology, see Schultz and Lavenda (2013).

hard to imagine a more intimate view of the character and functioning of monarchical France and contemporary French society.

Over the last decades any number of wonderful microhistories and histories of mentalities have come forth. Emmanuel Le Roy Ladurie's *Montailleu, village occitan de 1294 à 1324* (1975, appearing in English in 1978 as *Montaillou: The Promised Land of Error*), Carlo Ginzburg's *Il formaggio e i vermi. Il cosmo di un mugnaio del '500* (1976; in English as *The Cheese and the Worms*, 1980), Davis' *Martin Guerre*, and Darnton's study of eighteenth-century print shop workers are all classics in the genre. But microhistories have come to be criticized as merely evocative, of limited scope and vision, and representing a kind of neo-antiquarianism.[100] And so, today, the enterprise of historians needs to go beyond microhistories as snapshots for their own sake and to be more chronologically encompassing and engaging of more far-reaching change and historical dynamics.

The present book is decidedly not a microhistory, unless considered as a microhistory covering 600 years. Rather, the ambition here is to combine microhistory with macrohistory and Fernand Braudel's *longue durée*—combining the vignettes and detailed insight of sharply focused instances that jetons provide with the larger course of historical change extending over long periods. In this case we face the expanse of the 600-year period, 1200–1800. Indeed, the jetons under study here first appeared in the late twelfth century, and they have a continuous history that spills into the nineteenth century and down to today. Therefore, jetons offer the rich possibility of combining microhistories with the larger sweep of French history from the Middle Ages through the French Revolution. The author's effort here is thus as a pointillist painter, and the result we might call pointillist history, seeking to depict the whole of French society and culture as it developed over the centuries using jetons as microdots to do so.[101] The image is an illusion, of course, but perhaps a not uninteresting one.

Jetons indeed do hold a mirror to their times. They were woven into the fabric of contemporary society, and they reveal telling instances of social practice. The chapters that follow lead us through the details of the story of jetons as made objects in institutional and cultural contexts and their use in practice. Their materiality is a primary starting point for all that follows and distinguishes the present work a bit further. The aim is to grasp something of the ineffable, but no less real and distinctive character of life in those times. We venture some general conclusions at the end, but there are no surprises here. Like Blake's grains of sand, what jetons provide is a window, a narrow one to be sure, but one that looks out on a wide world we already know well: The dynamic, complex, and in many ways stunning, often astounding world of early modern and Old Regime France.

---

100. On this point and a defense of microhistory, see Cohen (2004), pp. 75, 202, 224–27, 241, 282; see also his "The Larger Uses of Microhistory," www.microhistory.eu/index.php/2017/02/10/the-larger-uses-of-microhistory-thomas-cohen/; accessed September 23, 2018.

101. See further James E. McClellan III, "Pointillist History," https://www.youtube.com/watch?v=P5Z8mQM9qRc; accessed May 14, 2019

FIGURE 1.1: Medieval school token (*jeton d'école*).

1.

An Historical Overview of Early Jetons

Jetons are numismatically distinct and historically distinctive. Their origins date to the Middle Ages, and with a special efflorescence in the seventeenth and eighteenth centuries, they continue on to today in various numismatic guises.[1] Jetons form a particular class of tokens to be distinguished from other kinds of tokens, coins and medals. For one thing, the jetons under consideration here had no formal monetary value and did not circulate as money. For all kinds of reasons jetons deserve our attention, but at the outset they need to be differentiated more precisely from these other categories of struck coinage, notably coins with monetary value, historical medals, and church or mass tokens known as *méreaux*.

Money, Medals, and Méreaux

The right to mint money is a powerful privilege, one claimed from the early Middle Ages by all French kings.[2] Charlemagne issued an edict concerning minting in 804, and the establishment of the venerable royal mint in Rouen in 864 also indicates early efforts at royal control over minting money.[3] The contemporary reality of producing coinage was a different matter, however. Major bishoprics stamped their own currency, for example, and at other times, particularly at periods of political instability and with or without royal sanction, feudal lords turned out their own specie, too. In the early stages royal control was most often leased out to what became strictly family businesses, but centralized oversight strengthened in 1216 with the creation of royal superintendents of the mints (*généraux-maîtres des monnoies*). Thereafter, a hodgepodge of legislation governing minting operations amalgamated diverse regions. Centralization of royal control was well underway in the seventeenth century, and by the eighteenth century the crown had established royal minting operations in fully seventeen principal cities in the kingdom.[4] These mints

1. La Tour (1897), somewhat against the grain of other commentators, thinks the sixteenth century is the golden age of jetons, but admits just the same, p. xxvi, "Le XVIIIe siècle est une époque privilégiée pour le jeton."
2. At least from the Carolingians onward. On these points, see Darnis (1996/1999), pp. 639-45; Barnard (1916), p. 59. On coin production during the Merovingian period, see Grierson (1991), pp. 19-21 and Grierson and Blackburn (2006), pp. 95-96. For further background on medieval coins, jetons, *méreaux*, and accounting see Labrot (1989).
3. Darnis (1996/1999), p. 14.
4. Darnis (1996/1999), p. 16 and list of royal mints, n. 31; see also pp. 61-84 and vol. 2, pp. 647-55 for case study of "Les divisions financières de la France sous l'Ancien Régime." For a survey of the various French mints and their coming into being, see also Droulers (1989) who discusses fourteen separate royal mints in France. Spooner (1954) provides a comprehensive history of French mint activity in the sixteenth and seventeenth centuries. Stahl (2000) offers a point of comparison with the medieval mint in Venice, the Zecca.

FIGURE 1.2: An *Écu*, 1790.

reflected the town's and region's economic importance and were organized administratively subservient to royal intendancies (the intendant) which reported back to the crown.[5] The silver *écu*—fixed at six *livres* (monetary "pounds") in 1726—depicted here (Fig. 1.2) is an entirely typical product of these currency mints.[6] Jetons, by contrast were more finely crafted than circulating coinage, and whereas French currency from roughly 1500 to 1800 was more or less standardized, jetons were more varied in their design.[7] Close-up studies of French mints, particularly by Jambu on the provincial mint at Caen and Blanchet on Bayonne, but other ones concerning La Rochelle, Nancy, and Paris show the pressing need to produce small coinage to ease economic exchanges, and they reveal these mints to be complicated technological operations that continued unabated over centuries.[8] They required lots of equipment such as forges, mills, rollers, counters, weights, balances, and other tools of various sorts.[9] They entailed large staffs, including masters, royal agents, guards, controllers, engravers, clerks, assayers, and workers.[10] They became seats of advanced technology and technological practice. And, certain limited exceptions aside, none of them struck jetons. A separate and very special mint in Paris, the Monnaie des Médailles, did that as will be shown below.

It turns out that the French monarchy was not the only authority to mint money. From the Middle Ages powerful counts held that right.[11] Given their role in organizing and governing towns and regions after the collapse of Roman government, local Catholic bishops did as well, perhaps extending as far back as the Merovingians. The bishop of Metz, for example, independently had money struck at least from 963, and the Metz bishopric continued to do so into the seventeenth century even as Lorraine fell fully under the sway of the French monarchy.[12] The right of bishops and authorities in outlying principalities to strike money was taken away as provinces gradually became incorporated into France proper. Coincident with the ascendancy of the Bourbon monarchs, bishop of Metz from 1612 to 1652, Henri de Verneuil, was the last bishop to strike his own

5. On French minting in general, see previous note and Darnis (1996/1999), pp. 11–33; Belaubre and Collin (1992); Rachline (1992) on the Paris mint.
6. Royal French gold and silver *écu* date back to the thirteenth century; the major monetary reform of 1640-1641 standardized the gold French *louis d'or* and the silver *écu*; on these points see Droulers (1989), (1991), and Glossary.
7. La Tour (1897), pp. xxix–xxx.
8. Jambu (2010) and (2013b); Blanchet (1920c), especially pp. 200–01, describing rooms and equipment; see also Tricou (1973), p. 380 for the Lyon mint and the organization of minting in other towns in the kingdom; and Blanchet (1924) for Nancy. A study of the Rouen mint seems in order. Sargent and Velde (2002) explore the "big problem of small change" from the point of view of general economic theory.
9. Jambu (2010), pp. 517–20, provides relevant details concerning the mint at Caen.
10. Darnis (1996/1999), p. 17, notes that in 1789 the Paris mint employed a staff of 64 actual minters, not counting other employees.
11. D. Barthélemy (2012), p. 44.
12. Robert (1890), pp. 235–37.

FIGURE 1.3: Medal: Treaty of Ryswick, 1697.

money in France. The dukes of Lorraine struck coinage until the final assimilation of that province into France in 1766.

Jetons and historical medals are both exonumia or paranumismatica, that is, numismatic items that are not government-issued coins or money. Nonetheless, historical medals are a category of minted objects that need to be carefully distinguished from jetons. Numismatically and historiographically, medals constitute a separate subject beyond the scope of the present focus on jetons, but a few words of comparison are in order.[13] As a quaint archival reference expressed it in 1744, "Medals are a particular kind of coinage that the king has struck to commemorate events and perpetuate their memory."[14] The history of medals extends back to early antiquity. The practice of fashioning medals strictly to commemorate events arose in Europe ca. 1400, and historical medals became a defining feature of princely courts and the humanist movement of the Renaissance. As the culture of historical medals spread northward from Italy, French kings began to produce medals commemorating landmark events in the early 1500s. While notable and often of high artistic and technical caliber, this production remained relatively sporadic, but with the Bourbons again, particularly Louis XIV, the fabrication of historical medals increased dramatically. One contemporary list enumerates 283 different royal medals struck between 1644 and 1700.[15] Another source reports more than 550 medals issued in the personal reign of Louis XIV (1661–1715).[16] The tradition continued under Louis XV and Louis XVI, but to a lesser extent.[17] The historical medals of Old Regime France are remarkable artistically and numismatically (Fig. 1.3), even if the golden age of art medals and commemorative medals lay ahead in more modern times.

13. On historical medals, see works by Blanchet (1930), pp. 1–201; Guiffry (various 1884–1888); and especially Jacquiot (1968–1970), Jones (1979a), de Turckheim-Pey (2004), and Sarmant (2003). See also Valin (2016), vol. 1, p. 7; Dillemann (1997), pp. 6–7; Pastoureau (1984), pp. 40–43; and Rachline (1992), pp. 63–76 for a nice treatment.

14. "La Medaille est une monnoye particulière que le Roy fait frapper, pour marquer les evenemens et en perpetuer la connoissance," AN O[1] 2064, "1744, Memoire concernant les Medailles."

15. AN, O[1] 2064, Ms. "Inscriptions des Médailles de l'histoire du Roy rangées par ordre alphabétique."

16. Rachine (1992), p. 64; see also Burke (1992), p. 207; J.-J. Guiffrey (1884), p. 478, and de Turckheim-Pey (2004), pp. 21, 27.

17. Rachine (1992), p. 64 who notes only 150 medals under Louis XV; see also p. 66.

Medals and jetons enjoy complementary histories. At least from the mid-1500s French medals, like all jetons, were minted and issued through the same bureaucratic processes and from the same mint, the Monnaie des Médailles in Paris mentioned above. Medals were cast through the early part of the seventeenth century, but thereafter they were struck in coin presses in elaborate production processes and with an exceptional degree of skill and artistry. Historical medals and jetons carried no monetary value, and, although a subsidiary market existed for collectors, neither were technically available for sale.

But the differences between medals and jetons are categorical.[18] Medals always commemorated something to do with the reign of the king and notable events in the kingdom. Jetons were not commemorative and connected to historical events only indirectly as per their issuing organizations or, at certain moments, obliquely referring to an event in the previous year, but jetons never served directly to commemorate an event. Then too, medals were one-offs, as it were; that is, although the technology was in place to create many different medals, each medal issue was unique, while whole series of jetons, for the Royal Navy, for example, or the King's Council, continued for decades and decades as a series. Particularly for the more elaborate and exquisite medals, one suspects that medals were struck in much smaller numbers, but it is impossible to be specific about the runs.[19] Jetons, by comparison, were minted by the thousands. The Bourbon medals were always much larger than jetons, twice the diameter or more (40–60 mm) of the typical jeton (28 mm) and on very thick bronze or silver planchets that were two, three or four times the thickness of the disk used in striking jetons. Medals were always cut in high relief, while jetons were always stamped in low relief. Medals were distributed as gifts by the king and on his behalf, whereas the social functions of jetons were more complex.

The broader context for considering medals involves the notion of what is known as medallic history, *histoire métallique*, or documenting and narrating history via a collection of medals, what Robert Wellington has labelled a "physical encyclopedia."[20] It was not crazy for early modern amateurs and savants to focus on medals as a medium for recording and retelling history because medals are enduring in ways that other sources are not. Metallic history became fashionable in the reigns of Louis XIV and Louis XV to the point where Thierry Sarmant refers to the "science of medals" and a contemporary Republic of Metals centered on Paris.[21] Louis XIV was, of course, responsible for producing many medals and was a genius at exploiting the propaganda and narrative potential involved, but he himself also was an ardent student and promoter of metallic history.[22]

18. Burke (1992), p. 206.
19. Daniel Renaud, private communication.
20. Wellington (2015b), p. 20. On contemporary medallic history, see further Wellington (2015a), pp. 44–51; Wellington (2016), p. 289; Sarmant (1994); Sarmant (2003); de Turckheim-Pey (2004); Veillon (1994), pp. 23–26. The tradition of medallic history continued into the twentieth century. The Le Maistre collection of history medals and jetons commemorating peace treaties, sold in 1912, shows what can be done with medallic history. Regarding the 1697 Treaty of Ryswick, illustrated above, that ended the nine-year War of the League of Augsburg against the French, Schulman (1912), pp. 85–96, inventories 47 different medals and jetons that give a rounded picture of this important treaty. Jean Babelon and Jacquiot's 1951 presentation of the history of Paris via its medals is another excellent example, as is Jacquiot et al. (1970) on the medals of Louis XIV. The history of medicine via its medals is a notable subset of metallic history as evidenced by Faludi (1929) and Freeman (1946) and (1964).
21. "République des médailles," Sarmant (2003), p. 363; see also Veillon (2008), especially chapter 3. Collections for *histoires métalliques* continued to be built through the eighteenth century; see Nocq (1910); AI, AIBL archives, A:68, minutes of the meetings of August 22 and September 4, 1786, relative to a medal to commemorate the king's visit to Cherbourg on June 11, 1786.
22. On the general point see especially Burke (1992); see also Sarmant (2003), pp. 97–99;113ff; (1999), pp. 83–88; Mainguy (1999), p. 69; J.-J. Guiffrey (1884), p. 478. On Louis XIV as a collector, see also Wellington (2015a), ch. 3 and

Bureaucratically, the royal Cabinet des Médailles draws attention here as the official repository of the king's jetons and historical medals.[23] The origins of this institution go back to the reign of Charles IX (1560–1574) and the creation of a *Garde particulier des médailles et antiques du roi* (Special guardian of the king's medals and antiquities). The Cabinet des Médailles existed as a department within the royal library (the Bibliothèque du Roi) along with other departments (manuscripts, imprints, genealogies, and engravings). All historical medals (and jetons formally struck on royal account) were required to be deposited at the royal Cabinet des Médailles, which independently held a collection of over 9,000 ancient medals in gold, silver, and bronze.[24] The Cabinet kept records and employed researchers to catalog and explicate all royal medals.[25] Louis XIV moved the Cabinet des Médailles to Versailles, but it returned to Paris under Louis XV and new quarters on the rue Richelieu where it continues today as the Département des Monnaies, Médailles et Antiques of the Bibliothèque Nationale de France (BnF). In the 1780s one Abbé Barthelemy directed the Cabinet des Médailles for which he received a handsome 5,000 *livres* a year and another 2,000 for a housing subsidy; the Cabinet employed an assistant and two clerks for a total of 9,000 in salaries, not including housing or gratifications.[26] The Cabinet des Médailles purchased elaborate and expensive velvet-lined cases (*médalliers*) in which to keep medals and jetons, and the artisanship involved in making these cases is to be noted.[27] The Cabinet bought medals from time to time, and on one occasion at least sold duplicates out of its collection.[28]

Finally in this connection we need to differentiate jetons from another type of contemporary token, church tokens or *méreaux*.[29] Fontenay (1854) defines them as "cathedral chapter tokens and the common money of men of the church."[30] *Méreaux* (L. *merellus, i*) may be related to Roman counters and gaming pieces, such as those used in the ancient board game Nine Men's Morris (*merelles* in French), but the etymology is not solidly established. The distinction was not always drawn between jetons and *méreaux* in the Middle Ages, but soon the ecclesiastical context for *méreaux* as distinct from jetons established itself.[31] *Méreaux* were small (20–25 mm) hammered coins, using common metals such as lead or tin; they were imprinted on thin planchets with only elementary designs. Minted locally and issued by local cathedral chapters, *méreaux* formed part of

especially fig. 1, and (2015b), pp. 12–13.

23. On the Cabinet des Médailles, see especially Sarmant (1994); Wellington (2015a), pp. 79–97; Wellington (2015b); Valin (2016), vol. 1, pp. 3–4; Veillon (2008), pp. 49–51. For its later history, see Villela-Petit (2012a).
24. Nocq (1908), pp. 426–27. Wellington (2015b), p. 12, puts the total number of coins and medals held in the Cabinet des Médailles at 27,000.
25. See Wellington (2015b), p. 19, and folder labeled "Cabinet des Medailles...de 1685 à 1690," AN O[1] 2064.
26. On these points see AN, O[1] 609, #390–433 for accounting details of the Bibliothèque du Roi.
27. See previous note. In 1785 the Cabinet spent 400 *livres* for velvet alone. For an example, see the royal *médailler*, circa 1750; Musée des Arts Décoratifs, Paris, Legs Bernier, 1920, inv. 21986.
28. See sale of 1690 noted in Guiffrey (1881–1901), vol. 3, col. 389.
29. Many sources take up *méreaux*, some in passing, some substantially. See Labrot (1989) on medieval jetons and *méreaux*. See also Fontenay (1847), pp. 124–42; Loir (1856), pp. 12–17, especially on the town of Mantes; Blanchet (1890), vol. 2, pp. 406–09, 436–45; Blanchet (1917–1918); Blanchet (1930), pp 511–78; Mainguy (1999), pp. 19–21; Pastoureau (1984), pp. 38–40; Lerouge and Lerouge (1977), pp. 27–28; de Turckheim-Pey (1997), pp. 46–48; Roelandt et al. (2004), p. 20; Dillemann (1997), p. 6; Barnard (1916), pp. 81–82; Roman (1880), pp. 376–77.
30. "Le véritable sens attaché aujourd'hui au mot *méreau*, celui, du moins, sous lequel nous le considérerons exclusivement ici, est le sens de *Pièce de Chapitre* et de *monnaie conventionnelle des hommes d'église*," Fontenay, (1854), p. 68; on Fontenay's extended treatment of *méreaux*, see pp. 64–101.
31. Blanchet (1930), pp. 512–13.

FIGURE 1.4: A *méreau* of the Église de Saint-Omer, 1526.[32]

the urban landscape in larger cathedral towns and religious communities across France from the later Middle Ages and into the eighteenth century.[33]

Méreaux functioned in several ways. They possessed a literally token monetary value and could be exchanged at appointed times for cash at the canon treasury or used as a voucher for a meal in the church commissary or bread at the local baker, and merchants could redeem them back to the chapter as well. As mass tokens, *méreaux* were handed out to choir members as they entered the choir and given to priests and brothers to encourage attendance at services. Priests received *méreaux* for performing anniversary masses or other special masses, such as for the cloth merchants of Paris in 1527.[34] Those attending formal chapter meetings involving the bishop, the dean and chapter canon received *méreaux* as well.[35] This represents the first function of *jetons de présence* or jetons for attendance later taken over entirely from *méreaux* by jetons. In addition, *méreaux* were used for charitable purposes and given out to the deserving poor by the confraternity (*confrérie*) associated with the cathedral or local parish. Workmen such as masons received *méreaux* from cathedral chapters as payment for services. And *méreaux* may have played a part in the upside-down world of popular festivals.[36] The sums involved could be substantial, as evidenced by the 13,618 *livres* budgeted for *méreaux* by the cathedral at Rouen in 1657.[37] In effect, as bishops lost the power to mint their own money and with the trustworthiness of other coinage sometimes in doubt, *méreaux* became a substitute currency that circulated locally and supported local commerce, often with the tacit approval of the chapter.[38]

An incident from 1557 sheds light on *méreaux* and their place in the world, notably that they did circulate as currency, and this practice inevitably brought conflict with central, royal authority that otherwise minted and controlled the circulation of money. The crown had established a special monetary court in Paris in 1522 to superintend these matters, the Cour des Monnaies. The Cour des Monnaies sent out commissioners to the hinterlands to stamp out the illicit circulation of *méreaux* (Fig. 1.4). At the time in question, two judges (*conseillers*) from this court passed through the town of Mâcon. They observed the circulation of *méreaux* not only "as payment for choir members and other priests serving in the said church, but even more for the whole of the said town of Maçon, some worth six *deniers* [pence], the others for double that and as Tournois *deniers*." The outraged judges then noted with satisfaction that "this abuse was corrected by a ruling of the said commissioners, notwithstanding letters of protest regarding it sent in by the aforesaid dean, canons and chapter" of the church.[39] We can only imagine what those letters said.

32. On the jeton and *méreaux* of Saint-Omer, see Harmand (1834), pp. 53–94.
33. *Méreaux* were used not just by French Catholic churches; Blanchet (1890), vol. 2, pp. 446–48 notes *méreaux* issued by French Protestant churches prior to the Revocation of the Edict of Nantes in 1685.
34. Mazerolle (1902–1904), vol. 1, p. 509.
35. Loach (1995), p. 25.
36. de Tuckheim-Pey (1997), p. 48.
37. Blanchet (1930), p. 514.
38. Roman (1880), p. 377; Blanchet (1930), pp. 516–17.
39. "Pour le payement des choristes et autres prestres servans en la dite église...lesquelles pièces avoient abusivement

In this connection we should point out merchant tokens as another distinct class of tokens. Common from the thirteenth century, these were pieces of metal or pottery marked with the insignia of a merchant or merchant house that licensed traders used to claim lots at wholesale markets such as *Les Halles*.[40] A regulation of 1698 standardized how these jetons had to look, i.e., marked with the merchant's initials. Fruiterers would put down their marked jetons on baskets of butter, eggs, cheese, and whatnot to signal a purchase. Merchants' tokens are documented to have been used in wholesale markets into the late nineteenth century, and doubtless the practice continued into the twentieth century.[41] Merchant tokens were used as jetons for in-house accounting on counting tables, and in addition they served as a form of advertising as they circulated hand-to-hand in markets and elsewhere.[42]

But jetons as understood here and in the world of numismatics generally were none of the above: They weren't coined money. They weren't historical medals, and, although they came to circulate in church circles, they were not *méreaux*. They are not to be confused with tokens used by merchants. No, early modern and Old Regime French jetons constitute a distinct class of coinage with distinct historical origins and fates. Our jetons first appear innocuously enough in medieval Europe as little tokens moved around on counting tables by those keeping the books. Jetons and the exchequer system continued to function as an accounting tool for many centuries, but over time, especially in France and especially in the Old Regime, the nature and social significance of jetons took on wildly expanded meanings.

Medieval Jetons and the Casting Counter

The exchequer system was a practice for manually performing arithmetical calculations and keeping accounts that arose in medieval Europe. So-called manual arithmetic involved an operator, a casting counter (a table or a cloth) and a bag of jeton tokens.[43] Manipulating jetons on a properly incised table or marked cloth, an operator could reliably perform the simple arithmetic functions of addition, subtraction, multiplication and division. The later medieval school jeton pictured at the head of this chapter (Fig. 1.1) shows the casting counter with jetons on it and, although hard to decipher in this image, standing behind the table we see the Rechenmeister himself with his puffy sleeves moving around his jetons. Jetons and the table or cloth thus functioned something like an abacus and was known as the line abacus, and historically the first use of jetons was in this accounting function as counters, *jetons de compte*. The exchequer system of calculating originates in the twelfth century in Europe, and its practice spread and continued through the eighteenth century, a period of 600 years. The system has disappeared, but it left many imprints, notably, for example, in the treasury post of Chancellor of the Exchequer in England.[44]

The connection is not clear between the exchequer system that arose in medieval Europe and ancient accounting traditions. All kinds of tokens of bone, horn, ivory, lead, clay, and stone were used in all kinds of ways in classical antiquity: for voting, calculating, establishing a queue, as game

---

cours non seulement parmy lesdits prestres et choristes de ladite église, mais encore par toute la dite ville de Mascon, les unes pour six deniers, les autres pour doubles et deniers tournois...Lequel abus fut corrigé par l'ordonnance desdits commissaires, nonobstant les lettres en opposition envers ielle impetrées par lesdits doyen, chanoines et chapitre...," contemporary document quoted in Robert (1860), pp. 28–29.

40. de Turckheim-Pey (1997), p. 43; Blanchet (1930b).
41. d'Affry de la Monnoye (1878), p. xxvii.
42. Chautard (s.d.), p. 7.
43. Berry (1977), p. 26; Berry (1968).
44. On these points see Lerouge and Lerouge (1977), p. 26; Schärlig (2003), pp. 129–30; Schodt (1873), p. 11; Haskins (1924), pp. 156, 333–35; Haskins links the exchequer system to the abacus.

pieces, tickets to the theater, and so on.[45] The Romans did develop an abacus for computational purposes similar to the more familiar Chinese abacus.[46] Regardless, ancient accounting practices did not pass directly to Europe after the fall of the West Roman Empire. Rather, as Europe revived after the year 1000 and more particularly in the twelfth century, a new form of computational practice using jetons and a counting table arose in Europe. Pope Sylvester II (999–1003) reintroduced the abacus to Europe from the Islamic world, a move indicative of burgeoning conditions in Europe.[47] Yet, it seems that it was Crusaders who brought the technology of the casting counter (or something like it) back to Europe from Constantinople in the twelfth century.[48] The use of counting tables and jetons first became implanted in northern and central Italy, whence it spread to France, the Low Countries, England, and the whole of the West.[49]

But then, in 1202 the Italian mathematician, Leonardo of Pisa (aka Fibonacci), let out into the Western world (one hesitates to say published) his influential book, *Liber abaci*, a work famous for introducing Arabic/Indian numerals into Europe. Fibonacci's book was not about the abacus, as its title might suggest, but about how to calculate *without* using the abacus or the casting counter. In this, Fibonacci introduced into Europe a new form of arithmetic and new techniques for calculating requiring only pen and paper. The techniques were simple and reliable—essentially what is taught to us in elementary school—and their use spread from the first quarter of the thirteenth century.[50] Calculating by quill, as it were, and calculating with jetons were two different ways of performing arithmetic functions, and the techniques jutted up against one another in the thirteenth and fourteenth centuries. Italian regions were quick to switch to Arabic numerals and calculating by quill, and as a result we find no jeton or exchequer tradition there. The same story in Portugal and seemingly Spain, too, where jetons never made it or were soon replaced with Fibonacci's pen and paper techniques.[51] That left the regions of France, England, the Low Countries, and parts of Germany as the heartland of the exchequer system and the enduring culture of accounting by means of counting tables and jetons.[52] In these areas only in the sixteenth century did reckoning with numerals more commonly begin, and not before the nineteenth century did Fibonacci's methods win out entirely over the exchequer, a process that took 600 years overall.[53]

Jetons and the casting counter came to England, if not in 1066 with William the Conqueror himself, then within the next few decades. The medieval treatise, *Dialogus de Scaccario* (Dialogue on the Office of the Exchequer) written around 1180 by Richard Fitzneale, Lord Treasurer to English King Henry II, is an early reference, as is another of 1190 for the, "purchase of jetons destined for the counting house of the queen."[54] Numismatists date the earliest Italian and French jetons to

45. The 1724 paper by the Montpellier physician, priest, antiquary and academician Nicolas Mahudel (1673–1747) is surprisingly informative on this point; see Mahudel (1729). For a discussion of Roman calculi, see Berchem (1936) and further de Turckheim-Pey (1997), p. 14; Rossignol (1851), p. 33; on unrelated Chinese tokens, see Amandry (2001), p. 361.
46. For this background, see Mainguy (1999), pp. 22–25; de Turckheim-Pey (1997), pp. 13–16; Roelandt et al. (2004); Blanchet (1930), p. 203; van Beek (1986), p. 195; Fontenay (1847), pp. 10–13; Müller (1984), p. 808 and figure 11; Loach (1995), p. 26.
47. Provinciale Dienst (1981), p. 31; Anonymous (1982), p. 98.
48. van Beek (1986), pp. 199–201; Schlinder (1952).
49. Pastoureau (1984), p. 18.
50. Pastoureau (1984), p. 19; see also Devlin (2011); Swetz (1987).
51. Veillon (1994), p. 23; van Beek (1986), p. 201; Schärlig (2003), p. 37, suggests more continuities with reckoning with jetons in Italy than is normally assumed.
52. This is not to say that calculating by quill did not penetrate these regions at all; Barnard (1916), pp. 90–91, notes resistance to using Arabic numerals and arithmetic methods.
53. On shift from Roman numerals to Arabic ones and continuities of the former, see Swetz (1987); Schärlig (2003), pp. 29–33, 37–45. See also Blanchet (1890), vol. 2, p. 400; Barnard (1916), p. 26.
54. Schärlig (2003), p. 33, quoting Karl Menninger (1958).

the last decade of the twelfth century and the first decades of the thirteenth, notably around the court of the French regent queen, Blanche de Castille, in the 1220s and 1230s, and jetons were struck in the reign of every French king thereafter.[55] The earliest documented date for jetons in the Low Countries, used by merchants and others in Bruges, is 1284, although they were doubtless present long before that.[56]

The word, jeton, derives from the Latin *jacere*, meaning to throw away or to launch. The related French verb, *jeter*, also means to throw, and so the word jeton evolved to mean the little coins thrown down onto counting tables. In this way, in medieval and early modern French the verb *jeter* or *jetter* explicitly came to mean to compute or to perform calculations.[57] *Jeton, jetton, gettons, gectons, gets, gectouers* were just some of the names by which jetons and calculating in this sense were known. Schodt (1873) gives an astonishing 47 variants of the words *jeton* and *jeter* and their etymological connections, and, most notably for us, this large variety of names for essentially the same word and act is indicative of the depth of the social penetration of jetons and the exchequer system.[58]

FIGURE 1.5: A typical medieval jeton.

At the outset all jetons were just jetons, that is, ordinary, common tokens or *jetons banaux* (from the plural of *banal* in French) (Fig. 1.5).[59] These would be jetons used on counting tables and soon, if not simultaneously as gaming pieces (whence the ultimate origin of the word jeton to mean poker chips in modern French). The point is that across the centuries private individuals who had need of jetons to run their accounts or play their games could obtain them from merchants and manufacturers of various sorts. On the other hand, as jetons became a perk of office and as royal and other bodies began issuing them for ceremonial and other purposes, a new class of jetons soon emerged distinct from *jetons banaux*. Put simply, the later could be purchased, but the former, issuing from an institution, could not, and this, following Rouyer (1879), is the easiest way to distinguish between *jetons banaux* and the vast world of institutional jetons otherwise to be explored in the pages that follow.[60] First, however, how jetons and the exchequer system worked at the counting table must be examined.

55. On these early jetons, see van Beek (1986), p. 200, Schärlig (2003), pp. 29–30, 34; de Turckheim-Pey (1997), p. 14; Roelandt et al. (2004), p. 11; Pastoureau (1984), pp. 17–18; Gadoury (1986), p. 3; Blanchet (1890), vol. 2, p. 401.
56. Provinciale Dienst (1981), p. 31; Anonymous (1982), p. 97; Schärlig (2003), p. 36; Schodt (1873), p. 9.
57. On names and etymologies, see Pastoureau (1984), p 15; Blanchet (1890), vol. 2, p. 401; Blanchet (1930), p. 205; Josset (1978), p. 337; Boutin (1966), p. 240; La Tour (1897), p. vi; de Turckheim-Pey (1997), p. 13; see also Barnard (1916), pp. 26–28.
58. Schodt (1873), pp. 10–11.
59. On *jetons banaux*, see Valin (2016), vol. 1, pp. 126–27 and pp. 246–47, where she indicates that *jeton banaux* were sometimes struck in the provinces using dies manufactured at the Monnaie des Médailles in Paris. See also Pastoureau (1984), p. 20, n.; Jaumonet (1985), p. 19; Barnard (1916), pp. 66–67; Blanchet (1930b); Rouyer (1893), pp. 185, 515; Blanchet (1909c) for gaming pieces of the fourteenth century; Blanchet (1930), p. 237, reports gaming pieces continuing to be manufactured through the eighteenth century.
60. Rouyer (1879), p. 14 n.

FIGURE 1.6: Gregor Reisch's *Margarita philosophica*, 1503.

Reckoning with Jetons

In 1503 at Freiburg the German Carthusian humanist, Gregor Reisch, published his knowledge compendium, *Margarita philosophica* (The Pearl of Philosophy) (Fig. 1.6). Accompanied by a now-famous woodcut, chapter four concerned arithmetic (TYPUS ARITHMETICAE).[61] The image depicts the muse along with Boethius seated on the left and Pythagoras on the right.[62] The two are doing arithmetic. Boethius performs his calculations at a desk with pen and ink. Pythagoras performs his on a counting table using jetons.

Modern historians and numismatists have thoroughly documented the techniques involved in "manual arithmetic" and using the casting counter and jetons. Barnard (1916) devotes over 50 pages to the methods.[63] Smith (1921) is given over entirely to such operating instructions, and more recently Alain Schärlig (2003) has produced a comprehensive manual on the exchequer system, *Compter avec des jetons: Tables à calculer et tables de compte du Moyen-Age à la Révolution* (Computing with Jetons: Counting Tables from the Middle Ages to the Revolution). Schärlig devotes almost 100 pages to the details of how to manually perform arithmetic calculations with jetons on the counting table.[64] Suffice it to speak here generally and give only an example of what was involved.

Operators began at the table, and craftsmen built and sold special tables with horizontally and vertically incised lines (the exchequer or checkerboard) for this purpose.[65] Alternatively, a marked cloth would be spread out and jetons put down.[66] Operators would then manually position and move jetons around on the board in adding, subtracting, multiplying or dividing sums. The casting counter is known as the line abacus, and, although the techniques varied somewhat, the numerical values at any point depended on the position of the jetons relative to the lines on the board. Horizontal lines represented ascending powers of ten (1, 10, 100, 1,000, 10,000, 100,000), although these were not marked explicitly on the board, and for keeping moneyed accounts the lines would represent pence, shillings, and pounds to the English or *deniers, sols,* and *livres* to the French. To simplify matters, jetons placed in the spaces between lines counted for five of the lower unit (or otherwise when dealing with money). The 1,000 line was often marked as a 'holding line' where calculations above the line were performed first and then the lesser numbers below added in. Calculations proceeded across columns (Fig. 1.7), so wishing to add 21,968 and 5,632, the calculator using jetons puts those two numbers in the first two columns, combines them in a first result, and then reduces them to the sum indicated in the fourth column, 27,600. Subtraction works by removing jetons appropriately line by line and adjusting lower lines as necessary. Addition and subtraction could easily be performed using the first two columns alone or multi-columns for multiple numbers. But multiplication and division required more complicated operations, operating

---

61. Schärlig (2003), pp. 24–26 on this and other early visual images of reckoning with jetons; see also Veillon (1997), p. 12.
62. Newsome (2011) offers a provocative interpretation of the role of the Muse in this image.
63. Barnard (1916), especially Part III, "The Methods of Casting with Jettons," pp. 254–319. See informative review by Smith (1920), who notes that Barnard had a personal collection of 7,000 jetons.
64. Schärlig (2003), pp. 47–145. On instructions for reckoning with jetons and counting tables, see also de Tuckheim-Pey (1997), pp. 12–16; Pastoureau (1984), pp. 16–17; Roelandt et al. (2004), pp. 11–14; Boutin (1966), p. 240; Fontenay (1854), pp. 135–37; La Tour (1897), p. xiv; Schodt (1873), pp. 24–27; Berry (1968); Ch. Florange (1933), pp. iii–iv; Provinciale Dienst (1981), pp. 32–36, and Anonymous (1982), pp. 99–104. Pastoureau (1984), pp. 35–37, noted that much remains to be learned about the actual practices of medieval Rechenmeisters, a lacuna now more than filled by Schärlig's exhaustive study of 2003.
65. Schärlig (2003), pp. 147–246. Only a few of these have survived. Schärlig (2003), p. 14 notes the existence of fewer than thirty tables and cloths. See also Barnard (1916), p. 8 and Part II, "Counting Boards and Counter-Cloths," pp. 230–53, for his inventories.
66. Schärlig (2003), pp. 247–64.

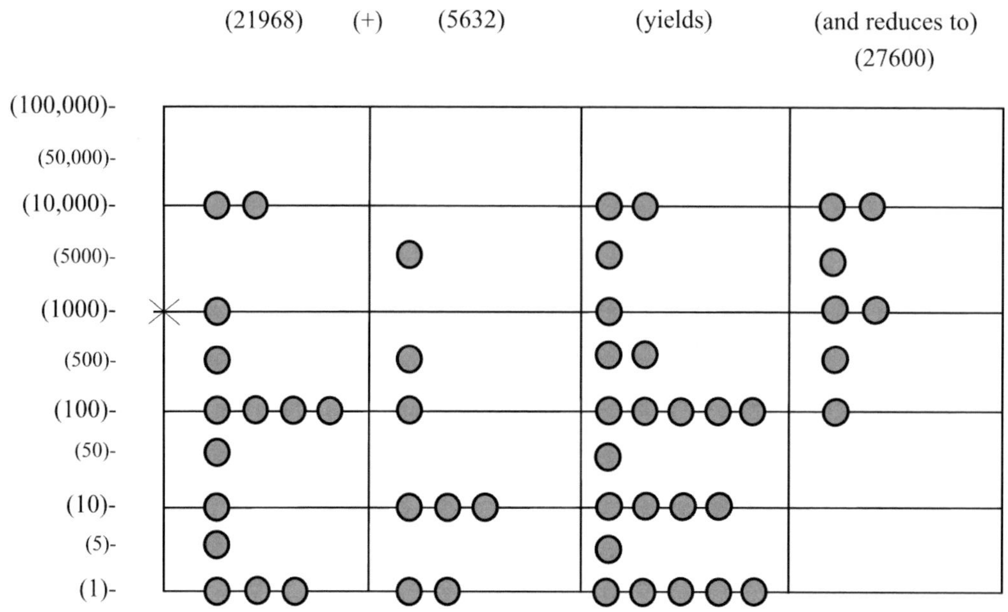

FIGURE 1.7: The casting counter.

with each line of the multiplier or divisor in turn on every line of the multiplicand or dividend through a series of intermediary steps involving temporary place holders. The principles were not difficult to grasp, but skill in calculating with jetons was something to be gained only through practice. Obviously, considerable experience was required to make these calculations quickly and accurately, and a deft operator was someone with a valuable skill, a skill which is completely lost today.[67]

A centuries-long tradition of contemporary manuals and how-to books provided instruction to the uninitiated on the art of calculating with jetons on the counting table. Barnard (1916) examined some 120 different treatises on reckoning with jetons, and he gives specific details of a dozen of them.[68] Johann Widman's *Algorithmus Linealis* of 1488 is an early example, as is Robert Recorde's *The Ground of Artes Teaching the worke and pratise of Arithmetike* (London, 1542). The 1608 *Arithmétique de Jean Trenchant, avec l'art de calculer aux jetons* is one of many French how-to manuals, but François Le Gendre's, *L'arithmétique en sa perfection* topped them all. With its requisite chapter on reckoning with jetons, the first edition of this book appeared in 1646. Le Gendre himself published at least six editions through 1673, and publishers in Paris, Rouen and Lyon reprinted the work every few years throughout the eighteenth century, with upwards of two dozen editions and reprints in all, the last appearing in 1812.[69] Le Gendre and company, too, were part of the world of jetons not to be overlooked.

The techniques of reckoning with jetons were taught in colleges and universities where arithmetic was one of the seven undergraduate liberal arts. Such instruction also took place in grammar

67. Guiffry (1881–1901), vol. 1, p. ii, reminds us that accounts in the eighteenth century were not kept with the regularity to which we are accustomed today and that, despite the serviceability of jetons and counting tables, errors were still made.

68. Barnard (1916), p. 254 ff.; Mainguy (1999), pp. 23, 316–19, lists several other contemporary sources on how to calculate with jetons.

69. The BnF catalog lists 22 editions. On Legendre, see also Pastoureau (1984), p. 20; van Beek (1986), pp. 196–99; Jaumonet (1985), p. 18; Barnard (1916), p. 89.

schools along with reading, writing, and a touch of Latin grammar. School masters employed special jetons, "school jetons" or *Schützen-Jetons* to instruct students in the techniques.[70] The school jeton (Fig. 1.1) is an example, showing again the Rechenmeister at the table on the obverse and the alphabet on the reverse, functioning as a mini-hornbook. In addition, such jetons were used as admission tokens for paying-students to these grammar schools. One school in 1697 boasted of teaching students to reckon "with quill *and* with jetons."[71]

Performing arithmetical operations with Roman numerals is not easy, and so long as Roman numerals remained in use, jetons and the counting table offered obvious advantages and were thought to be more convenient. But do note that one can just as easily use Arabic numerals to record and jetons to calculate, so that the coming of Arabic numerals itself was not a direct challenge to the technology of the counting table. And then, too, it is not necessary even to read or write to handle jetons on a counting board.[72] Furthermore, although the decimal system was in place for ordinary calculations, the decimal system was not used when it came to money or keeping financial accounts.[73] Rather, money accounts were recorded in pounds, shillings, and pence or, in French, *livres, sols, and deniers*, where 12 *deniers* made one *sol* (or *sou*) and 20 *sols* (or *sous*) made one *livre* When making calculations involving money the value of the lines would be adjusted accordingly to *deniers*/pence (6 pence in the space), *sols*/shillings (10 *sol*/shillings in the space), and *livres*/pounds.

From the thirteenth through the eighteenth centuries, a great variety of people who needed to keep accounts employed the casting counter and jetons to do so. This included anyone who had funds to maintain or dispense, budgets to live by, or transactions to conduct, i.e., a large swath of society involved in economic activity: merchants and every mercantile or commercial enterprise, bankers, money changers, artisans in their shops, innkeepers, gaggles of government financial officers and agents (including receivers, payers, treasurers, collectors, controllers, verifiers, etc.), clerks in various capacities in royal, princely and bourgeois houses; managers of seigniorial properties; agents in any number of bureaus in towns and associated with regional estates, not to mention operatives in religious communities, guilds, and domestically at home. The exchequer was a tool, and the learned and unlearned alike used it in their calculations.[74] Incidentally, the use and spread of the casting counter is an indicator of growing economic activity in Europe over the period in question.

Because of its direct connection with political and military power, the French royal counting house, the Chambre des Comptes or CAMERAE COMPUTORUM REGIORUM, was a notable center of calculation with a long history.[75] Ordinances establishing the formal institution date to the 1250s and 1260s, and evidence from the 1350s shows the Chambre des Comptes allocated 1,000 jetons a year for the purposes of its operation.[76] Given the feudal (dis)organization of France in the thirteenth and fourteenth centuries, one is not surprised to find a number of other sovereign counting houses complementing the one personally associated with the king in the region of Île-de-France. The Chambre des Comptes of Burgundy dates from 1317, the one for the Dauphiné from 1367, and others appeared for Lorraine, Languedoc, Brittany, Picardy and ultimately 13 sites spread

70. Provinciale Dienst (1981), p. 31; Anonymous ((1982), p. 97.
71. "à jetter à la plume et aux jettons," mentioned in Gadoury (1986), p. 3; author's emphasis.
72. Amandry (2001), p. 297.
73. Schärlig (2003), p. 44.
74. Pastoureau (1984), p. 20; Provinciale Dienst (1981), p. 30; Anonymous (1982), p. 96.
75. See especially Sarmant et al. (2007); see also Blanchet (1890), vol. 2, pp. 416–17; Blanchet (1930), pp. 260–62; Fontenay (1847), pp. 81–83. See also Mousnier (1990–1992), vol. 2, pp. 275–76, 338–40.
76. A. Barthélemy (1875), p. 229: de Turckheim-Pey (1997), p. 42.

across the kingdom.[77] Royal and related counting houses obviously mirrored the larger political and economic structure of the growing kingdom. A more minor point to be emphasized here is that internally counting houses developed elaborate formal procedures for the purposes of royal bookkeeping. Official sessions were announced; observers were present and minutes taken. A calculator would perform his calculations on the casting counter, another would check the calculations. A reporter announced the results aloud, which an auditor would then record using jetons on another exchequer.[78] The checking of accounts became a function within the treasury for which extra jetons were given as payment.[79]

Some residue of these practices from the Middle Ages is preserved in the old English nursery rhyme, "The king is in his counting house, counting out his money." A close-up view of 1474 involves the Duke of Burgundy, Charles the Bold (1433–1477), who was decidedly in charge of the Burgundian counting house and himself worked the counting table. According to the report:

> The duke came to the counting room very often, and no account was closed without him being present or without his seal. He himself would sit at the head of the table, throw his jetons and reckon like the others, and there was no difference among them in this activity, except that the duke reckoned with gold jetons and the others with silver jetons. The master of accounts, the master of the payroll, and the custodian of accounts all operated and went over these accounts and summed them up together. For this, each of the latter every year receives a half a pound of silver jetons stamped with the arms and watchwords of this prince.[80]

Women—many women, in fact, and perhaps more women than men—knew the exchequer system and how to manipulate jetons on the counting board. Women used the apparatus in business or to keep household accounts, and skill in using jetons was considered an advantage for girls of marrying age.[81] With jetons so ingrained in domesticity, women apparently continued reckoning with jetons after the practice had been abandoned by men.[82]

In France and elsewhere to the end of the eighteenth century jetons and the casting counter remained prominent accounting tools. In terms of any cultural history of the Old Regime the counting table and jetons need to be seen as one constant in that firmament. Although jetons continued to be moved around on counting tables decade after decade and century after century, the historical interest in jetons stems more from what else jetons became, how else they functioned in society, and what else they can tell us about their world.

---

77. On these counting houses, see Sarmant et al. (2007), pp. 31–34; Vallier (1879), p. 8; Robert (1890), pp. 235–37; Mazerolle (1902–1904), vol. 1, pp. 507, 510; Mainguy (1999), p. 71; Josset (1978), p. 337; Rossignol (1851), p. 38, n.
78. On these points, see Ch. Florange (1933), p. i; d'Affry de la Monnoye (1878), p. xxii.
79. d'Affry de la Monnoye (1878), p. xxii.
80. "Là vient le duc en la chambre des finances bien souvent, et ne se cloent nuls comptes sans luy ou sans son secu. — Luy mesmes il sied au bout du bureau, ject et calcue comme les autres, et n'y a différence en eux, en iceluy exercice, sinon que le duc ject en jets d'or et les autres de jets d'argent — Le maistre des comptes, le maistre de la chambre aux deniers, le contrerolleur ject et calculent icelles parties et sur ce sont mises les sommes, et pour ce faict, ont tous les ans chascun d'eux, pour un marcq de jets d'argent, aux arms et devises du prince." This, from the *Estat de la Maison de Charles le Hardy* included in the memoirs of Olivier de la Marche written in 1474 and published in 1562; different editions and modernizations have transformed the quote somewhat in different published sources; see de la Marche (1825), vol. 2, p. 487; on Charles the Bold at the counting table, see also A. Barthélemy (1875), p. 241; Barnard (1916), p. 28; Schodt (1873), p. 16 n.; Rossignol (1851), p. 38, n; citing Mahudel (1729), Valin (2016), vol. 1, p. 8, suggests that Louis XIV himself also used jetons at the counting table.
81. Provinciale Dienst (1981), p. 31; Anonymous (1982), p. 98. Schodt (1873), p. 23.
82. Schärlig (2003), pp. 6, 22.

The Manufacture and Aesthetics of Medieval Jetons

Some medieval jetons and *méreaux* were cast, as evidenced in the permission given by the royal monetary court, the Cour des Monnaies in 1530 "to Jean Auzoul, founder in lead and tin, to be able to cast tin jetons so long as by their form they be not confused with hard currency."[83] Exceptionally, in the case of sieges, for example, jetons or *méreaux* of paper, leather, or stone would be used.[84] But by far, medieval jetons were small stamped circular metal disks. Medieval jetons (and all contemporary coinage) were hammered, not pressed.[85] That is, a prepared blank or flan was set on one die tooled by the engraver and secured on a base. The other die was set in a metal tube and placed over the first die and its flan. A presumably swarthy worker would then strike the top of the tube with a hammer, imprinting the flan/planchet and the process repeated.[86] The earliest jetons were always of common metals: tin, copper, bronze, and various alloys.

Official permission to strike jetons was required in the kingdom of France from the fifteenth century, and they could be struck only in royal mints. Most medieval French jetons were struck in Paris, although exceptions arose from time to time.[87] The master of the mint received a fee (the *droict de brassaige*) regardless of who was striking them.[88]

Medieval jetons were typically stuck on thin planchets with minimal attention to design. They were utilitarian objects, not works of art. They might be decorated with simple quatrefoils or simple images: a cross, a Pascal lamb, a ship, a rose; a crude crown. Fleurs-de-lys were sometimes present. The image of a balance on early jetons of the *Chambre des Monnaies*, the office in charge of overseeing royal mints, already indicates some specialization of jeton design, but overall the first jetons were plain and largely uninteresting aesthetically.[89]

Because they were destined for the counting table, medieval jetons were often stamped with edifying and amusing legends: "Don't make a mistake," "Count well, count well," "Your count will be your end," "Wait, count, and tally well," "He who counts well, his tally will find," "Attend well and loyally to [your] accounts," and so on.[90]

But in the end the actual practice of ordering, engraving, making or distributing jetons at this early period remains unclear.[91]

Jetons as a Perk of Office

Jetons started out innocently enough as tokens moved around on medieval counting tables all across Europe. But then a shift occurs in the thirteenth and fourteenth centuries as they move off of the counting table and into the French royal court to become a perk owed to officeholders.

83. "Permission donnée à Jean Auzoul, fondeur en plomb et étain, de pouvoir mouler des jetons d'étain pourvu qu'ils ne puissent, par leur forme, être confondus avec la monnaie courante," quoted in A. Barthélemy (1875), p. 255.
84. Blanchet (1890), p. 14.
85. On the manufacture of medieval jetons, see de Turckheim-Pey (1997), p. 42; Pastoureau (1984), p. 25; Blanchet (1890), vol. 1, pp. 1–4. On general background to medieval coinage, see Labrot (1989) and the handsome volume by Belaubre and Collin (1992).
86. The coin press is not introduced until the 1550s, and the hammering of coinage did not stop until 1645.
87. de Turckheim-Pey (1997), p. 42. On jetons minted by the Monnaie de Paris and not the Monnaie des Médailles, see Delcamp (2015b), pp. 220, 222.
88. Mazerolle (1902–1904), vol. 1, p. 516.
89. Barnard (1916), p. 34–38. On the iconography of medieval jetons, see also de Turckheim-Pey (1997), pp. 37–40; Blanchet (1890), vol. 2, p. 402.
90. "gardez vous de mescompter," quoted in Boutin (1966), p. 240; "gettes bien gettes bien," Schodt (1873), p. 12 n.; "car la fin fera vre compte," "attendez, comptez et gettez bien," A. Barthélemy (1875), p. 260; "Entendez bien & loyaument aux Comptes," Barnard (1916), p. 47; "Qui bien jettera, son compte trouvera," Mahudel (1729), p. 263. Blanchet (1930), p. 503–10, gives an extensive list of legends of medieval *jetons banaux*, as does Barnard (1916), pp. 45–51; see also Blanchet (1890), vol. 2, pp. 402, 409–11; Berry (1968); Mainguy (1999), p. 31.
91. Pastoureau (1984), p. 25.

So much of the history of jetons at this point centers on the royal court and its operational extensions. In the French case, as the wealth and complexity of the royal court and the king's household and treasury grew in the thirteenth and fourteenth centuries, offices were created and accounts established that used jetons and the exchequer system. The key and for us trailblazing office for royal accounts, the Chambre des Comptes, as mentioned, grew out of the earlier counting house and dates formally from the 1250s. The formal institution of the king's household (the Maison du Roi) was in place from 1261; the lesser office of the king's silverware (the Argenterie) was an early one from 1317, as were the royal stables. The arm concerned with coinage and minting (the Chambre des Monnaies) is from 1358. The paying office for the king's household, the Chambre aux Deniers du Roi, soon joined these departments supporting the functioning of the court and the trappings of government. In 1380/1381, for example, we see Charles VI distributing jetons to the "Six Ancient Offices" and officers tending his bakery, wine cellar, personal and common kitchens, his fruits, furs, and stable.[92] The queen's household soon plied its own exchequers, and still in the thirteenth century the court supplied jetons for households and accounts of royal princes and princesses. The king through his agents provided the required jetons.[93]

At the outset this royal distribution was undoubtedly and simply for the purposes of calculating for the various bureaus, but soon packages of 100 silver jetons became a perk, an entitlement coming to those holding one of these offices. In the example mentioned above of Charles the Bold at his counting table in 1474, it is not clear whether his officers received their full half-pound of silver jetons in order to do their calculations or as a function of the positions they held (Master of Accounts, Master of the Payroll, Comptroller). Jetons continued to be used as counters, probably at the outset by these same office holders, but regardless, the novelty from the fifteenth century is that they now went to individuals qua office holders. Increasingly jetons came to be associated with the office itself and the bureaucratic exercise of the functions of an office.[94] A recognized "right to jeton" or "right of jeton" (*droit au jeton* or *droit de jeton*) came into being, which made jetons a symbolic form of remuneration and an honorific for the office holder.[95] For the Chambre des Comptes this right was codified in 1505 and reaffirmed in letters patent in 1613 and 1614.[96]

Indicative of these novelties, silver jetons appear only at the beginning of the fourteenth century and are rare before the fifteenth century.[97] The first gold jetons were struck in the 1460s and 1470s, but probably quickly melted down.[98] The first jeton with someone's name on it dates from 1345, a jeton struck for one Clarin de Paumier, master of accounts in the Chambre des Comptes.[99] Indicative of these changes, the earliest dated jeton is 1474.[100] Jetons had moved off of the counting table and on to become a perk of office. Here, they formed part of what was known as the *épices*

92. "Panneterie, echançonnerie, cuisine, fruiterie, écurie, fourrière," as indicated in d'Affry de la Monnoye (1878), p. xx. See also Mazerolle (1888), pp. 355–56 for the same offices mentioned a century later. On the royal stables, see Fontenay (1854), p. 128.

93. On these points see de Turckheim-Pey (1997), pp. 18–36; Préau (1890b). On the bureaucratic organization of the French court and Old Regime society generally, see Marion (1923), Mousnier (1990–1992), Bély et al. (2003), Barbiche (1999).

94. de Turckheim-Pey (1997), pp. 16, 41, 49. On jetons as perks of office, see also d'Affry de la Monnoye (1878), pp. xviii–xix; Barnard (1916), pp. 73–75; Müller (1984), p. 809; La Tour (1897), pp. viii–ix.

95. Mainguy (1999), p. 26; Fontenay (1854), p. 174.

96. Mainguy (1999), p. 26.

97. Mainguy (1999), p. 27; Blanchet (1890), vol. 2, p. 401; Barnard (1916), p. 30; d'Affry de la Monnoye (1878), p. xvii, dates the first silver jeton at 1345.

98. d'Affry de la Monnoye (1878), p. xvi.

99. Boutin (1966), p. 240; de Turckheim-Pey (1997), pp. 18–19, 42.

100. Blanchet (1930), p. 204; de Turckheim-Pey (1997) puts the earliest dated jeton at 1498; Barnard (1916), p. 53, puts the earliest dated jeton at 1468.

(spices), a package of incidentals that came with an office. These incidentals usually included paper, pens, penknives, inkstands, candles, matches, real spices, and jetons.[101] Jetons of common metals would be given for any actual calculation that an office or bureau might do and a purse of silver ones for the office holder himself. Rarely, but on occasion these jetons were budgeted and paid in cash. Thus, for example, in addition to emoluments on the order of 8,500 *livres*, masters of the king's paying office, the Chambre aux Deniers, and its two controllers-general each received an additional 64 *livres* as jetons.[102] Lest we think this a wholly trifling sum, in the case of the Cour des Monnaies and other institutions it came part and parcel with a range of other privileges, such as freedom from the salt tax. These privileges added up and bespoke an individual's state and status in the complex society of the day.

Soon the practice of receiving jetons as a function of an office migrated from the royal court and court offices into offices of the city administration of Paris and to provincial estates. For example, officers of the Cour des Aides in Paris, a sovereign tax court, soon began receiving jetons, and sometime between 1557 and 1578 attendees of the triennial meeting of the Estates of Burgundy began to receive their packages of 100 silver jetons.[103] The fact that more and more jetons began to be struck in silver is one indication of their new function.[104] Copper and brass jetons still served on the counting table, but silver ones served as part of a package of perks of office.[105] These were generally not cash payments, but rolls of 100 actual silver jetons. Again, jetons never possessed monetary value, although they could be converted to cash by selling them or redeeming the silver.

Indicative of the office and status of the beneficiary, contemporaries valued the receipt of a purse (*bourse*) of jetons as a point of honor, and the privilege went beyond the intrinsic value of the metal. A revealing court case of 1520 evidences this sensibility. The ruling speaks for itself:

> The Paris engraver, Le Gaule, took 60 *sols* for crafting dies with the arms of M. Claude Delacroix, royal councilor and master of the Chambre des Comptes. The latter complained of this to the court [the Cour des Monnaies], given that he had nothing to pay since it was a question of the 100 silver jetons [*gectouers*] due him by the king on account of his office. These dies should have been paid to the said Le Gaule by the king as is the custom. The court orders that these 60 *sols* be reimbursed this very day [to Delacroix] by the guard of the secretariat of the court, under penalty of prison. Furthermore, it is henceforth forbidden to exact anything from the officers to whom the king owes silver jetons by dint of their office. When an engraver will have made or engraved any of these dies he will present his invoice to the Chambre [aux Deniers] to be paid.[106]

101. For these details see earlier note and Bouclier (1924), p. 60–61; Préau (1890a), p. 125; Blanchet (1930), p. 204; de Turckheim-Pey (1997), p. 16. For an example from 1775 of *épices* at the Cour des Monnaies, see "Menues nécessités des officiers de la d. Cour [des Monnaies]," CAEF, Ms. A-1, 9 (18–18c) 1775. See also comment, Delcamp and Jambu (2018), p. 66.

102. Fontenay (1854), p. 174.

103. Rossignol (1851), p. 37.

104. Josset (1978), p. 337; Boutin (1966), p. 241; La Tour (1897), p. xii, all note an expansion of silver jetons beginning in mid-sixteenth century.

105. Lippens (1969), p. 34; Amardel (1901), pp. 10–11 on copper used more for accounting, silver for more symbolic uses, yet he notes that copper jetons still have remunerative value. Mainguy (1999), p. 308, points out that much less is known about distribution and social function of copper and brass jetons.

106. "Le Gaule, tailleur de la monnaie de Paris, avait pris 60 sols, pour avoir gravé une pile et un trousseau aux arms de M. Claude Delacroix, conseiller du roi et maître des Comptes; celui-ci s'en était plaint aux généraux, attendu qu'il ne devait rien payer, puisqu'il s'agissait de 100 gectouers d'argent à lui dus par le roi à cause de son office; ces fers devaient êtres payés audit Le Gaule par le roi comme c'était la coutume. Il est ordonné que ces 60 sols seront le jour même remboursés par le garde au greffe de la Cour, sous peine de prison; de plus il est défendu de prendre

The Sixteenth-Century Transition

If by the fourteenth and fifteenth centuries jetons had become a small perk of office, along with quills and candles, in the sixteenth century jetons took another historical turn. No longer simply tokens used on counting tables nor even a handsome perk of 100 struck pieces of silver, jetons took on the character of the institutions and the offices themselves. They came to represent and be the face of administrative units and particular bureaucratic entities and less the office holder. Furthermore, within offices and institutions these administrative/bureaucratic jetons came to be handed out for attendance at official meetings, so-called *jetons de présence*. These institutional jetons, as it were, were issued annually in long-running series that spanned decades. They thus became one outward face and a symbol of a bureau or an institution. More than that, jetons were no longer confined to the personal offices of the king and court, but in the sixteenth century they expanded to include the larger offices of government administration and various departments of the emerging French state. In this way, as Sylvie de Turckheim-Pey has noted, with the outset of the sixteenth century jetons came to occupy "an important place in social relations."[107]

The rise of the nation-state is recognized by all as portentous, and as an indicator and agent, the nation-state is synonymous with modern history.[108] In the sixteenth century Spain, Portugal, Holland, France, and England all transitioned from feudal polities and the relatively weak central authority of the medieval king, to laying the foundations for much stronger, more centralized states. France recovered slowly from the costly, but victorious Hundred Years' War that ended in 1453. The Renaissance and a new era came to France, and the French state began to solidify under François I who reigned from 1515 to 1547. Not coincidentally, with François I jetons took on a new role inside of government and the institutions of state.

The king's privy council, the Conseil du Roi, is an early and representative example of this trend in formalizing jetons as emblems of bureaucratic government (Fig. 1.8).[109] French government per se prior to Louis XIV centered on the royal privy council, where weighty political matters were debated. The council struck its own jetons from 1514 at the onset of the reign of François I. From 1550 the jetons of the Conseil du Roi were dated, issued annually, and took on a standardized appearance. Crown, shield and fleurs-de-lys firmly stamped royal authority on the obverse along with the inscription, on every jeton since introduced by Henri II in the 1540s, NIL NISI CONSILIO (Nothing without counsel). The jeton series of the Conseil du Roi was the first of its kind and one of the longest running, ending only in 1662, when Louis XIV assumed complete personal rule.

In the Conseil du Roi jetons came to be used in a new way. They were given out as *jetons de présence* to those who attended sessions. In 1537 François I himself authorized 400 silver ones "for each of the princes and participants in the privy council."[110] For 1549, for example, one reads that

---

dorénavant aucune chose des officiers auxquels le roi doit des gectouers d'argent, à cause de leurs offices; mais que lorsque le tailleur aura fait ou gravé quelques-uns de ces fers, il baille sa requête à la Chambre, pour avoir taxation," quoted in A. Barthélemy (1875), p. 253.

107. "En plus de sa vocation « arithmétique », le jeton va acquérir, dès les premières années du XVIe siècle, une place importante dans les relations sociales. Après avoir été utilisé pour la vérification des comptes, les jetons vont bientôt entrer dans le circuit des honneurs et des gratifications s'éloignant peu à peu de leur utilisation première," de Turckheim-Pey (1997), p. 49.

108. For an entrée into this topic, see Tilly (1992).

109. On the Conseil du Roi and its jetons, see Mousnier (1990–1992), vol. 2, pp. 132 ff.; Bély et al. (2003), pp. 320–25; Barbiche (1999), pp. 279–310; Mazerolle (1907b), p. 5; Blanchet (1930), pp. 256–57; Mainguy (1999), p. 308; A. Barthélemy (1875), pp. 262–63. Jambu (2014), p. 18, outlines the structure of various Conseils that succeeded after the personal rule of Louis XIV.

110. "A Chastillon, le XVme de septembre...A François Perdriel, maister de la monnoye de Paris, pour argent et façon de quatre cens gectons d'argent pour aucuns des princes et gens du Conseil prive. IIII^xx^ XVI^l^," see entry in J. Guiffrey (1877–1880), vol. 2, pp. 234–35.

FIGURE 1.8: Jeton: Conseil du Roi, 1585.

the jetons of the Conseil du Roi were "to be distributed to the king, our sire, and also to the honorable princes, prelates, lords and officers of the king who customarily attend the king in his private council."[111] Jetons thus became small tips or lagniappe associated with attending meetings and handling the business of an institution. This custom would expand considerably within French government and society in the two hundred years that followed. The language of the quotation suggests that the king himself received jetons for sessions of the Conseil du Roi.

Significant changes in the scale in the production of jetons for the Conseil du Roi indicate their new place and role within the bureaus of government that developed in the sixteenth century. The first striking for the Conseil du Roi in 1514 was a run of 700 jetons in silver, but the numbers quickly grew, particularly from the second half of the century, as the following table indicates, reaching 18,000 per annum by the 1630s just for the Conseil du Roi.[112] The sheer mass of silver and gold distributed in the form of artfully crafted jetons within the Conseil du Roi speaks volumes about changes underway for jetons and the bourgeoning French state beginning in the sixteenth century.

Year	No. of Jetons
1514	700 in silver
1528	1,800 (in silver)
1534	2,500 in silver
1543	2,600 (in silver)
1549	3,700 in silver
1561	5,400 in silver
1571	11,700 in silver
1573	14,100 in silver
1576	4,800 (in silver)
1594	1,200 in silver
1603	14,500 in silver; 314 in gold
1608	(variously) 16,000 (in silver) 12,300 in silver; 314 in gold
1635	18,000 in silver; 208 in gold

TABLE 1.1: Production of jetons of the Conseil du Roi, 1514–1635

111. "... pour les distribuer tant au Roy nostre dict seigneur que à Messeigneurs les princes, prelatz, sieurs et officiers d'icelluy seigneur qui ont acoustume d'assister en son Conseil privé." From the extant records of the Cour des Monnaies, quoted in Mazerolle (1902–1904), vol. 1, p. 537. For an almost identically worded reference of 1552, see A. Barthélemy (1875), quoted pp. 262–63.

112. Data extracted from Mazerolle (1902–1904), vol. 1, pp. 508–16, 578–80; Mazerolle (1907b), pp. 5–9; Blanchet (1930), p. 207; Barnard (1916), p. 30. Presumptive information in brackets. In 1545 the royal monetary court granted permission to strike 10,000 "gettons d'argent pour Messeigneurs du Conseil privé du Roy," but undoubtedly fewer than that number were actually minted; see Mazerolle (1902–1904), vol. 1, p. 516.

FIGURE 1.9: Jeton: The Chambre des Comptes du Roi, 1603.

The same standardization and "institutionalization," as it were, is evident in the jetons of the venerable Chambre des Comptes, the king's old counting house with institutional roots deep in the Middle Ages.[113] By the sixteenth century the Chambre des Comptes in Paris and its ilk across the rest of France had long since coalesced into major royal institutions and institutions of state. Jetons were struck, and each president, master, royal agent, other officer, and subaltern received his packet or half-packet of 100 silver ones annually.[114] But now, like those of the Conseil du Roi, in the sixteenth century the jetons of the Chambre des Comptes took on a standardized appearance and became part of an institutional series. The obverse of the jeton of the Chambre des Comptes of 1608 (Fig. 1.9), for example, is essentially identical to those depicted of the Conseil du Roi above or the Cour des Monnaies below: the royal crown and shield occupy the center with the particular institution identified (CAMERAE COMPUTORUM REGIORUM).

A remarkable continuity in design and appearance applies to these and like jetons of government and court from the sixteenth through the first half of the seventeenth century, a period of a solid century. Overall, sixteenth- and seventeenth-century jetons are relatively bland. The reverses are dated, and a Latin inscription encircles an image. We will have more to say about these inscriptions and the iconography involved in a later chapter. Suffice it to indicate here that the images and inscriptions connected to each other and carried meanings artfully associated with the office or unit in question. Engraving and striking these images is where the world of the fine arts enters the picture, much after the case of historical medals, and such artistry as is now evident in jetons is what distinguishes sixteenth- and seventeenth-century jetons from their cruder, stamped medieval antecedents. It was at this point that they became *petits monuments*. Other offices adopted standard images, such as the rake for the royal stables, where the image and the jeton itself serve as a kind of logo for the office or bureau in question.[115] In this way jetons became incorporated as part of a larger pattern of government and state development in contemporary France.

The Cour des Monnaies (CURIA MONETARIA FRANCIÆ)

The French royal monetary court, the Cour des Monnaies, is an institution of particular note for the story of jetons, not least because its own jetons likewise manifest this new institutional character that emerged in the sixteenth century.[116]

113. On the early history of the Chambre des Comptes, see de Tuckheim-Pey (1997), pp. 18–19.
114. For such arrangements for the Chambre des Comptes of the Dauphiné in 1628 and the Chambre des Comptes of Burgundy in Dijon circa 1647, see Vallier (1879), pp. 13–14.
115. Boutin (1966), p. 240.
116. On the Cour des Monnaies, see Parsons (2014), pp. 17–59; Bluche (1966); Bouclier (1924); Préau (1890a). See also Valin (2016), vol. 1, pp. 33–37; Darnis (1996/1999), pp. 11, 13; de Turckheim-Pey (1997), pp. 19, 41; Schärlig (2003), p. 34; Blanchet (1890), vol. 1, pp. 12–13; Blanchet (1890), vol. 2, p. 417; Blanchet (1930), pp. 262–64; Mousnier (1990–1992), vol. 2, p. 277.

The Cour des Monnaies was a powerful sovereign court with authority over all matters monetary. It grew out of earlier antecedents, notably the Chambre des Monnaies of 1358. The Cour des Monnaies formally dates to 1522, and it was chartered as a full-fledged royal court in 1551. The composition of the court varied somewhat over the years, but by 1716 it was composed of a first president and four other presidents, 36 associate judges (*conseillers*), a royal prosecutor and royal attorneys, a secretariat, and a sergeant at arms. The presidents wore the long robe made of velvet; associate members of the court and the king's representatives wore the short robe of satin; the first sergeant was decked out in taffeta. The position of president was hereditary and conferred nobility on its holder after a certain period (if he wasn't already a nobleman) and his heirs thereafter.[117] These offices could be purchased from the king, and in the 1770s the office of first president of the Cour des Monnaies cost an astounding 180,000 *livres*.[118]

The Cour des Monnaies issued its own series of jetons (CURIA MONETARIA FRANCIÆ), evidencing the sixteenth-century transition in jetons and the making of the French state just discussed, i.e., that jetons came to represent the institution, a nontrivial one in this case. Internally, jetons functioned within the Cour des Monnaies as they did elsewhere, i.e., as a perk of office and part of a package of *épices*. Notably, letters patent of 1613 confirmed the *droit aux jetons* as a perk of office for the officers and clerks of the Cours des Monnaies, a right reaffirmed in another royal edict of 1645.[119] The importance of the Cour des Monnaies to the present account extends further, however, in that for 150 years the Cour des Monnaies absolutely controlled the production of jetons (Fig. 1.10).

FIGURE 1.10: Jeton of the Cour des Monnaies, 1577.

The Cour des Monnaies had wide and great authority. All money struck in the kingdom passed through the Cour des Monnaies, and its most important function was to verify and certify the production of royal mints in the "judgment of the boxes" (*jugement des boîtes*), a formal session/ceremony involving assaying samples and tallying accounts.[120] The court had authority over the guilds of engravers and mint workers, and only members of these guilds could work in the mints or otherwise practice their craft.[121] In addition, the Cour des Monnaies oversaw silversmiths, jewelers, smelters, molders and everyone involved in the sale, purchase, or manufacture of anything in silver or gold. For some reason the court also had distilleries under its jurisdiction.[122] The Cour des Monnaies had police authority, and in one instance 1572 it paid to have an unlicensed engraver tracked down and arrested, an Italian, Jérôme Zoucha, "for an unauthorized die for a jeton."[123]

117. On these points, see Préau (1890a), pp. 122, 124; Rouyer (1892), p. 7.
118. See reference, CAEF, Ms. A-1, 9 (15).
119. Préau (1890a), p. 124.
120. On these points, see Bluche (1966), pp. 17, 18; Jambu (2014), p. 18, likewise discusses these processes.
121. Blanchet (1890), vol. 1, pp. 7–8.
122. Rouyer (1892), p. 3.
123. "...pour la fourniture d'un coin de jetons à la Cour des Monnaies," Mazerolle (1902–1904), vol. 1, p. 573. On another occasion the Cour des Monnaies investigated and then fined a smelter for providing the Monnaie des Médailles with inferior ingots; Mazerolle (1932), vol. 2, pp. 66 ff.

From the outset, the manufacture of jetons was tightly controlled by royal authority and actively supervised by the Cour des Monnaies. A fire in 1776 destroyed most of its archives.[124] Remaining, sixteenth-century archival records published by Fernand Mazerolle show the workings of the Cour des Monnaies in this regard.[125]

Court authorization was necessary to strike any jeton, whether for a public institution, a private individual, or as common jetons or gaming pieces (*jetons banaux*).[126] This requirement extended to restrikes or any second strikings. Approval was usually conditioned on submission of designs for obverses and reverses; requests for permission often came with the intended designs, the court approving "the aforesaid portraits, legends, and emblems.[127] Striking without permission brought fines, and in a revealing case in 1554 an entrepreneur named Jean Cousin got a royal order protecting him from any action by the Cour des Monnaies for his having struck, "or so it is said," 3,900 jetons in his own atelier without its prior permission.[128] Dies (*fers*) had to be brought to the court for prior approval, and changes to dies not were allowed after approval, as the engraver Pierre Turpin found out in 1601.[129] Flans or blanks could be prepared outside of the Paris mint, but silver had to come from the Paris mint and jetons struck at the mint. The case of the 1,000 jetons struck in 1570 for the Duc d'Anjou for distribution to the Gendarmerie seems typical, the Cour des Monnaies granting permission "under the condition that casting and alloying be done at the regular mint in Paris in the presence of the guards of the same mint who will keep registers and make reports of trials and their results to our said Court."[130] Once struck, jetons were sealed at the mint and brought back to the Cour des Monnaies for weighing trials. In the presence of judges, the essayer weighed and tallied up the finished jetons against the silver used and issued from the mint. Thus, for example, 2,600 jetons made of 40 *marc* of silver should weigh 11 *gros* each. This checking acted as a control on fraud and insured that the proper number were struck. After all this, the Cour des Monnaies would issue a certificate for the lot, and only then could jetons be distributed.

The Cour des Monnaies thus tightly controlled the production of jetons. Remarkably, as part of that process court inspectors took a portion of each run not only for inspection and testing for quality, but also to line the pockets of the court. An inspector's report of 1657 regarding the jetons for the Conseil du Roi makes plain this new role for jetons: "...from these jetons, whether of gold or silver, we have had a certain number taken, as much to make the necessary trials and counter trials as to distribute to the members of the Court, as is the custom."[131] This bonus was in addition to jetons received annually as a perk of office.

---

124. de Turckheim-Pey (1997), p. 41; A. Barthélemy (1875), p. 227. On the other hand, recent work by Marion Delcamp (2015a), Jérôme Jambu (2012, 2013b, 2014), and Sabrina Valin (2016) reveals that the AN still holds valuable archival materials related to the Cour des Monnaies in several series under the rubric Z^{1b}.

125. Mazerolle (1902–1904), vol. 1, pp. 501–82. On Mazerolle, see Darnis (2005), pp. 9–11.

126. See previous note and La Tour (1897), pp. x, xiii. It was not thought necessary to cite particular page numbers from Mazerolle (1902–1904) to support every point here. On *jetons banaux* see further Valin (2016), vol. 1, pp. 125–27.

127. "...lesdicts pourtraictz, legendes et devises," Mazerolle (1902–1904), vol. 1, p. 552.

128. "...ainsi que dict est...," Mazerolle (1902–1904), vol. 1, p. 542.

129. "[Pierre Turpin, graveur]...a esté defendu de graver lesdicts carrez sans permission de ladicte Cour...," Mazerolle (1902–1904), vol. 1, p. 577.

130. " à la charge de faire la fonte et alleaigne en la Monnoye ordinaire de Paris en la presence des gardes de ladicte Monnoye, lesquelz en tiendront registres, pour, iceulx rapportés en ladicte Court, en estre faict essay ainsi que de raison," Mazerolle (1902–1904), vol. 1, p. 569.

131. "...desquel jettons tant d'or que d'argent nous avons fait prendre certain nombre, tant pour faire les essays et contre essays nécessaires que pour distribuer à Messieurs de la Cour, ainsi qu'il est accoustumé," Mazerolle (1932), vol. 2, p. 81; see also related archival material reprinted in Mazerolle (1902–1904), vol. 1, p. 502; Jambu (2014), pp. 18–19, gives a further example of this traditional hustle.

The Cour des Monnaies flourished for over a century reaching the apogee of its power in the 1640s.[132] After that, however, its jurisdiction and authority diminished, and it was eclipsed largely as a result of expanding central authority of the Bourbon monarchs and the French crown itself. Royal legislation of 1645 took away control over some provincial mints, and in 1704 Louis XIV created a second monetary court in Lyon with jurisdiction over other mints.[133] Later changes in the 1760s and 1770s further undermined and humiliated the Cour des Monnaies, placing all matters concerning royal accounts, currency, and the mints firmly in the hands of the royal controller-general of finances.[134] More to the point here, legislation issued by Louis XIV in 1696 took away all authority of the Cour des Monnaies over jetons and their production.[135] From that moment and across the eighteenth century, the Cour des Monnaies remains largely in the background as far as the story of jetons is concerned. Nevertheless, the Cour des Monnaies continued as a large, august and expensive royal court down to the French Revolution, powerful enough to engineer the suppression of the monetary court in Lyon in 1771. Annual salaries for the top officers of the court in 1786 amounted to 93,701 *livres* paid for by the king.[136] The Cour des Monnaies ceased to exist in 1791.

Jetons in Contemporary Culture

Jetons and the exchequer system of performing arithmetic calculations were commonplace in early modern Europe. One report from 1729 labeled jetons, "these pieces of metal that are so widespread today," so one is not surprised to find that, as a metaphor or an image or tool to think with, they penetrated contemporary society and culture.[137] In language, for example, the French expression "*faux comme un jeton*" (false as a jeton, i.e., a liar), meaning not real money or not the real thing, dates from the fourteenth century and endures today.[138]

In the world of law and jurisprudence, contemporaries knew passing off jetons as real money as "theft by jetons" (*vol aux jetons*), a recognized crime that carried stiff penalties.[139] Two soldiers were arrested and thrown into the Châtelet for this in 1430, and in 1786 the Cour des Monnaies adjudicated a case concerning one Élisabeth Ambach, who was condemned to be pilloried for three consecutive days in front of the Hôtel de Ville in Metz from 11 a.m. to midnight with a sign: "Scammer, passing off copper jetons for gold crowns."[140] Ambach was banished from the Metz jurisdiction for five years, was assessed a fine of 12 *livres*, and the 12 jetons at issue were confiscated by the king, with the orders of the court published and posted in Paris, Metz, and elsewhere. In a related connection counterfeiters sometimes had jetons sown onto their clothes on the day of their execution. Conversely, confiscated counterfeit coins were sometimes used as jetons, but jetons were never counterfeited.[141]

132. Bluche (1966), pp. 14–15.
133. On the monetary court in Lyon, see Jambu (2012), p. 84; Préau (1890a), p. 124; Darnis (1996/1999), pp. 12, 38–39; Droulers (1989), p. 186; and materials in Mazerolle (1932), vol. 2, pp. 66–99. The Lyon monetary court had jurisdiction over mints in Lyon, Grenoble, Aix, Riom, Montpellier, Toulouse, Bayonne, and (in 1710) Perpignan.
134. Darnis (1988), p. 104 n; Darnis (1996/1999), p. 11.
135. Bouclier (1924), p. 107 and further below, chapter 7.
136. "Etat des Gages et autres droits des officiers de la cour des Monnoies de Paris." CAEF, Ms. A-1, 10 (7d), 1786.
137. "...ces piéces de métal qui sont aujourd'huy si comunes," AILB, H&M, vol. 5 (1729), p. 259.
138. Lerouge and Lerouge (1977), p. 26; de Turckheim-Pey (1997), p. 15; Pastoureau (1984), p. 25; Schärlig (2003), p. 16; Serrure (1897); Barnard (1916), pp. 78–79.
139. Roelandt et al. (2004), pp. 15–19. On the larger background to crimes involving currency, usually resulting in the death penalty, see Parsons (2014), pp. 193–236; Jambu (2012); Valin (2016), vol. 1, pp. 127–30.
140. "Escroqueuse, distributrice de Jetons de cuivre pour des Louis d'or," quoted in Serrure (1897), pp. 55–56; for similar cases see Jambu (2012), pp. 90–91.
141. Pastoureau (1984), p. 26. Barthélemy (1875), p. 229; Barnard (1916), pp. 8, 32. On various forms of counterfeiting (including the illicit use of genuine dies) and their prosecution in the Old Regime, see Jambu (2012).

At a more refined social level, on June 10, 1671, the literary *précieuse*, the Marquise de Sévigné, could write proudly to her daughter, "We found using these wonderful jetons that, counting all my small possessions, my wealth will have totaled 530,000 *livres*."[142] Montaigne lets us know that he doesn't know how to count either by jeton or by quill.[143] While in 1673 Molière's Argon in the *Malade imaginaire* verifies M. Burgon's bill "with jetons."[144] In Rabelais, Panurge suggests his importance, "by rattling around a big purse of common coins and jetons as if they were gold."[145] Jetons even show up in the pornographic classic of 1775, the *Paysan perverti* by Rétif de la Bretonne, where the hero, Edmond, complains to his brother that, "when we were at school under master Jacques I had the misfortune to learn to read, write, and count [jeter] faster than you."[146]

These resonances are not limited to the francophone world. Shakespeare is full of jetons, as in *Cymbeline*, for example, where the hangman declaims "Your neck, sir, is pen, book, and counters." Iago in the opening scene of *Othello* derides an opponent, "a great arithmetician...This counter caster." The clown in *A Winter's Tale* wonders how much his wool is worth, "I cannot do't without counters," and Troilus in *Troilus and Cressida* asks his brother:

> Weigh you the worth and honour of a king
> So great as our dread father in a scale
> Of common ounces? Will you with counters sum
> The vast proportion of his infinite?

Chaucer, Luther, Hobbes, Leibnitz, and Goethe all make oblique references to jetons because jetons were part of their world. A common trope was to see people as jetons, meaning pawns in power games. For the royal tutor, François Fénelon (1651–1715), for example, "Courtiers resemble jetons used for counting and count for more or less according to the whims of the prince." For the dramatist, Edmé Boursault (1638–1701), "We are jetons animated by the king," and Friedrich II of Prussia is quoted as saying, "When kings play with provinces, men are the jetons that pay."[147]

Non-French Jeton Traditions

Numismatically and historiographically, what one finds in France in the seventeenth and eighteenth centuries far outshines anything going on elsewhere in Europe with regard to jetons. No nation more than France came to embrace jetons and incorporate them into the common culture. But France was not the only nation to adopt them. In England jetons and the exchequer system were used for centuries, jetons being known locally as "abbey counters." But for some reason English manufacture of jetons collapses by the end of the fourteenth century with demand in England then being met by production from Germany.[148]

142. "Nous avons trouvé, avec ces jetons qui sont si bons, que j'aurai eu cinq cent trente mille livres de bien, en comptant toutes mes petites successions," quoted in Jaumonet (1985), p. 18.

143. On these and related references, see especially Schärlig (2003), pp. 18–24; Schodt (1873), p. 23; Barnard (1916), pp. 85–89. The ARTFL database of works published in French is to be noted in this connection; https://artfl-project.uchicago.edu/; accessed September 23, 2018.

144. "...avec des jetons...," quoted in Boutin (1966), p. 240; Pastoureau (1984), p. 20.

145. "...faisant sonner une grande bourse pleine d'escus du Palais et de gettons comme si ce feussent escus au Soleil," *Pantegruel*, I, ch. xx, quoted in Mowat, p. 403; Serrure (1897).

146. Restif de la Bretonne (1977), p. 22; Vallier (1879), p. 7 n.

147. "...les gens de cour ressemblaient aux jetons dont on se sert pour compter: qu'ils représentaient plus ou moins selon la fantaisie du prince," "Nous sommes des jetons que le roi fait valoir," quoted in Lhôte (1994), p. 273. "Quant les rois jouent des provinces, les hommes sont les jetons qui les payent," quoted by Schodt (1873), p. 25 n.

148. Delgrange (1987); Thompson (1990); van Beek (1986), p. 201; Barnard (1916), p. 5. On the earlier flourishing of English jetons, see Berry (1968); see also Blanchet (1890), vol. 2, pp. 449–50.

The Low Countries (modern Belgium and the Netherlands) stand out, too, and need to be included alongside France in thinking about jetons on a European-wide scale.[149] The practice of reckoning with jetons had long since established itself in the Low Countries, a pattern reinforced by government under Burgundian dukes. Jetons were manufactured there (and in Germany for use in the Low Countries), and in the sixteen and seventeenth centuries expanded social usages of jeton in these areas began to mirror the complexities of the French case. The scale was comparable. J.-F. Dugniolle's four volumes of 1880, for example, inventory nearly five thousand seventeenth- and eighteenth-century jetons of the Low Countries.[150] Feuardent counts over 1,500.[151] But Dutch jetons stand out for another, very particular reason, that is, their role as propaganda pieces in the long Dutch revolt against Spain across the years 1568–1648. More than a thousand different Dutch jetons were struck as propaganda pieces in this period, and, incredibly, they circulated to mark events and as insignia to reinforce Dutch morale. The independent Dutch and other provinces of the Low Countries continued to make and use them across the rest of the seventeenth and eighteenth centuries, and, as noted in our introduction, jetons of the Low Countries and jetons of the Dutch revolt in particular sparked the first numismatic interest in jetons among seventeenth- and eighteenth-century antiquarians and collectors. We previously mentioned the very substantial eighteenth-century works by van Loon and van Mieris, giving us, in the words of Victor Jaumonet, "an exceptional documentation on the jetons and medals of the Low Countries from 1405 to 1716" (Fig. 1.11).[152] But, alas, a larger cultural and institutional history of Low-Country jetons remains to be written.

FIGURE 1.11: Dutch jeton, 1578: David and Goliath.

Finally in this connection, attention needs to be drawn to Germany and the mining and metallurgical region around Nuremberg where stock jetons by the thousands were hammered out for European-wide distribution.[153] Across three centuries extending to the end of the eighteenth century whole families of Nuremberg manufacturers churned out jetons for sale: the Krauwinckels, the Lauffers, the Schulteses. The most famous producer, Hans Krauwinckel, was active between 1586 and 1635 and had agents in Paris and Amsterdam. Nuremberg *Rechenpfennige* were required to be stamped with the word "Rechenpfennig" and the name of the manufacturer (Fig. 1.12). They were cheap and of poor quality struck on thin planchets, with simple iconography and often

149. For entrées into Low-Country jetons and medals see Jambu and de Oliveira (2014); on their connection to the Dutch revolt, see Paris (1887), Blanchet (1890), vol. 2, pp. 451–53; Rulau (1964); Drewing (1970); van Beek (1986), pp. 201–04.

150. See Dugniolle (1880) and comment on the plethora of Low-Country jetons in Blanchet (1890), vol. 2, p. 451.

151. F.13550–15026.

152. Jaumonet (1985/1986), p. 13, van Mieris: "...constitue avec Van Loon une documentation exceptionnelle sur les jetons et médailles des Pays-Bays de 1405 à 1716."

153. On Nuremberg and these Rechenpfennigs and Rechenmeisters, see Schärlig (2003), pp. 27–28; Levinson (1998); van Beek (1986), pp. 201–07; Fontenay (1854), pp. 47–50; Blanchet (1930), pp. 216–17; Blanchet (1930), pp. 216–17; La Tour catalog (1910), pp. 373–433 for Rouyer's Nuremberg jetons.

FIGURE 1.12: Jeton: A Nuremberg *Rechenpfennig.*

embellished with Latin barbarisms. Many were routinely made especially for the large French market and featured the requisite royal insignia and portraits. Nuremberg jetons were cheaper than the better-crafted French ones, and the need for jetons for routine accounting purposes and gaming in France seems to have been met largely by these cheaper German jetons.[154] Contemporary French jetons, as we have seen, struck at the Monnaie des Médailles in Paris and under the supervision of the Cour des Monnaies, were more finely crafted and also more expensive, so there were always pressures for lower quality German jetons to make their way into France and, conversely, for German jetons to be kept out of France. Legislation increased over time to prevent contraband jetons, and the Cour des Monnaies issued edict after edict proscribing their import into France.[155]

There seems a paucity of jetons in other cultures outside of England, France, the Low Countries, and Nuremberg. Rulers in Spain, Austria, Italy, Russia, and Scandinavia did strike commemorative medals on the occasion of weddings or coronations, but not on a regular basis. Small exceptions aside, these regions of Europe seem to lack any indigenous tradition of jetons apart from whatever was the case regarding keeping accounts with the manual arithmetic of the exchequer.[156] Elsewhere in Europe through the eighteenth century we see nothing like the scale or social complexity of jeton production and use as seen in France, the Low Countries, and the extension in Nuremberg...France above all.

The reign of the Bourbon monarchs in France ran continuously from 1589 to 1792. It began with Henri IV (r. 1589–1610) and ended with the instauration of the French Republic in 1792[157] For almost exactly two centuries of Bourbon rule jetons continued to play their part as counters on counting tables. They remained small perks for those holding certain offices, and more and more, especially under the burgeoning bureaucracy of Louis XIV, they came to be appropriated by the state to symbolize and represent various administrative bureaus.[158] By the time of Henri IV and the turn of the seventeenth century jetons had already come a long way from their origins hundreds of years earlier on medieval counting tables. Yet, the explosion and efflorescence of French jetons after that and their ramification even further into contemporary French society and culture makes Old Regime jetons a distinct subset of jetons worthy of particular historical attention. The rest of this book is devoted to examining these jetons and what they have to tell us about the organization and character of France in the Old Regime.

154. Blanchet (1890), vol. 2, p. 405. See snide comments by Tricou (1972), p. 197, about eighteenth-century France being "infested" ("infesté") with German jetons displacing "our good French jetons" ("nos bons jetons français").

155. Blanchet (1930), p. 217, and archival material, CAEF. The distribution and use of Nuremberg jetons further to the east and south of German lands is a topic worthy of further exploration.

156. Compare Blanchet (1890), vol. 2, pp. 449–60, for his short list of jetons outside of France and the Low Countries.

157. The Bourbon monarchy was restored and continued 1814–1830.

158. Mainguy (1999), pp. 74–77.

FIGURE 2.1: A silver jeton of Louis XIV.

2.

Old and New in the Old Regime

"...nous dirons seulement que créés à l'origine comme instruments de calcul, ils reçurent ensuite des destinations nombreuses, et les occasions d'en émettre se multiplièrent à l'infini sous l'ancien régime."

—Jules Florange (1902), p. i.

The history of jetons truly covers the *longue durée*, in this case 1200–1800, and thus, by capturing the little snapshots of their time and culture that jetons offer here and there, we can discern the outlines of larger changes over longer periods. In the manner of a pointillist painting, again, we can connect the dots and bring into renewed focus the better known transitions from the European Middle Ages to an early modern period and the coming of the French state and on to the distinctive forms of social, political and economic organization that came to characterize France in the Old Regime or the *Âge classique*, as it is known to French historians. From this broad brush perspective, by 1500, as we have seen, jetons already had a 300-year history as tokens shunted around on the counting tables of medieval Europe, a function that would go on for another 300 years. Nevertheless, the three centuries that lay ahead from 1500 to 1800 represent the golden age of jetons (French jetons in particular), and they coincide with other developments in their world. After 1500 jetons took on a number of new and revealing functions in society and culture. Already in the sixteenth-century transition discussed in the previous chapter, jetons became evermore perks of office and institutional symbols of a burgeoning bureaucracy and the growing complexities of state and government. By 1600 jetons had engrained themselves into the workings of French society in these ways and in others we will explore further in the present chapter.

The heyday of jetons came in the next 200 years, or, again, coincident roughly with the reign of the Bourbon monarchs, 1589–1792. We can zero in even further on the century 1650–1750 as the ultimate apogee and true efflorescence of jetons and their importance in French society. This period covered the reign of Louis XIV and the better part of the reign of Louis XV through the late 1750s.[1] After that, jetons continued as before in a variety of social settings and social practices, but for reasons to be explored, by the later 1750s the golden age of jetons, if not of the Old Regime itself was fading. Then, of course, more than just practices surrounding jetons came crashing down with the French Revolution.

In the epigraph that heads this chapter Jules Florange in 1902 saw the extent to which jetons ramified throughout contemporary society in new and unexpected ways in the Old Regime. Sylvie de Turckheim-Pey echoed this thought in 1997 when she wrote that from the outset of the sixteenth century, jetons came to have "an important place in social relations."[2] Indeed, after 1500

1. Boutin (1966), p. 241, makes something like this point.
2. "...une place importante dans les relations sociales," de Turckheim-Pey (1997), p. 49; on these changes, see also Pastoureau (1984), p. 20.

and even more so after 1600 jetons came to be used in new ways and penetrated various levels of society and cultural practice. The present chapter is devoted to exploring some of these new ways jetons permeated society. Suffice it to say again that what started out as little counting tokens to a considerable degree ended up reflecting the structure and functioning of the whole of Old Regime France. Here we can use numismatics, or more specifically the numismatics of jetons, to shed light on Old Regime France and its history as a whole. An otherwise well-known, but very particular world becomes even more brightly illuminated.

Jeton Production and Costs, 1500–1800

The sheer scale of jeton production is to be remarked upon. Jetons were minted by the thousands every year from the Middle Ages onward.[3] Henri de La Tour (1897) speaks of "the enormous quantity of jetons that were struck in every epoch."[4] Michel Pastoureau counts half a million jetons from the early Middle Ages in public collections today.[5] Thousands of different jetons were struck over the decades, sometimes tens of thousands at one minting. If that scale was always large, it became even larger in the seventeenth and eighteenth centuries. We have already seen the explosion in jeton production after 1500 in the case of the Conseil du Roi, where the number of its jetons jumped two orders of magnitude from 700 in 1514 to 16,000 in 1638. But the Conseil du Roi was only one of many royal departments and offices, not to mention other bodies in society that struck jetons in the seventeenth and eighteenth centuries, and the totals are staggering.[6] Exceptionally, for example, 30,000 jetons were struck in Sedan in 1639.[7] For 1683, the Trésor Royal alone produced 26,000 jetons in silver and 700 in gold.[8] Jan Lippens notes that 20,000 and sometimes even 40,000 were struck a year for the Bureau des Finances in Flanders.[9] Claude Rossignol estimates that the Estates of Burgundy struck 100,000 silver jetons and 170,000 copper ones from 1742 to 1793.[10] And so on and so forth. Marie Veillon's 1994 estimate of 50,000 jetons minted annually pertains only to those produced and paid for by the crown, but even with that as a conservative estimate, the totality of jeton production comes to 10 million jetons struck in France over the 200-year period, 1589–1789.[11] A better overall estimate would probably be 15–20 million. Literally millions of these little things touched the lives of contemporaries, scattered in bureaus or in purses or bouncing around in the back of drawers. They were indeed Blake's grains of sand.

The jetons issued and paid for by the crown constitute a nontrivial subset of all of these jetons and are labeled administrative jetons.[12] Although the precise units varied from time to time, at least from the 1680s and growing out of ancient obligations of the king to supply jetons for his counting tables or otherwise provide jetons as perks to his officeholders, the crown annually underwrote the production of jetons for a standard set of court and administrative offices. These usually included the queen's household, the royal treasury, royal paying accounts, the office of venal offices, the army and artillery, the navy and the galleys, and the key administrative ministry, the Bâtiments

3. Gadoury (1986), p. 3.
4. "...l'énorme quantité des jetons frappés à toutes les époques," La Tour (1897), p. xxvii.
5. Pastoureau (1984), p. 27
6. Lippens (1969), p. 36, similarly notes the larger scale of jeton production in seventeenth and eighteenth centuries than earlier.
7. "au prix de 45 sols le cent," Blanchet (1930), p. 216.
8. Mainguy (1999), p. 72; Veillon (1994), p. 31; Blanchet (1930), p. 210.
9. Lippens (1969), p. 36.
10. Rossignol (1851), p. 10.
11. Veillon (1994), p. 28.
12. This set of government-sponsored jetons is studied in particular by Mainguy (1999); see also Josset (1978), pp. 337, 340.

du Roi. Many other units made up the contemporary court, administrative, or royally-sponsored bodies (such as the academies), and they annually or otherwise issued their own jetons, as did a wide range of yet other institutions, such as, for example, various guilds or the city of Paris. We will turn to all of these in due course, but for now it will repay to have identified this core set of administrative jetons and court and ministerial bodies with annual standing orders for jetons. The withdrawal of government funding for administrative jetons in 1758 marked the onset of the decline of jetons in French society.

The accompanying table summarizes the production of jetons for this core group of royal administrative units in the 1680s.[13] It becomes apparent that every year, decade after decade through the late 1750s, the royal Monnaie des Médailles in Paris banged out 50,000 or so jetons in silver and gold ordered and paid for directly by the crown for these administrative jetons. The Trésor Royal issued over half of the total for these administrative offices, and the leading role of the treasury at the center of jeton culture within the halls of government and court is both apparent and understandable, given the finances involved in running the French state.[14] The status of the august royal navy, the Marine Royale, and the associated department of the galleys likewise stand out in the production and ceremonial use of jetons within circles of royal officialdom and administration (Table 2.1).

Unit	Silver	Gold
Trésor Royal	26,000	700
Marine	4,500	
Parties Casuelles	3,500	100
Galleys	2,800	
Chambre aux Deniers	2,000	
Ordinaire des Guerres	2,000	
Extraordinaire des Guerres	2,000	
Bâtiments du Roi	1,600	
Artillerie	1,500	
Maison de la Reine	1,500	200
Maison de Mme la Dauphine	1,000	100
Total:	**48,400**	**1,100**
Grand total: 49,500		

TABLE 2.1: Administrative jetons for 1682

These are the so-called administrative jetons for the various departments of government and administration paid for by the king. New jetons were issued annually, and institutional series of these jetons continued essentially uninterrupted over very long periods as indicated in Table 2.2.[15]

13. Jacquiot (1968), p. lxxviii. Different authors report slightly different numbers from the Ur-sources of the *Mercure galant*, 1682(1), pp. 53–54, and 1683(1), pp. 258–72; see Mainguy (1999), pp. 72, 154–56; Veillon (1994), p. 28; Depeyrot (1980), p. 8. Note that the figures for these units varied over the years, too; for the Trésor Royal, for example, the 1683 *Mercure galant* reports 800 gold jetons, but (a possible misprint?) only 2,600 silver ones; for 1683 the number of silver jetons for the Maison de la Reine jumps to 6,100.

14. Veillon (1994), p. 31.

15. Mainguy lists many of these, p. 70. The longest dated series (230 years) was of the *Bureau des finances* in Flanders that went from Charles the Bold in the 1470s down to 1697; on this Low-Country series, see Lippens (1969).

Name of Unit	Period of Series	Duration of Series (in years)
Chambre aux Deniers	medieval to 1758	[centuries]
Chambre des Comptes	medieval to1552–1608	[centuries]
Conseil du Roi	1550–1662	112
Chambre des Monnaies	1554–1657	103
Artillerie	1563–1756	193
Marine	1593–1758	165
Extraordinaire des Guerres	1615–1778	163
Parties Casuelles	1616–1758	142
Bâtiments du Roi	1617–1758	141
Ordinaire des Guerres	1634–1757	123
Trésor Royal	1672–1758	86
Galleys	1680–1748	68
Maison de la Reine	1727–1758	31
Maison de la Dauphine	1746–1758	12

TABLE 2.2: Central royal administration jeton series and their dates

The production of government jetons represented only one portion of the total production of jetons. The city of Paris, for example, itself came to strike 24,000 jetons annually in silver and in brass, and the jetons issued by other bodies only added to the mounting number manufactured every year.[16] The production of so many jetons inside and outside of royal administration entailed considerable expenditures, and jetons were a decidedly nontrivial drain on budgets of units that ordered and distributed jetons.[17] Henri de La Tour (1897) understood the "enormous expense" involved in producing jetons.[18] In 1899 Émile Bonnet similarly remarked on the budgetary strain imposed on the Estates of Languedoc by the increasing number and expense of the jetons it issued, and Adrien Blanchet (1930) noted the same stresses on the budget of the city of Paris caused by the large number of purses of jetons it distributed.[19]

The budgetary parameters depended on the number of jetons ordered, the cost of silver (or gold, set officially by the Cour des Monnaies or royal edict), and the thickness of the planchets or blanks used to mint the jetons. In 1608, for example, the Conseil du Roi spent 6,567 *livres* for 12,300 silver jetons using 250 *marc* of silver (the *marc* being half a *livre* in weight); at 26 *livres*, 6 *sols* per *marc*, a roll of 100 jetons thus cost a little over fifty *livres*.[20] The Conseil spent an additional 2,936 *livres* for 314 *gold* jetons using 11¼ *marcs* of gold at 260 *livres* per *marc*, or something like 930 *livres* for a roll of gold jetons, roughly 20 times the cost of silver ones, gold being both heavier and more expensive than silver. As another example, in the same year, 1608, the city of Paris spent 2,235 *livres* for 3,400 silver jetons using 85 *marcs* of silver at the same 26 *livres* 6 *sols*

16. Veillon (1994), p. 28.
17. Lerouge and Lerouge (1977), p. 28; Gadoury (1986), p. 3.
18. "...des sommes énormes que [les jetons] avaient fait dépenser chaque année...," La Tour (1897), p. xxi.
19. Bonnet (1899) in Depeyrot (2007), p. 78; Blanchet (1930), p. 211 ("à cause du grand nombre de bourses distribuées"). On the expenses involved, see also Barnard (1916), pp. 76–77.
20. On these and related examples, see Cour des Monnaies data reprinted in Mazerolle (1902–1904), vol. 1, pp. 537, 567, 578–79, 582; the figures have been rounded off. Note that the French *livre* (like the English pound) was a unit of currency as well as a unit of weight; see Glossary.

per *marc* of silver, with the resulting roll of 100 costing a little over 60 *livres*. For the city of Paris forty years later in 1647 the number of jetons produced and the associated costs had gone up considerably: that year it spent 8,400 *livres* for 9,550 silver jetons struck using 254 *marcs* of silver at 33 *livres* per *marc* for a total cost of almost 90 *livres* for a roll of 100, in addition to the 106 *livres* the city had to pay for engraving the dies.[21] Fifty or 60 or 100 *livres* for a roll of silver jetons was a not insignificant sum, being about 10 to 20 times the average day's wages for a skilled artisan, but the sums really added up when it came to providing dozens of such rolls of silver jetons, not to mention gold ones. And, these expenses only increased as the price of silver rose from 20 *livres* per *marc* at the beginning of the seventeenth century to 33 *livres* in 1677 to 57 or even 69 *livres* by the end of the eighteenth century.[22] Thus, the Brittany estates spent 10,359 *livres* on jetons for its biennial session in 1782–1784.[23] In the 1770s, to pick another example, the Estates of Languedoc were spending almost 11,000 *livres* a year on jetons, 13,000 *livres* in 1787.[24] In 1754 the queen's household, the Maison de la Reine, set aside 12,000 *livres* for jetons to be distributed as New Year's gifts.[25] And so on and so forth. Individually and collectively these constituted extraordinary sums spent on jetons.

Continuations of Jetons as Accounting Tools and Perks of Office

As much as jetons and practices surrounding jetons came to expand and diversify in the seventeenth and eighteenth centuries, we must not lose sight of the fact that the exchequer system had not gone away and that jetons continued to be used on counting tables everywhere to the end of that period.[26] The jetons of the royal administration just seen are only one locus and indicator of where such accounting activities continued to take place. In addition to silver and gold jetons which served more ceremonial and remunerative functions, millions of other tokens were struck in common metals for accounting purposes and continued to be used year in and year out by expert manipulators operating at the casting counter. Thus, for example, again back in 1608, in addition to the 3,800 jetons it had struck in silver, the city of Paris ordered an additional 4,400 in brass (*laiton*) for lesser exchanges and more ordinary accounting purposes.[27] Jetons continued to serve as an accounting tool on the counting tables of the king, for the burgeoning bureaucracy, and elsewhere in increasingly vibrant realms of economic activity across the board in society... from the tavern keeper, to manorial managers, to city councils, regional authorities, the church, and elsewhere through the top ranks of society and government and everywhere accounts needed to be kept and routines established to keep them.

Then, too, packages of 100 silver jetons continued to be given to various office holders as part of a package of *épices*, or an annual perk or right of office. We saw this development in the fifteenth and sixteenth centuries in the Chambre des Comptes and in the Cour des Monnaies, and the tradi-

21. Order reprinted in d'Affry de la Monnoye (1878), p. 244. Blanchet (1930), p. 244, and Lerouge and Lerouge (1977), p. 43, give additional examples of the costs involved. Lacronique (1902b), p. 39, provides further data for 1699. For 1671, Jambu (2014), p. 19, reckons a roll of 100 gold jetons were worth 1,300 *livres*; a roll of 100 in silver, 65.
22. See CAEF, Ms. 8, passim and Ms. A-1, 11 (unnumbered): "Prix du Marc D'argent"; see also Bonnet (1899) in Depeyrot (2007), p. 84; Mainguy (1999), pp. 8, 168–70, who indicates prices as high as 93 *livres* per *marc* at some points in the eighteenth century; Delcamp (2015a), p. 501, puts the price of a *marc* of silver in 1771 at an unbelievable 491 *livres*. Costs may have been more stable given coinage debasements.
23. Daniel (1980), p. 148.
24. Bonnet (1899) in Depeyrot (2007), p. 129 n.
25. AN, KK 225b, 1754 entry for the Maison de la Reine "pour *les Étrennes* de Sa Majesté."
26. Thierry Sarmant, private communication. Blanchet (1930), p. 204 and La Tour (1897), p. xiii, also note that the primary use of jetons through the end of the eighteenth century was still on counting tables; see also Lippens (1969), p. 36.
27. Mazerolle (1902–1904), vol. 1, pp. 578–79.

tion and the "right" to jetons continued through the eighteenth century.[28] The principal officers of the Estates of Brittany, for example, started to receive their statutory *Gits de jectons* in 1641.[29] This practice continued uninterruptedly in the background and frames the expanded uses of jetons elsewhere in contemporary society.

Cult of the Monarchy

The French crown tightly controlled the production of money, medals, and jetons, and already in the sixteenth century the imprint of the French state began to show itself in standardized jetons issued for the Conseil du Roi, the Chambre des Comptes, the Cour des Monnaies, and other bodies of state, as we have seen, and this development continued into the seventeenth century. Nevertheless, by dint of his power and personality, Louis XIV turned his medals and jetons ever more into instruments of propaganda and tokens to promote the veneration of the monarch and the monarchy.[30] From 1650 onward essentially all jetons glorify the king, notably given the royal bust that almost universally graced the obverse of these coins along with other emblems of national, regional, and local authority.[31] Jetons became more oriented to celebrating the monarchy as the seventeenth century wore on into the eighteenth century.[32] This newer propagandistic and ideological function is seen in virtually every French jeton struck from Louis XIV forward.

Stylistically, French jetons of the era of Louis XIV, Louis XV, and Louis XVI are distinctive and represent a break with earlier jetons. This break is evident in the 1770 jeton for the Estates of Brittany (Fig. 2.2).[33] The obverse depicts an august Louis XV in the manner of a Roman emperor, crowned with laurels and the honorific of Most Christian King. The reverse, tellingly, offers the arms of Brittany, already adorned with lilies, alongside the arms of France and now encircled with a cloak of royal authority and set beneath the royal crown.[34] The irony here and indicative of the effectiveness of such royal propaganda is the fact that the Estates of Brittany were the most centrifugal of all the provincial estates making up Old Regime France that resisted central authority.[35] But they showed their obeisance in their jetons.

FIGURE 2.2: Jeton of the Estates of Brittany, 1770.

28. Mainguy (1999), pp. 160–64. Note that distribution is by function of office, i.e., jetons went not to individuals, but to people occupying positions. This understanding reinforces what Mainguy (1999), p. 164, sees as "une structure hiérarchisé où chaque département devient un corps à part entière."
29. Daniel (1980), p. 7.
30. On this point see especially Burke (1992); Wellington (2015a).
31. See especially Jones (1979b) and (1983) for medals as propaganda and Veillon (1994); for a detailed special study of propaganda in the consular jetons of Lyon, see Loach (1995). See also Rossignol (1851), chapter 16; d'Affry de la Monnoye (1878), pp. x–xi. For a larger take on this same subject that emphasizes the nineteenth and twentieth centuries, see van Wie (1999).
32. Veillon (1994), p. 28; Sarmant (1994), pp. 88–89.
33. The Estates were the issuing authority for this jeton, not the king.
34. See also the visually comparable 1746 Burgundy Estates jeton pictured above, introduction, ln. 5.
35. The so-called Brittany Affair of 1763 involved the regional parlement; see Rothney, ed. (1969).

Among scholars, wide agreement exists regarding this propagandistic and ideological dimension to contemporary jetons. Anne Mainguy, for example, sees "a language of the elite par excellence" in the political, symbolic, and ideological messages embedded in jetons.[36] In particular, we can speak of the cult of the monarchy and a veneration of the person of the king and the radiant authority of the sovereign as bestowed by God.[37] With its jetons the army accounting department (the Extraordinaire des Guerres) seems to have gone overboard in its adulation; Marie Veillon comments that their design unashamedly "extolled the heroic gesture of the sovereign via his Herculean powers."[38] In 1994 she summed up the situation nicely concerning propaganda and the cult of the monarchy as evident in late seventeenth- and eighteenth-century jetons:

> Under the cover of witty entertainment, the designs of jetons, were put to the service of monarchical glory. They exulted devotion to the King and had as their principal element the insignia of the King, the King as the subject of the design, or devotion to the royal order; for themes, they conveyed the power of the State or the subordination of other bodies and officials.[39]

But we can ask how much were these propagandistic impulses directed toward the person of the king as a royal being or to the king as the embodiment of the emerging state? Paul van Wie in his 1999 study of imagery in the coinage of modern Europe cautions us not to go too far in conflating veneration of the person of the king with demands for loyalty to the state. He writes, "The denominations and iconography were products of a polity focused on the person of the monarch rather than on a nation in the modern sense, while established religion played a conspicuous and important role."[40] Van Wie's position goes too far, however, for in contemporary jetons we see both simultaneously, the state to which allegiance is owed and the person of the king to be revered. The editor of the monthly gazette, the *Mercure galant*, in 1682 struck a better balance between jetons as part of a personality cult and as emblems of state power and authority:

> The surprising diversity of jetons made in the kingdom every year signals the power and greatness of the State. The royal Treasury alone distributes twenty-six thousand of them in silver and seven hundred in gold. All the others [departments] make generous gifts in proportion, and given that several sorts of jetons are made, one can say that their number mounts almost to infinity. France alone is capable of sustaining this so often recurring expense, and she is today the first of all nations, just as her King is the source of their admiration and astonishment.[41]

36. "...un language d'élite par excellence," Mainguy (1999), p. 302.
37. On the king as a divine personage, see Mousnier (1990–1992), vol. 1, pp. 505–25; vol. 2, pp. 15–31; Barbiche (1999), part one.
38. "Les devises de l'Extraordinaire des Guerres chantaient via les prouesses herculéennes la geste héroïque du souverain," Veillon (1994), p. 35.
39. "Sous couvert d'un agréable divertissement de l'esprit, les devises de jetons, mises au service de la gloire monarchique, exaltaient la dévotion au Roi, avec la devise du Roi pour corps principal ou le Roi pour sujet de devises, ou, la dévotion à l'ordre royal avec pour thèmes de devises la puissance de l'Etat ou la sujétion des corps et des officiers," Veillon (1994), p. 39.
40. van Wei (1999), p. 3.
41. "La surprenante diversité de jetons qu'on fait tous les ans dans le royaume marque la puissance et la grandeur de l'État. Le seul Trésor royale en distribue vingt-six mille d'argent et sept cens d'or. On fait largesse de tous les autres à proportion; et, comme il s'en fabrique de plusieurs sortes, on peut dire que le nombre monte presque à l'infiny. La France seule est capable de soutenir une dépense si souvent réitérée, et elle est aujourd'huy la première de toutes les nations, comme son Roy en est l'admiration et l'étonnement," *Mercure galant*, 1682(1), pp. 53–54; also quoted in d'Affry de la Monnoye (1878), pp. 299–300 and Guichon de Grandpont (1862), p. 16.

At Versailles Louis XIV and his successors would occasionally organize large entertainments (*festes*) for the pleasure and amusement of the royal court and invited guests. These would be replete with costumed courtiers, lavish decorations, ample food and drink, water displays, music, and fireworks. As further evidence of his beneficence and largesse, on these occasions the king would hand out jetons and medals as presents to guests and visiting officialdom.[42] These entertainments are another revealing locus where jetons intersected with, in the minds of some, the essentially divine royal person of the king. We will see a related instance below in the case of jetons used in royal coronations.

New Year's and Les Étrennes

Perhaps the most surprising new function for jetons that developed from the sixteenth century was the custom of offering little purses of jetons as presents on New Year's Day as part of seasonal greetings.[43] The custom began at the French court and expanded elsewhere into the administration and further in society. These ceremonial and seasonal greetings shared at the start of the new year were (and still are) known in French as *les Étrennes*. Such rituals harked back to Roman times; they were common throughout Europe, and seasonal greetings like these continue in one form or another widely today. The remarkable thing for us is that from the sixteenth century in France and the Low Countries little packages of not just anything, but of *jetons* came to be handed out and exchanged as trinkets on the new year. In other words, the jetons of the counting table became transmogrified into New Year's Day presents (*jetons d'étrennes*). In France the custom of distributing these *jetons d'étrennes* began with François I, reigning in 1515. On the first of January every year he handed out jetons for the Étrennes to members of the queen's household and the households of the dauphin and other princes, as well as the Conseil du Roi.[44] The practice may have begun slightly earlier in 1510 in the Low Countries; the first real evidence of *jetons d'étrennes* there dates from 1528, but by that time their use was already seen as customary.[45] Back in France, evidence from the 1560s shows the Cour des Monnaies approving the striking of *jetons d'étrennes*.[46] One way or another, by the end of the sixteenth century, the custom of exchanging New Year's *jetons d'étrennes* had become well rooted in society or at least in court and administrative circles to the point where we might speak of the madness of the Étrennes as evermore luxurious purses of 100 silver jetons came to change hands.[47]

However much it was already installed, the tradition of *jetons d'étrennes* seems to have taken off after 1589 and the ascension of Henri IV, the first of the Bourbon kings.[48] Henri IV took a direct personal interest in the jetons issued under his authority, and he and his first minister, the Duc de Sully (1560–1641), beyond other weighty matters, consulted over the jetons for the upcoming year

---

42. For example, those handed out to "M. l'Ambassadeur d'Espagne et autres Seigneurs" in 1728; AN, O¹ 2064, "Registre des ordres de Paiment, 1711–1736." See also, Wellington (2016), pp. 291–92, 296.
43. On *jetons d'étrennes* and these essential details, see Valin (2016), vol. 1, p. 9; Mainguy (1999), pp. 157–60; Gadoury (1986), p. 3; Jaumonet (1985), pp. 18–19; Josset (1978), p. 337; Blanchet (1930), p. 208; Barnard (1916), pp. 79–80; Blanchet (1890), vol. 2, p. 401. For a revealing study of *jetons d'étrennes* struck by Austrian authorities in the Low Countries, see Hoc (1950).
44. Mazerolle (1907b), p. 4.
45. Lippens (1969), p. 34; see also Schodt (1873), p. 19; de Witte (1893), p. 6. A. Barthélemy (1875), p. 252, provides a 1519 reference to *jetons d'étrennes*.
46. Mazerolle (1902–1904), vol. 1, pp. 549–50.
47. Josset (1978), p. 337; Lerouge and Lerouge (1977), p. 28; Blanchet (1930), p. 211. In the Low Countries *jetons d'étrennes* sometimes came in small cylindrical tins made of silver; see illustration in Schärlig (2003), p. 15; van Beek (1986), figure 33; Hoc (1947), p. 106, mentions jetons distributed in paper rolls.
48. Mainguy (1999), p. 305; Mazerolle (1888), p. 360, notes this trend in the case of the Chambre aux Deniers.

and together devised Henri's jetons.[49] Sully's private journal affords us a close-up look at what took place when these were distributed at the Étrennes.[50] Sully wrote:

> Early morning, go to wish the King and the Queen good day and a happy new year and to bring them their customary purses of jetons, the design of which consisted of a golden breastplate or shield set against branches of green laurel with these words around: NIMI PLEBIS AMOR (Unbounded love of the people).[51]

Sully brought three sacks with him: one contained three purses of gold jetons and 10 purses of silver jetons; the second, carried by a page, held 25 purses of silver jetons; 30 rolls of new *demi-francs* (i.e., cash) to be distributed to the ladies-in-waiting filled the third. The king, queen and dauphin each received a purse of gold jetons.[52] The one for the dauphin was supposed to go to his governess, unless the queen kept it, as she had in the past. Of the eight purses of silver jetons, the king received two with the rest distributed to the queen and the queen's household that included the dauphin, ladies-in-waiting, wet nurses, and a variety of other attendants. Twenty-five purses of silver jetons went to others in the royal household and suite.[53] Sully brought with him two other big sacks containing 12,000 *sols* (600 *livres*) for distribution to the poor. The occasion of the Étrennes at the court of Henri IV—repeated and recorded in Sully's journal from 1600 to 1609—occasioned banter and celebratory goodwill all around.[54]

In part these *jetons d'étrennes* were given out as tokens of appreciation for royal service in circumstances where passing on cash might appear unseemly for the giver or recipient. For those who received them silver *jetons d'étrennes* had real monetary value and were doubtless swapped for hard currency. The silver trade and the guild of silversmiths were highly regulated, and so authorities always knew the quality of silver items, and anything in silver, including jetons, could be redeemed at royal mints for cash at a set rate.[55] Gold jetons were undoubtedly quickly exchanged or melted down. The gold jetons offered to the king were made into gold dinner plates for the royal service; Louis XV had six plates made every year from gold jetons he received at the Étrennes.[56]

The presentation of *jetons d'étrennes* took on another character as the custom migrated beyond the court per se and into the central institutions of government that by the seventeenth century had achieved an independent bureaucratic existence. In these and related instances the treasurers of various royal offices and departments, i.e., the various bureaus included among the abovementioned administrative jetons (the Treasury, the Army, Navy, etc.), would exchange *jetons d'étrennes* among themselves in an act of institutional solidarity that happened to fatten the pockets of recipients. As the *Mercure galant* put it in 1707: "This year the Monnaie des Médailles struck the eleven

49. On these points see especially Longpérier (1863); also, Barnard (1916), p. 67; Mainguy (1999), pp. 34–36; Jacquiot (1980), p. 857.
50. Longpérier (1863), pp. 443–45.
51. "...dès le matin donner le bon jour et le bon an au Roi et à la Reine, et leur porter leurs bourses de gettons accoustumées, desquels les deuises estoient une targe ou bouclier d'or sur une touffe de lauriers verdoyans avec ces paroles alentour: NIMI PLEBIS AMOR," quoted in Longpérier (1863), p. 443.
52. It was already the custom for French kings to receive little purses of gold jetons on the first of the year; Blanchet (1909a) p. 399.
53. Longpérier (1863), pp. 444–45.
54. Longpérier (1863), p. 427; see also Boutin (1966), p. 240. For a close-up view of the continuation of this tradition and the distribution of jetons within the royal court in 1671, see Jambu (2014), passim and especially pp. 17, 19.
55. See CAEF, Ms. 8 passim, where every kind of item in silver is specified to be marked in a certain way: snuff boxes, chocolate boxes, etc. Weil (1988), p. 252, notes that many silver jetons were melted down in the Revolution. The rule, per Delcamp (2015b), p. 219, seems to be that for every 1,000 jetons struck only one survives today.
56. d'Affry de la Monnoye (1878), p. xvi; Barnard (1916), p. 29. Schodt (1873), p. 15 n.

jetons that it customarily strikes every year. Many treasuries distribute purses on New Year's Day filled with one hundred gold or silver jetons, according to the rank and the right of those to whom they are given."[57] Within institutions, receiving *jetons d'étrennes* became ever more explicitly a right of royal office holders, and this development in large measure only reinforced the longstanding practice of receiving jetons as part of a package of *épices* and a perk of office.[58] The novelty here is the *jetons d'étrennes* came to be exchanged externally to institutions and between and among the accounting bureaus of government in particular. In this way the court, government and then other elements of society (notably the city of Paris) became involved in a larger pattern of institutional self-reinforcement and collective identity expressed through the seasonal offering of little purses of jetons at the Étrennes.

The Trésor Royal, for example, distributed its 26,000 jetons to 260 recipients of various ranks and stations.[59] The archival record provides several close-up examples of how jetons were exchanged in this way and how this social and institutional affirmation took place. In 1651 the army account, the Ordinaire des Guerres, for example, had 82 purses of 100 jetons to distribute. The king got one, presumably gold, as did the all-powerful chief minister, Cardinal Mazarin; other purses went to other royalty and nobility, high government officials and functionaries, and to masters of accounts at other exchequeries; the rest were distributed internally within the Ordinaire des Guerres.[60] The cost of these jetons was a substantial 6,932 *livres*. About this distribution Adrien Blanchet remarked in 1950, "This custom of purses of jetons was a real institution that events did not modify. From the historical point of view this realization is of real interest."[61]

The Chambre aux Deniers offers another example. Documented from 1688 to 1714 50 people received purses of jetons from the Chambre aux Deniers at the first of the year: The king, queen, other royal family, the controller general of finances, the chief officer of the Conseil du Roi, the state secretary for the King's Household (the Maison du Roi), the head of the Trésor Royal and its first officer, in addition to thirty-six other recipients within the Chambre aux Deniers itself.[62]

This set of seasonal exchanges of purses of jetons at the Étrennes came to involve the Académie Royale des Inscriptions et Belles-Lettres in a particular way. The Académie des Inscriptions et Belles-Lettres was the royal society devoted to history, numismatics and epigraphy founded by Colbert in 1663. We will have more to say about this academy and its ongoing role in the story of jetons, but suffice it to remark here that the Académie was largely responsible for creating the numismatic designs for administrative jetons, and hence the treasurers of those units would customarily offer the Académie a purse of their jetons every year in return. In this way the Académie des Inscriptions became caught up institutionally with other bureaus of government in the madness of the Étrennes.

Thus, in 1757, for example, that Académie des Inscriptions received 11 purses of jetons from various offices of government.[63] In 1695 for some reason the treasurers of the various government accounts did not send their purses of 100 silver jetons as usual to the Académie des Inscriptions,

57. "On a frappé cette anneé à la Monnoye des Médailles les onze jettons que l'on a coûtume d'y fraper tous les ans et dont plusieurs tresorieres distribuent le premier jour de l'année des bourses remplies de cent jettons chacune, d'or ou d'argent, selon le rang et le droit de ceux à qui ces jettons sont donnez," *Mercure galant*, 1707(1), pp. 162–63; also quoted in d'Affry de la Monnoye (1878), p. 319.

58. On this point see Boutin (1966), p. 240; Jaumonet (1985), p. 19.

59. Mainguy (1999), pp. 154–56.

60. Blanchet (1950).

61. "Cette coutume des bourses de jetons était véritablement une institution que les événements ne modifiaient pas. Et cette constatation présente un réel intérêt du point de vue historique," Blanchet (1950), p. 200.

62. Veillon (1994), p. 28; see also Mazerolle (1888).

63. AI, AIBL archives, A:63, minutes of the meeting of January 11, 1757.

and state minister Pontchartrain had to remind payers of the departments to return to their former usage and do so.[64] The scramble within the Académie des Inscriptions de Belles-Lettres over not receiving purses of *jetons d'étrennes* in 1737 is particularly revealing of how matters worked.[65] As noted in the minutes of its meeting of January 8, 1737, M. Bertin, treasurer of the Parties Casuelles, sent over a purse of 100 jetons to the Académie which was divided up on the spot among the academicians present. Claude Gros de Boze (1680–1753), permanent secretary of the Académie, remarked that he knew that the majority of treasurers were feigning not to give jetons that year because the Académie had not furnished them with any new designs and that the administrators of each department were content to choose the ones they thought best from among the old ones. Upon which, the minutes went on to note, it seemed to the Académie that it would be inappropriate to make any solicitation to get them, but to be content with jetons offered in good faith. Nevertheless, as it turned out, other treasurers did eventually come forth with the traditional purses: Marmontel of the Trésor Royal, Le Pestre of the Ordinaire des Guerres, du Sauroy of the Extraordinaires des Guerres, Denys of the Bâtiments du Roi, Le Fèvre of the Galleys, and Bertin of the Parties Casuelles, as mentioned, all sent over to the Academy purses of 100 silver jetons which were eagerly divvied up among the academicians over the course of January 1737.[66] The purse itself that the particular jetons came in went to the academician who furnished the design for the jeton.[67] Individual academicians who suggested *devises* while the academy was on vacation received the whole purse from the soliciting institution.

The custom of handing out purses of jetons at the Étrennes extended beyond ceremonies acted out at court or inter institutionally among government accountants, and these additional exchanges at the Étrennes further reinforced the existing power structure and existing institutional arrangements. The mayor and aldermen of the city of Paris (the *prévôt des marchands* and *échevins*), for example, participated, annually traipsing out to Versailles at the first of the year to hand out gold and silver jetons and other presents to the king and other dignitaries at court, as we will see further.[68]

The custom of jetons and the Étrennes was not limited to the Paris-Versailles axis or institutions of central government alone. Rather, the practice spread across the whole of France and contemporary French institutions in a pattern that reflected the established order of the Old Regime. The Estates of Burgundy provide a revealing example. In 1573 it distributed to its own officers a modest 11 purses containing 50 jetons for a total of 550 jetons.[69] By 1679 the total number had risen to 12,000 (200 gold, 2,700 silver and 9,000 copper), and that number only increased over the course of the eighteenth century, such that by 1789 the Estates of Burgundy handed out almost 17,000 jetons (200 in gold, 6,650 silver jetons in 66 purses, and 10,000 in copper). Gold jetons went to the Prince de Condé, head of the cadet branch of the Bourbon clan and an important person at court, and to his eldest son the Duc d'Enghien. The royal provincial governor and the intendant each received purses, as did the mayor and vice-mayors of Dijon, jurists within the Estates (the *Alcades*), representatives who paid compliments to the king on the part of the Estates, and a host of others, to the point where the historian of the Burgundian estates, Claude Rossignol, can speak of "abuses" in the distribution of jetons not actually authorized by statute.

64. Per Guichon de Grandpont (1854), p. 7.
65. On this episode, see Josset (1978).
66. Paraphrase of AIBL minutes as given in Josset (1978), pp. 339–40; also Blanchet (1930), p. 231.
67. Blanchet (1930), p. 211; Veillon (1994), p. 31.
68. Blanchet (1930), p. 211, and below, next chapter.
69. On these points see Rossignol (1851), pp. 39–45, 234, 281–83; also, La Tour (1897), p. xii; Barnard (1916), p. 29.

The same pattern can be seen in the packages of jetons handed out by the Estates of Brittany. From their origin in 1641 the right to jeton (*git de jeton*) went to the principal officers within the estates as a perk, but soon the distribution extended more widely. By 1655 the Brittany estates distributed 1,300 jetons in purses of 50 and 100 jetons, but the number of such purses skyrocketed in the second half of the seventeenth century.[70] Over the course of the years the Estates of Brittany issued 151 different jetons and came to distribute some 6,200 silver jetons in 62 purses. As well as an expanded number of officers and employees within the estates, recipients came to include the royal intendant and other royal agents and their wives, a representative of the local Chambre des Comptes, local dignitaries who attended sessions, select members of the three estates constituting the Brittany estates, and persons not part of the estates but who rendered it some service. Other French provincial estates did the same, as we will see further.

From the seventeenth century ruling elites and institutions in towns also seasonally handed out jetons. Judi Loach's remarkable 1995 study of consular jetons issued by the Lyon town council shows how this took place and what it meant in the urban context.[71] Lyon started issuing jetons in 1624, and Loach points to the "civic function" they served and how much they were a "ritual ceremony."[72] Recipients included not only a number of officers within and serving the town council in the older manner as a perk or salary supplement, but in addition to a wide range of individuals outside of the council itself, again reinforcing networks of power and the status quo. The happy few included the royal governor and intendant, the local archbishop, others holding local administrative or police authority, officers at the mint, notable local bankers and doctors, its own chargé d'affaires in Paris, other treasurers and exchequeries in Paris, and persons it wished to thank, recompense for services, or otherwise protect or curry favor with, such as the local fraternity of archers, whose goodwill and expertise the town council might need in a pinch. In an instance suggesting that purses of jetons became less seasonal gifts than required gestures as part of the order of things, in 1646 the town council of Lyon decided not to issue jetons, but its representative in Paris and the administrative treasuries in Paris insisted, "that they should be offered a comparable gift to that which they had received in 1643, namely a purse of jetons."[73] One way or another the jetons of the Lyon town council, as elsewhere, continued to be handed out year after year through 1789.[74]

Jetons entered the culture of the great French trading company, the Compagnie des Indes, which similarly sprinkled its purses of *jetons d'étrennes* here and there, its purse of gold jetons going to the Duc d'Orléans during his regency.[75] Jan Lippens in his 1969 study of the bureau of finances at Ghent indicates that as many as 200 functionaries and 500 people in total may have received *jetons d'étrennes*.[76] Lippens is certainly right in saying that if "1470 is incontestably the era of jetons as accounting tokens [*jetons de compte*], 1697 is undoubtedly that of the *jetons d'étrennes*."[77] We do not pretend to have documented the full range of estates and towns that participated in exchanging jetons at the Étrennes, but the tradition was widespread. As gratuities and gratifications distributed on the part of French institutions at the first of the year, *jetons d'étrennes*

70. On these points see Daniel (1980), pp. 7–8, 148; also Jaumonet (1985), p. 19. On the jetons of the Estates of Brittany, see also Feuardent (F.8704–8793), jetons struck from 1594 to 1788.
71. Loach (1995); see also Blanchet (1930), p. 366.
72. Loach (1995), p. 28.
73. Per Loach (1995), p. 31.
74. On the case of the town of Moulins, for example, see Fontenay (1854), p. 143; further on civic jetons, see next chapter.
75. Blanchet (1909a), pp. 398–99; Blanchet (1930), p. 271.
76. Lippens (1969), p. 36.
77. "...1470 est incontestablement l'époque du jeton de compte, 1697 celle du jeton d'étrennes," Lippens (1969), p. 33.

continued to reinforce the existing order across the eighteenth century, ceasing in only November 1789 when the Constituent Assembly prohibited public officials from receiving gifts at the Étrennes or budgeting of public funds for the Étrennes.

Parlor Games and Jetons in the Mercure galant

The *Mercure galant* was a breezy periodical gazette launched in 1672 by the writer and royal historiographer, Jean Donneau de Visé (1638–1710). Published monthly in Paris in-8° with a royal imprimatur, the *Mercure galant* reported gossipy news of court life, interesting events, ceremonies and appointments, deaths and weddings, the general comings and goings of the great, happenings in the world of theater, music, the fine arts, and fashion, and it featured little stories, popular songs, bits of poetry and literary divertissements. Donneau de Visé's editorship ceased with his death in 1710, but the periodical continued after 1724 as the *Mercure de France* all the way through to the French Revolution. It was a huge success, and everyone who was anyone read it and hoped to see his or her name in it. It was the *Vanity Fair* of its day.

For us, the significance of the *Mercure galant* stems from the fact that every year from 1680 through 1758, it announced and provided an engraving of the principal jetons issued for the upcoming year.[78] The engraving of jetons that appeared in the January 1682 issue pictured here is entirely typical (Fig. 2.3.) The jetons mentioned in the *Mercure galant* were mainly the administrative jetons paid for by the Crown, but from time to time other new jetons for the year were announced to the public, including jetons for the Académie Française, the city of Paris, the Estates of Burgundy, notaries, the royal secretaries, etc. As d'Affry de la Monnaie remarked in 1878 of Louis Quatorze high society, "In this era everything became literate and refined, grist for intelligent curiosity. Jetons were the great novelty of the year, and they were part of the vogue of the Étrennes."[79]

Donneau de Visé's reasons for including notice of jetons in the *Mercure galant* were only partly to inform his public of gallant seasonal novelties cascading from the cornucopia of the king and played out in the ceremonies of the élite. No, he had another reason, too, and that was to feed the game among aficionados of decoding the meanings conveyed in the design of a jeton. To understand this activity, we need to grasp what is known in French as a *devise* (emblem). (Although the English word "device" does include this meaning, there is no good English equivalent; perhaps "motto" comes closest.) According to the fourth edition of the *Dictionnaire* of the Académie Française (1762), a *devise* is, "an image accompanied by words that express in a concise, allegorical manner, some thought or sentiment. The image portrayed in the *devise* is called the body of the *devise* and the words the soul of the *devise*."[80] A *devise* is thus the ensemble of a jeton's design that principally included a symbolic image or emblem (called its *type*) and a legend and that carried an implicit message.[81] To give an example of a more standardized jeton, the *devise* of the Académie des Sciences always featured Minerva (the Roman goddess of knowledge and wisdom), scientific

78. Regarding jetons in the *Mercure galant*, d'Affry de la Monnoye (1878), pp. 291–381, reprints all the excerpts from the *Mercure galant* from 1680 to 1758 dealing with jetons. See also Veillon (1994), p. 28; La Tour (1897), p. vi; listed with engravers in Mainguy (1999), pp. 156, 177–80, 314–15; Rouyer (1879), p. 6; d'Affry de la Monnoye (1878), p. iv.

79. "...tout devenait, à cette époque paisible et lettrée, matière à curiosité intelligente. Les jetons étaient la grande nouveauté de l'année; il partageaient la vogue des étrennes...," d'Affry de la Monnoye (1878), p. 291.

80. *Le Dictionnaire de l'Académie française. Quatrième Édition. T.1* [1762]: "DEVISE. s.f. Figure accompagnée de paroles, exprimant d'une manière allégorique & courte, quelque pensée, quelque sentiment. On appelle La figure représentée dans la devise, *Le corps de la devise*; & on appelle les paroles, *L'ame de la devise*."

81. The obverse of a jeton was essentially always the royal bust, and on the reverse at the bottom in the exergue would usually appear the year, if any, and the name of the issuing institution, but these features did not really form part of the jeton's *devise*.

FIGURE 2.3: Jetons for 1682 in the *Mercure galant*

instruments, and the legend, INVENIT ET PERFICIT (It discovers and perfects), all describing and summing up the institution and its mission. But in the seventeenth century, legends and images of administrative jetons changed annually, and so an appropriate new design had to be concocted essentially every year. Someone had to "devise" the *devise*. To create one required arch erudition and some subtle insight into an institution and its mission. That someone was usually a learned academician of the Académie des Inscriptions et Belles-Lettres. New jetons were issued in January and dated for the year ahead, and in most instances from the mid-sixteenth century through the mid-eighteenth century they slyly alluded to some event or circumstance of the year just past.[82] And thus developed what was essentially a parlor game for others to try and decode the sense or the reference that the creator of the *devise* had in mind. It became a popular entertainment to uncover its hidden message, the lesson or moral it embodied, and the little commentary it involved or insight intended.

Donneau de Visé and the *Mercure galant* played into this long-running fad by providing explications for the new jetons late in January after the Étrennes. Entries described a particular jeton, noted who created the *devise* and who engraved the jeton, and then deconstructed the symbolic meaning behind the jeton. The jeton of the Bâtiments du Roi for 1680 as described in the January, 1680 issue of the *Mercure galant* is a typical example:

> For the Bâtiments du roi. A diamond with these words: ARGVIT AUTHOREM SPLENDOR [The splendor bespeaks the author]. This *devise*, engraved by Mr. Rottier, who is this quality engraver brought from England, is from Mr. Tallemant the younger [of the Académie des Inscriptions et Belles-Lettres].[83] The Diamond is formed by the sun in the entrails of the earth, and it gives off such a

82. In this abstruse way jetons were commemorative, letting medals commemorate the truly commemorable; Josset (1978), p. 337.
83. Paul Tallemant the younger (1642–1712), member of the Académie française and the Académie des inscriptions et belles-lettres.

> beautiful light that one well recognizes that so much brilliance can only come from this beautiful star. The connection with the edifices that the King has had built is entirely justified. They are of a magnificence that can only come from as powerful and as lavish in all things as Louis the Great.[84]

And so, year after year jetons issued forth from each of these great administrations—the Bâtiments du Roi, the Treasury, the Army, the Navy, and so forth—each with little references cached in images and legends composed by erudite individuals, references to current events or some ineffable institutional trait that lay waiting to be uncovered and explicated in a game of wits by the learned and the clever. The jeton of the Conseil du Roi of 1645, for example, alluded to the peace treaty concluded between the Pope and the duke of Parma; that of the 1668 Estates of Lille referenced royal victories in Flanders, and that of the city of Paris for 1695 concealed an oblique reference to the abundant harvest of 1694, and so forth.[85]

Donneau de Visé and the *Mercure galant* encouraged and facilitated this parlor game in publicizing new jetons, but after 25 years he stopped giving his own answers in 1706, saying, "I include the *devises* of the newly struck jetons that were distributed at the first of the year. I won't tell you any more about them. You will have the pleasure of examining them with your friends."[86] He came back to the point in his presentation of jetons the following year:

> I am not giving any explication of these *devises*; their sense should be easy when the *devises* are apt. However, nothing is more difficult, and the creators of *devises* often form ideas that don't always occur to the imagination of those who seek to explain them, and the latter sometimes offer a prettier and more natural interpretation than the inventors themselves. Allusions are sometimes concocted that make *devises* difficult to interpret, and it's then that, intelligence having to work harder, those who are lucky to succeed acquire even more glory.[87]

The "glory" to be won in this game is to be noted, but Donneau de Visé had still more to say on the subject again the next year in 1708:

> I send along as usual an engraving of the jetons that were struck at the end of last year and that were distributed as customary at the beginning of the year which we have just entered. I won't tell you

84. "Pour les batimens. Un diamant, avec ces paroles: ARGVIT AUTHOREM SPLENDOR. Cette Devise, gravée par Mr Rottier, qui est ce bon Graveur tiré d'Angleterre, est de Mr. Tallemant le jeune. Le diamant est formé par le soleil dans les entrailles de la terre, et il jette un feu si beau, qu'on connaît bien que tant de brillant ne sçauroit venir que de ce bel Astre. Le rapport en est fort juste avec les Batimens que le Roy fait faire. Ils sont d'une magnificence qui ne peut partir d'un Prince aussi puissant & aussi somptueux en toutes choses que Loüis le Grand," *Mercure galant*, 1680(1), pp. 32–33; also quoted in d'Affry de la Monnoye (1878), p. 293. See F. 12681; this is probably Norbert Roettiers.

85. The 1696–1697 inventory of jeton dies held at the Monnaie des Médailles published by Fernand Mazerolle (1903–1907) often gives the background story or interpretation of the design, here #1562, #2051, and #2707. Chapter 6 below further takes up covert references in jetons.

86. "je vous envoye les devises des jettons nouvellement frappez et qui ont esté distribuez le premier jour de cette année; je ne vous en dis rien davantage, vous aurez le plaisir de les examiner avec vos amis," *Mercure galant*, 1706(1), p. 313; also quoted in d'Affry de la Monnoye (1878), p. 319.

87. "Je ne vous donne point l'explication de ces devises, l'application en doit estre aisée lorsque les devises sont justes. Cependant rien n'est si difficile, et les inventeurs se forment souvent des idées qui ne tombent pas toujours dans l'imagination de ceux qui cherchent à les expliquer, et ces derniers y donnent quelquefois un sens plus beau et plus naturelle que les inventeurs mêmes. Il se trouve quelquefois des situations d'affaires qui rendent ces devises difficiles à trouver, et c'est alors que l'esprit ayant beaucoup plus à travailler, ceux qui ont l'avantage de réüssir acquierent beaucoup plus de gloire," *Mercure galant*, 1707(1), p. 165; also quoted in d'Affry de la Monnoye (1878), p. 319.

anything about the *devises*, and I don't believe that I am obligated to give you an explication, since the *devises* are plain and conformable to the subjects for which they were fashioned. They will be easy to explicate, and all their charms will be known from the start. And yet, if, on the contrary, obscurity resides in some of them and if there is not sufficient rapport between the subject and the object of the *devise*, I could make mistakes myself in wishing to explicate them. Here's something for speculative minds to grapple with.[88]

Although Donneau de Visé didn't explicate jetons after 1706, the *Mercure galant* continued to announce the new jetons every year. From 1722, "for the ladies," the *Mercure* gave Latin legends in French.[89] This continued all the way to 1758, when, as noted, government funding for administrative jetons was withdrawn. At that point, although jetons continued to spew forth from other bodies in society, the *Mercure* had nothing special to announce to its refined public.

Furthermore, by mid-eighteenth century pressures to standardize jetons and cut out the esoteric game playing had more and more effect, and jetons became evermore standardized. All allegorical elements of the jetons of the city of Paris end in 1716, and those jetons remained more or less uniform from then until the last one struck in 1788 with the arms of the current *prévôt des marchands* on one side and the standard emblems or views of Paris on the other.[90] In writing to the Académie des Inscriptions et Belles-Lettres in 1736 to request *devises* for the administrative jetons for the year ahead, the then head of the king's household, the Comte de Maurepas (1701–1781), passed down from even higher authority (the king's chief minister, the Cardinal de Fleury) instructions that, "these jetons must only refer to the department for which they were made without looking to find in them allusions to any particular event."[91] The academicians only promised to think about it. By the time we get to some of the late jetons of Louis XVI, such as that of the Société Royale de Médecine (pictured elsewhere), jetons had become completely standardized and bland, but the parlor game of decoding jetons and their *devises* never completely went away—certainly not for those involved in creating *devises* or for those glorying in jetons and their exchange at the new year.

Largesse!

The usages and cultural practices surrounding jetons spread beyond even these wild expansions of what jetons were and what they meant to people at the Étrennes or in the pages of the *Mercure galant.* One among this panoply was the custom of throwing money and jetons to crowds on grand occasions. The elaborate rituals associated with the entry of Charles VIII into Paris in 1484, for

88. "Je vous envoye à l'ordinaire, une estampe des Jettons qui ont esté frapez à la fin de l'année derniere, qui ont esté distribuez, suivant qu'il se pratique tous les ans, au commencement de l'année où nous venons d'entrer. Je ne vous dis rien des devises dont je ne crois pas vous devoir donner d'explication, puisque, si elles sont naturelles et convenables aux sujets pour lesquels elles ont esté faites, elles seront faciles à expliquer, et que toutes les beautez en seront d'abord connuës; et que si, au contraire, il se trouve de l'obscurité dans quelques-unes et qu'elles nayent pas assez de rapport avec les sujets qui en sont esté l'objet, je pourrais me tromper en voulant les expliquer...Voilà de quoy exercer l'esprit des speculatifs," *Mercure galant*, 1708(1), pp. 144–45; also quoted in d'Affry de la Monnoye (1878), p. 320.

89. "Nous mettons ici une petite explication des sujets avec les Legendes en françois, en faveur des Dames," *Le Mercure de France*, 1722(1), pp. 73–74; also quoted in d'Affry de la Monnoye (1878), p. 324.

90. Rouyer (1879), pp. 26–27.

91. "...il fallait que ces jetons eussent seulement rapport aux départements pour lesquels il étaient faits sans y chercher des allusions à aucun évènement particulier. Messieurs on promis d'y songer," AIBL, minutes of the meeting of August 23, 1737, as quoted in Josset (1978), p. 340. On this episode see also Guichon de Grandpont (1854), p. 8; Mainguy (1999), p. 68. Bonnet (1899) in Depeyrot (2007), p. 91, notes similar standardization of various provincial estates.

example, included tossing pieces to the crowd.[92] Accompanied with shouts of "*Largesse!*"—generosity!—jetons and money were thrown to the people on the entry into the town of Cambrai of the Duc d'Alençon in 1581; visits of the Queen Mother, Catherine de Médicis, on one of her tours of France in 1584 occasioned the fabrication of commemorative silver pieces and jetons which were cast before the hoi polloi.[93] Similarly at big civic ceremonies, such as the inauguration in 1733 of an equestrian statue of Louis XV at Bordeaux, along with having a fireworks display, organizers threw money to the crowds.[94]

Weddings became related events where jetons and medals had a role.[95] On the one hand, already from the fifteenth century the rich and the great, particularly royalty, had medals and jetons struck on the occasion of notable weddings. Some of these were plainly historical medals, but others were for the marriage ceremony itself and can be considered jetons or small medalions, something halfway between a formal and well-struck medal and a more common jeton.[96] In addition, in a tradition going back to the early Middle Ages, at an engagement ceremony the prospective groom would give his betroved either a purse of 13 coins, symbolizing Christ and the 12 Apostles, or one coin, known as a *treizain*, worth 13 *deniers* or 1 *sol* and 1 *denier*).[97] These figured in the royal marriage of Henri IV and Marie de Médicis, for example, in Lyon on December 17, 1600.[98] Four thousand guests attended the elaborate ceremony. Special coins in gold and silver and jetons worth eight *sols* were struck for the occasion, and again with cries of "Largesse!," these were tossed with great fanfare.[99] The tradition today of throwing rice or confetti at weddings is one residue of such practices.

Royal coronations were things of splendor in which jetons also played their part. From the tenth century with few exceptions the coronations of French kings took place at the Cathedral at Reims.[100] By the mid-1300s an elaborate and prescribed ritual had come into being that remained essentially unchanged through the coronation of Charles X in 1825. At the heart of this ritual lay what was taken to be an actual sacrament of the French Catholic Church, that of anointing the king's body with holy oil from the sacred Holy Ampoule held at the nearby Abbey of Saint-Remi, a vessel originally and miraculously brought from heaven by the Holy Spirit. This anointing was what legitimated the king's rule by divine right.

92. Le Fur (2010), pp. 86–94 provides these details.
93. For the specifics, see Robert (1853), pp. 15n, 19; Villela-Petit (2012b). These customs derived from ancient Roman practices where new consuls threw money to the crowds; see Barnard (1916), p. 80; Blanchet (1930), p. 212; Loach, p. 26, and Blanchet (1892), p. 202 n.
94. Évrard de Fayolle (1902), pp. 88, 125–27; for jetons and medals for this event, see F.9212–9214. The reference to commemorative medals of the Saint Bartholomew's Day massacres of 1572 being thrown to the people to memorialize the mass murder of Protestants remains entirely enigmatic; see Mazerolle (1903), pp. 127–28: "on frappa des médailles relatives à la Saint-Barthélemy, qui furent jetées au peuple pour éterniser la mémoire des massacres du 24 août 1572."
95. On marriage medals and jetons, see primarily Chalon (1867, 1869); Fontenay (1854), pp. 101–09; J. Florange (1936); see also Blanchet (1890), vol. 2, p. 403, and Loach (1995), p. 25. See also AN O[1] 2064, letter from Gros de Boze, dated "A Paris le 27e 9bre 1744" to "Monseigneur," transmitting two proposals from the Académie des Inscriptions et Belles-Lettres for "Medailles proposées pour le mariage de M. Le Dauphin."
96. Barnard (1916), p. 39, mentions 'medallic' jetons serving the function of a medal, but smaller than medals and more like jetons of a sort. Rouyer (1893), p. 407, says that coronation jetons were not real jetons, but he doesn't say why he thinks this.
97. Blanchet (1892), pp. 192–93.
98. Blanchet (1892), pp. 207–08. Valin (2016), vol. 1, p. 10, similarly mentions jetons thrown to crowds on the occasions of the weddings of Louis XIII and Louis XIV.
99. Schodt (1873), p. 22 n., notes that even gold jetons were distributed to "the people" on grand occasions.
100. See especially Blanchet (1892); see also Valin (2016), vol. 1, pp. 223–26; Le Fur (2010), pp. 72–80 provides extensive details for the first "modern" coronation of Charles VIII in 1484.

The whole affair was highly orchestrated beginning with the king's entry into Reims. At one stage the king, on his knees at the cathedral's altar, handed over a purse embroidered with gold thread that contained 13 gold coins struck for the occasion, a *treizain* that symbolized the king's marriage to the church and the state. The archbishop took the requisite "loaves" of gold and silver for the Reims cathedral, and at another point red velvet gloves, spurs, and other regalia received blessings.[101] The king made various oaths. At the investiture itself the archbishop placed the crown of Charlemagne on the king's head, and he was girded with Charlemagne's sword, *Joyeuse*. After that, but before the Mass, the king was escorted from the altar to the royal "throne of glory" at the back of the cathedral and was so formally enthroned. The *Cérémonie des Offrandes* followed, and it is this segment of the proceedings that interests us.[102]

With the crowned king seated on his throne of glory, the doors of the cathedral were thrown open and the general public swelled in.[103] To cries of "*Viva Rex in aeternum*," doves were released; trumpets, fifes, drums, and bassoons sounded loudly, and three canon salutes were fired off. While this public acclamation proceeded, sergeants-at-arms circulated through the choir and nave of the cathedral tossing handfuls upon handfuls of specially struck gold and silver jetons to the assembled multitude with cries of "Largesse! Largesse!" The custom of throwing jetons was in place by the coronation of King Henri II in 1547, and it continued in the period under consideration here (Fig. 2.4).[104] Mass and further scenes played themselves out as part of the coronation, including,

FIGURE 2.4: The *Cérémonie des Offrandes* at the coronation of Louis XVI, 1775.

101. "pains" per Blanchet (1892), p. 191.

102. For virtually identical accounts of the coronations of Louis XIV, Louis XV, and Louis XVI in 1654, 1722 and 1775 respectively, see Chapitre de l'Eglise Metropolitaine de Reims (1654), Menin (1724), and Pichon (1775). On jetons thrown at the coronation of Henri IV, see Blanchet (1898), p. 205; on jetons thrown at the coronation of Louis XIII, see Valin (2016), vol. 1, p. 9.

103. The coronation was otherwise attended by invitation-only dignitaries of one sort or another.

104. Lippens (1969), p. 33; Feuardent (F.7865–7916) lists 51 different jetons struck for royal coronations; see also Sombart (1993).

for Louis XVI in 1775 as well as for his ancestors, pardoning 600 convicts and touching 2,400 of the scrofulous waiting in line outside for the royal touch.[105]

Henri Nocq in his 1908 study of the coronation of Louis XV adds further relevant details. The young king brought to Reims in 1722 13 gold medals to present to the relevant clerics and 13 more for the ambassadors accompanying the king to Reims.[106] Various princes, princesses, cardinals, other prelates and court nobles received smaller gold medals. For the *Cérémonie des Offrandes* 200 very small gold medals were mixed in with the 4,700 silver jetons tossed to the crowds by the sergeants-at-arms (Fig. 2.5). But it turns out that the sergeants-at-arms first set aside nearly 1,200 pieces for separate distribution to members of the king's household and stables, including royal musicians, chaplains, porters, barbers, pages, and the like.[107] The king's coronational largesse and jetons thus reached "downstairs" and lesser folk associated with running the court. The ceremonial jetons produced on the occasion of royal coronations were so popular for the general public that knock-offs were struck in Germany as souvenirs.[108]

FIGURE 2.5: Jeton/Medal for the coronation of Louis XIV, 1654.

The practice of "largesse" extended to more minor inaugurations as well. For example, new aldermen (*échevins*) at Metz regularly had jetons struck with the arms of the town on one side and their own on the other. After being sworn in, a new alderman and the city council together entered city hall, and from its high windows jetons were tossed to the crowd below.[109] Seemingly because of the expense, by the eighteenth century the habit became to throw candy instead of jetons.[110] The case of a baptism at Metz in 1754 offers an unusual opportunity to see jetons and "largesse" in yet other ceremonial occasions.[111] For the baptism of his son on September 19 of that year Antoine-Louis Lefebvre de Caumartin (1725–1803), then the royal intendant at Metz, spent an astonishing 10,048 *livres* for 12 gold, 3,000 silver, 5,000 copper, and 200 gold-plated jetons. Madame the Maréchale de Belle-Isle, wife of the governor-general of the province, was given a magnificent bouquet of flowers and two red velvet purses trimmed with gold, each containing eighty silver jetons. Madame de Caumartin received the same, and each of the eight ladies accompanying the Maréchale got a purse of forty silver jetons. As for the copper ones, some were sown to the public as the procession made its way to the church. Undoubtedly, other scenes now lost to us involving jetons and "largesse" took place elsewhere across France in the *Âge classique*.

105. Menin (1724), pp. 299, 460.
106. On these points see Nocq (1908), pp. 428–35.
107. Nocq (1908), p. 430.
108. Blanchet (1892), p. 212.
109. Robert (1853), p. 16n; Blanchet (1930), 377; Barnard (1916), p. 6.
110. See previous note. For inaugural tokens in Low Countries, which date from 1549, see Gilleman (1934).
111. Robert (1853), pp. 57–59 and sketches of these jetons; Barnard (1916), p. 84.

Gratifications and Beyond

Establishing masses is an area where jetons came to replace traditional Church *méreaux*. The last will and testament of 1681 of one M. Perraut, for example, president of the Chambre des Comptes, endowed an annual solemn mass and funeral oration in memory of Henri de Bourbon, the second Prince de Condé (1586–1646).[112] The will provided the necessary funds annually for four purses of jetons worth 80 *livres* each; three of the purses went to the administrators of the main hospital in Paris, the Hôtel-Dieu, attending the service and one went to the preacher pronouncing the funeral oration, and special jetons were struck for this foundation.

On other occasions jetons were handed out as gratifications pure and simple. For example, from 1577 the king awarded an annual bourse of 100 jetons to the chief of police.[113] From 1665 at least through the 1720s, to pick another example, the city of Paris annually gifted Théophraste Renaudot with 100 silver jetons for what was essentially a subscription to the weekly periodical established by his father, *La Gazette*.[114] A certain Duplessis also received 100 jetons for his *Gazette*. In 1722 the surgeon François Gigot de la Peyronie operated successfully on the Duke of Lorraine for a fistula, for which the town of Nancy awarded La Peyronie 200 silver jetons in thanks.[115] In 1769 a surgeon donated a set of surgical instruments to the company of Bordeaux surgeons, which, to express its gratitude, straightaway sent a gold jeton belonging to the community.[116] Then, in 1787 a bishop gave a number of books to the library of the city of Paris and 600 *livres* for the librarian to buy works of natural history. The city council sent a purse of jetons to express its gratitude, a gesture the bishop took to be very flattering.[117] Doubtless more examples of these sorts await discovery by a more diligent historian.

Then, jetons were offered as out-and-out prizes. In 1747, for example, the director general of the Bâtiments du Roi charged the history painters of the Académie de Peinture et de Sculpture to paint a picture for the king on a subject of their choice. He sent six gold medals and six purses of jetons intended for the artists judged to have best succeeded. Not wishing to make a choice, the academicians simply divided the booty.[118] Another prize in 1765 for this same academy was privately funded.[119]

A final, late example can show just how deeply ingrained jetons became in the social and institutional life of Old Regime France. The example concerns the public health officers of the port of Marseille (the *intendants de santé*). In 1784 there were 16 of them, all volunteers, who inspected arriving ships looking for lepers to be sent to the Marseille leper colony and unhealthy cargo for deposit and further inspection. These officers exercised a significant supervisory role in trade with the Levant, for which they each annually received a purse of 80 jetons for a total annual cost to the government of 6,600 *livres*. The leper colony was running into debt, however, and the government wanted to off-load it from its books. As part of this economy drive, the minister of the navy, the Marquis de Castries (1727–1801), ordered that, instead of annual purses of jetons, a medal

112. d'Affry de la Monnoye (1878), p. xxiv.

113. d'Affry de la Monnoye (1878), pp. xxiii–xiv. It is not clear how long this practice remained in effect.

114. Nota bene: This Théophraste Renaudot, sieur de Boissemé, a royal councilor at the Cour des Monnoies, is the *son* of the eponymous founder of the *Gazette*, Théophraste Renaudot (1586–1653). Chautard (1890), pp. 480–84, has it right; Rouyer (1892) and Ch. Florange (1933), p. ii get it wrong. See also La Tour (1897), p. ix, d'Affry de la Monnoye (1878), pp. 286–88, and CAEF, Ms. F° 82. "Collection Robert de Cotte (1656–1735)," p. 146.

115. Lacronique (1902a), p. 412. These were not struck especially for La Peyronie, although such has been claimed, but were only the regular jetons of the town.

116. Évrard de Fayolle (1901), p. 20.

117. d'Affry de la Monnoye (1878), p. xxiv.

118. d'Affry de la Monnoye (1878), p. xxiv.

119. Per d'Affry de la Monnoye (1878), p. 377.

be struck and given to each officer as a token of the king's appreciation on their retirement after their statutory 10 years of service.[120] In writing to his colleague at the Bâtiments du Roi about this medal, de Castries was sarcastic about these inspectors, saying that because they do this service as volunteers they are obviously not in it for the money. The *intendants de santé* protested the suppression of their jetons, saying that they were all the more sensitive to the matter because they considered their jetons as a distinction and any withdrawal of their jetons as a slight to their honor. De Castries replied to them through the intermediary of the local intendant that he was sure that, given that they were concerned about points of honor alone, they would have no regrets about receiving retirement medals that would be a precious indicator for them and their families of the satisfaction and benevolence of His Majesty.[121] In the end the inspectors settled for the medals, but in 1784 jetons and medals alike had little history left.

The new practices surrounding jetons seen in this chapter make plain that jetons became embedded in the fabric of French society and culture across the sixteenth, seventeenth, and eighteenth centuries in expanded, fascinating, and revealing ways. And note, to this point the full range of practices that came to involve jetons in one way or another has not been exhausted; subsequent chapters will examine those further usages in the academies, in the guilds, and in the creation and manufacture of jetons. Nevertheless, jetons became commonplace in contemporary society across a broad spectrum of cultural practices. Jetons were such a common part of their world that contemporaries probably did not even think of jetons as anything out of the ordinary or with anything like the comprehensive view the historian might bring to bear on the subject. But next, beyond the diverse functions already outlined for jetons, it will repay to step back and take a census of jetons and revisit some of the institutions of government and society in the Old Regime that issued jetons. This overview will allow us to better see the function of jetons as reinforcing institutions of power and social control.

120. On this episode, see AI, AIBL archives, A:68, minutes of meetings of May 18, June 18, July 6, and August 6, 1784.
121. de Castries letter dated "De Versailles le 9 Mai 1784," copied into AI, AIBL archives, A:69, minutes of the meeting of Tuesday, May 18, 1784.

FIGURE 3.1: Jeton: Secrétaires du Roi, 1705.

3.

Governing Bodies

Early modern French jetons collectively mirror the society and culture that produced them, but they do so in their own way. One sees essentially nothing of the peasantry, for example, in contemporary jetons, a point addressed at the end of this study. Jetons thus reflected and refracted a particular fraction of their world. This chapter underscores that jetons were associated with and rained down from centers of power and governance in the prevailing socio-political order. Jetons were a top-down phenomenon. They reflected the very complicated institutional structure of France in the Old Regime, and the institutions they reflected were institutions of power, governance, and social control.[1] Joseph-Étienne de Fontenay in 1854 already saw that surveying jetons collectively reveals "the whole system of government of the old monarchy."[2] But Fontenay did not go far enough. At issue is not just the government per se of Old Regime France, although notably that, but jetons and the whole structure of contemporary society. Jetons reveal the levers of social control and the operations of the institutions of government in the largest sense, those institutions being the crown, the bureaucracy, the military, regional estates, churches and church hierarchy, the nobility, towns, academies, guilds, chambers of commerce, and others. That being the case, the main point emerges that jetons taken collectively reveal the ruling centers of power and institutions of social control governing contemporary society.

Cliometrics

Félix Feuardent's monumental inventory of some 15,000 pieces, *Jetons et méreaux depuis Louis IX jusqu'à la fin du Consulat de Bonaparte* (3 vols., 1904–1915), provides a wonderful opportunity for a quantitative overview of the totality of jeton production and to zero in on the institutions and centers of power that issued jetons.[3] Feuardent doesn't quite exhaust French jetons and the people and institutions associated with jetons, but his catalog can stand as a reliable and valuable proxy. Table 3.1 tabulates what we can glean simply by counting the jetons in Feuardent. The table shows just how intertwined jetons became with the reigning institutions of life in early modern France.

Several caveats need to be introduced regarding Table 3.1 further. One pertains to Feuardent's classification system as presented in the left column. Because it so well mirrors institutional and

1. On the complicated institutional and organizational structure and functioning of early modern and Old Regime France, see again Bély et al. (2003), Barbiche (1999), Mousnier (1990–1992), and Marion (1923).
2. "...tout le système gouvernemental de l'ancienne monarchie," Fontenay (1854), p. 166.
3. On this work and Feuardent's importance in the historiography of jetons, see above, Ch. 1.

regional realities, Feuardent's classification scheme has been universally adopted.[4] Feuardent doesn't consider strictly royal jetons until volume three of his compendium, but he says he could have started with these.[5] His royal jetons have therefore been moved to the top of the list on Table 3.1 to better reflect the centrality of royal authority. The numbers need to be treated with some caution. Feuardent lists every striking. That includes jetons struck from the same dies, but on different metal blanks, e.g., gold, silver and alloys of copper, brass, and tin. Feuardent lists minor variants; alas, *méreaux* (church tokens) as well as jetons have been counted, although in making the tabulation the effort was made not to count reproductions coming out of Germany. Table 3.1 only sums the run of Feuardent numbers, not any variants or additions signaled by Feuardent with alphabetic superscripts, e.g., F.4385[a], a jeton of the Académie des Inscriptions et Belles-Lettres circa 1717. Then, too, Feuardent extends his inventory through the Napoleonic era; although we have tried to exclude jetons after 1789, there is undoubtedly some spillover. In many instances Table 3.1 follows Feuardent's categories exactly (e.g., the Trésor Royal), but in others (e.g., Paris city officialdom) it has amalgamated several different issuing institutions under one rubric. Even if useful and not inaccurate, these larger categories hide other information and will need to be broken down further. Feuaradent's classification of jetons and *méreaux* by province groups together various bodies (estates, towns, nobility, clergy), and these categories will need to be unpacked for a more accurate picture of jetons in the provinces. With the clergy, for example, Feuardent subsumed church jetons and *méreaux* beneath these regional classifications and subdivisions. The same can be said for jetons of provincial nobility and towns, which we will treat in due course.

The associated dates given in Table 3.1 need to be treated with great caution. The range of dates indicated are for dated jetons *in the Feuardent collection* and not for any series of jetons. For the administrative jetons listed, the range does approximate the run of the particular series, but for the more general categories, a number of different bodies are mixed together (e.g., the guilds or a province), many of them ephemeral or having jetons struck only a handful of times, and so the range of dates given is only indicative in a general way of jeton activity in that category. Furthermore, the range of dates indicated obviously excludes earlier, undated medieval and Renaissance jetons that are inventoried by Feuardent and counted in the totals. The historical beginning of dated jetons is a revealing indicator of the development of jetons, and to judge by the proportion of Bourbon jetons in Feuardent versus earlier medieval and undated Renaissance royal jetons, the bulk of jetons and *méreaux* cataloged by Feuardent date from the seventeenth and eighteenth centuries, the periods of concern here.

These qualifications notwithstanding, the data presented in Table 3.1 paint a revealing portrait of French jetons and associated institutions and the ruling power structure governing contemporary life. They indicate that on the order of magnitude of 10,000 distinct and different jetons and *méreaux* strikings were made through the seventeenth and eighteenth centuries (13,388 totaled here). The individuals and institutions for which they were struck, again, are those of ruling power and social control, starting with the king and court at roughly 15% of the total of jetons and *méreaux* in Feuardent. The bureaucracy and institutions of royal government account for approximately another 25%, meaning almost 40% of all jetons issued from the court and government bureaus. Paris institutions represent another 20% or so, including jetons distributed not only by city and police officials, but notably by the guilds as well, all of which indicate the multifaceted character of Parisian institutions and how embedded jetons became in the life of institutions in

4. On classification see Blanchet (1890), vol. 2, pp. 411–36; Fontenay (1854), pp. 143 ff. speaks about classification. See also Boutin, p. 241. Blanchet (1930) picks up from Feuardent.

5. Feuardent (1904), p. viii.

Issuing Institutions	Different Strikings			Range of Dates of Dated Jetons	Feuardant Nos.
	#	#	%		
KINGS AND QUEENS OF FRANCE		**1,845**	**14%**	1525–1789	11480–13392*
(Henri IV–Louis XVI) 12%	1,541			1589–1789	11851–13392
ROYAL COURT/GOVERNMENT/ ADMINISTRATION		**3,210**	**24%**		
Offices serving the Royal Court	108			1619–1758	2853–2961
Conseil du Roi	261			1550–1662	1–261
Ordinaire des Guerres	226			1634–1758	390–616
Extraordinaire des Guerres	287			1615–1778	617–904
Artillerie	189			1563–1756	953–1142
Navy	224			1593–1758	1178–1402
Galleys	253			1680–1748	1403–1550
Trésor Royal	273			1555–1758	1830–2103
Chambre des Comptes	157			1552–1708	1664–1821
Chambre aux Deniers	250			1598–1758	2293–2543
Parties Casuelles	237			1616–1758	2580–2817
Bâtiments du Roi	249			1617–1758	2962–3211
Other	368				
PARIS INSTITUTIONS		**2,868**	**22%**		
City of Paris and City Officials	590			1548–1789	3352–3924
Police and Judicial Offices	122			1583–1777	3212–3333
Clergy and Churches (includes *méreaux*)	426			1586–1792	3925–4351
Academies and Faculties	371			1638–1780	4352–4760**
Merchants and Guilds	749			1549–1802	4764–5513
Nobility	400			1550–1788	5513–5913
Towns: Île-de-France	166			1561–1799	5914–6080
PROVINCIAL ESTATES, TOWNS, NOBILITY		**5,398**	**41%**		
Burgundy	565			1525–1790	9654–10219
Brittany	421			1489–1792	8555–8976
Lorraine	398			1523–1780	7436–7834
Normandy	383			1543–1789	6081–6464
The Lyonnais	336			1601–1805	10494–10795
The Dauphiné	289			1524–1796	11111–11400
Other	3,006				
TOTAL:		**13,321**			

*Nuremberg reproductions (F.12958–13025) excluded

**Excluding F. 4454–4491

TABLE 3.1: French jetons and *méreaux* cataloged in F. Feuardent (1904–1915).

Paris. The big surprise here is that 40% of jetons and *méreaux* in Feuardent are associated with entities in the provinces. These provincial bodies need to be specified in more detail, too, but this finding indicates just how deeply rooted jetons became not only in Île-de-France but in the whole of Old Regime France and French society.

The present chapter and the two that follow survey the institutional panorama opened up to us by these metrics. This chapter looks more closely at the royal court, administrative bodies of government and the military, regional estates, churches and the clergy, the nobility, Paris city government, and other cities and towns. To round out this portrait of power and authority, Chapter 4 moves on to examine the academies, and Chapter 5 explores the world of the guilds and the place and role of jetons in that world.

The Royal Court and Its Jetons

Jetons arose in the Middle Ages as tokens moved around on counting tables, but early on jetons found a significant place in the royal court, if only because the king was responsible for providing the jetons for the counting tables of the offices that surrounded him and that rendered personal services to him and the royal household. Notwithstanding the expansion of jetons outward into the machinery of government and beyond, the king and court remained at the center of jeton culture. This would be particularly so in the king's royal cabinet and the workings of government that involved the king personally and directly. The centrality of the court to any story about jetons was cemented as the custom of the Étrennes took hold in the sixteenth century and jetons came to be offered and exchanged between and among august personages at court and in the bureaucracy at New Year's.

The royal court as it developed in the seventeenth and eighteenth centuries was a large, complex, and highly orchestrated affair. Louis XIV built up Versailles from a modest hunting lodge and moved the court and government there in 1682. Except for the period 1715–1722 under the regency of Philippe d'Orléans, the court remained there until the women of Paris brought Louis XVI and Marie-Antoinette back to Paris in October 1789, over a century later. Readers will be familiar with Versailles and the elaborate lifestyle of the court and courtiers there. Although there were other, competing centers of power in contemporary France and other organizations used jetons in one way or another, the court itself—the entourage of the king and various royal households—was the primary center from which radiated power and institutional authority. It embraced and fostered jetons and a culture that embodied jetons as no other single organization or entity.[6]

Feuardent counts over 1,900 different jetons struck for French kings and queens, although, again, that includes the same jeton struck in different metals. Although royal jetons stretch back to the thirteenth century, as indicated in Table 3.1, by far the majority of these royal jetons were struck for the Bourbon kings starting with Henri IV whose reign began in 1589. Not unexpectedly, Louis XIV, the Sun King, predominates with 577 different jeton designs, including German knock-offs, featuring him solely and personally.[7] That number averages eight strikings a year for the whole of his 72-year reign. Louis XV subsequently merited 121 different jetons, including German ones, an average of only two a year; the number of personal jetons for Louis XVI fell to only 15 in the period 1774–1789, or only one a year, indicating both a greater degree of royal modesty and the relative decline of jeton culture as the eighteenth century wore on. Other court jetons of this sort featured king and queen together, and some were issued for the queen only. Jetons for

6. Blanchet (1890), vol. 2, pp. 411–19; Blanchet (1950).
7. F.12448–13025.

the queen's household become more common under Louis XV; the dauphine ultimately got her own jetons, and the number of recognized princely households under the king's roof multiplied in the course of the eighteenth century.[8] The dauphin, next in line for the throne, had his own jetons struck for him, but the personage seems strangely absent from the record of the jetons of the court.[9] In this context we should not overlook the king's gaming jetons struck especially for him and for his pleasure at the gaming table.[10]

Elaborate troops of chaplains, chamberlains, masters of ceremonies, grand masters of ceremonies, seneschals, maître-d'hôtels, various noble gentlemen attendants, sergeants at arms, body guards, escorts, squires, pages, valets and miscellaneous other personal servants surrounded the king, queen and other royal personages and met their needs in the most luxurious and comfortable ways possible.[11] These little services required officials and departments, many of them venal, and most at various points issued jetons. So, jetons coming from within the court are revealing not only of the spread of jetons deep within the court but also of the complex character of the machinery of court. And thus came into being the elaborate bureaucracy of the king's household, the Maison du Roi—the bureaucracy of the court and associated offices at Versailles.[12] At the king's service and on call were a number of separate departments within the Maison du Roi in addition to the ancient ones of the king's kitchens, bakery, cellar, furs, fruit, stables and the hunt (variously for hunting stag and wolves and hunting with falcons) all of which continued. The department of the king's silverware, the Argenterie, for example, was responsible for the king's tableware, to be sure, but in addition it superintended the royal wardrobe and the royal chambers. The department of the king's lesser pleasures (the Menus-Plaisirs du Roi) sought the king's comfort and was responsible for organizing balls, ceremonies, and the like. The Menus-Plaisirs itself subsumed royal violinists and other royal musicians.[13] At hand and as needed stood various military detachments and Swiss Guards, masters of the king's horses, masters of the hounds, keepers of the king's tennis, royal doctors, royal apothecaries, royal painters, the king's theater troupe (the Comédie Française from 1680), and others, all devoted to personal service to the king. These lesser departments fell under the aforementioned bureaucratic structure of the Maison du Roi, which was headed by a major state secretary or like high official. At one point or another virtually all of these departments in the Maison du Roi issued jetons, and jetons as perks and as packages to be exchanged at the Étrennes became part of the life and functioning of the court. The Chief Gentleman of the Chamber (the *premier gentilhomme de la chambre*) was a high-ranking noble court official who had a role in deciding on subjects for jetons and medals and in distributing jetons at court.[14]

---

8. On the queen's household and its jetons, see Mainguy (1999), pp. 219–25; Blanchet (1930), pp. 249–55. See also Pradel (1936) and Sarmant and Ploton-Nicollet (2010) on other princely households and their jetons.
9. By the same token, Feuardent (F.11111–11235) lists 124 different jetons struck for the series of dauphins and dauphines from 1517, classified under the province of the Dauphiné and not as part of the royal household.
10. These are not listed in Feuardent's volumes. An undated archival document, "Etat des Jettons pour le jeu de Votre Majesté" (AN K 506 #20 (9)), shows the king's gaming jetons to include 150 gold jetons, 200 square jetons in silver, and 281 round jetons in silver. See also Roston (1993), p. 189.
11. This, despite contemporary Versailles being an otherwise foul and unsanitary place; on this point see Newton (2014).
12. Bély et al. (2003), pp. 777–84; Mousnier (1990–1992), vol. 2, pp. 115 ff.; Mazerolle (1888), pp. 466–73, 480–82, gives a breakdown of the constituent elements of the Maison du Roi that issued jetons; see also Blanchet (1930), pp. 240–48.
13. Mazerolle (1888), pp. 353–54. On the jeton budget for 1747 and 1748 within the Argenterie, see AN K 506 #2 (Dépenses de la Maison du Roi).
14. Guiffrey (1885), p. 209. For these jetons see F.223–39.

Within the Maison du Roi came the disbursement office of the Chambre aux Deniers.[15] The Chambre aux Deniers was not comparable to the larger and more directly administrative Trésor Royal or the accounting office of the Chambre des Comptes. It was a smaller, if still substantial and staffed bureau within the Maison du Roi that covered the direct expenses of the king and the king's household, the Hôtel du Roi. Funds came from the Trésor Royal. The Chambre aux Deniers dates from 1316 and soon struck jetons for its counting tables. These are few in number until its first dated jeton in 1598 under Henri IV. Thereafter jetons for the Chambre were stuck essentially every year until 1758 to make one of the longest series of jeton strikings.[16]

Still within the orbit of the court, although more strictly part of government officialdom, one finds a number of bureaus and offices associated with the king acting in his capacities as ruler and as active head of state and government. In this regard we have already treated the royal privy council (the Conseil du Roi) and its important series of jetons.[17] Here we might mention the chancellery (the Chancellerie), which kept the seals and expedited documents from the king. Not to be omitted from this list are official royal secretaries, the Secrétaires du Roi, batteries of whom scribbled and copied royal documents. They had their own jetons, too (Fig. 3.1). Every year on the sixth of May, the saint day of Saint Jean Porte Latine, in something like a feudal offering, the Secrétaires du Roi would present the Chancellor with a purse of jetons. The Secrétaires du Roi gave this purse directly to Louis XV when he kept the seals personally.[18]

The Avocats du Conseil du Roi, a related group of lawyers accredited before the royal council and able to represent clients before it, broke off from royal secretaries in 1643.[19] Fernand Mazerolle's 1907 study allows us to see this group up close and the role played by jetons within it. As many as 160 lawyers held these positions at any one time through the end of the eighteenth century. These *avocats* had to be lawyers accredited before the Paris Parlement, and they were subject to a "severe" inspection of morals. They were organized as a guild, and jetons played a role within the Avocats du Conseil that we will see many times over in other, more pedestrian guilds. On the one hand, *jetons de présence* were handed out at guild meetings, officers receiving two jetons per meeting, ordinary *avocats* receiving one. Candidates for admission owed the common till an acknowledgment of 100 *livres*; the person writing the report on the candidate received three jetons; the person playing devil's advocate and arguing against admission collected two. It is not simply ironic that this group continues in existence today as lawyers to the French national appeals court, the Cour de Cassation, or that new members receive a medal upon entry.

Before leaving the august realm of the royal court and the people and parties hovering around it, mention should be made of military and chivalric orders and the jetons regularly struck for same.[20] The Order of the Holy Spirit (Ordre du Saint Esprit), for example, began in 1578 under Henri III and continued through 1789; Louis XIV instituted the Ordre de Saint-Louis in 1693 to reward outstanding military service. The king headed these orders, of course. They struck their own jetons, and at least from the Ordre de Saint-Louis the king received a purse of jetons when it submitted its accounts. In 1753 the dauphin got one, too.[21]

15. On the Chambre aux Deniers and its jetons, see Mainguy (1999), pp. 225–29; Fontenay (1847), pp. 88–89; Mazerolle (1888), pp. 352–53, 358–74; Préau (1890b), pp. 51–52.
16. F.2293–2543 for a total of 250 different jetons.
17. See above, chapter 1, and F.1–233. NB: Although the Conseil du Roi series stopped in 1662 and although Louis XIV governed essentially single-handedly, royal privy and ministerial councils continued down thorough 1789.
18. d'Affry de la Monnoye (1878), p. xxv; Nivaille (1987), passim.
19. Mazerolle (1907b), pp. 10–20; F.348–376.
20. See Fontenay (1854), pp. 159–63; Feuardent (1904), p. 122 n.; F.1605–1656.
21. d'Affry de la Monnoye (1878), p. xxv.

Bâtiments du Roi

The institutional structure of royal government extended beyond the court per se, and growing outward from the central hub and power of the king and court, several other large bureaucracies of state came into being to run France. Still within the overarching Maison du Roi, another large department, the Bâtiments du Roi—or department of the king's buildings—stands out as a separate bureaucratic entity requiring closer attention. Headed by a high-ranking noble as administrator—director general or superintendant, the Bâtiments du Roi came to have 16 departments, a staff varying between 200 to 300, and a budget of 2.5 million *livres*.[22] One of its major functions was indeed the construction and maintenance of royal palaces, and its budget mushroomed as Louis XIV built up Versailles. But the Bâtiments du Roi was also an umbrella institution that sheltered the arts and oversaw royal patronage for a multifaceted set of cultural institutions subventioned by the state. The royal academies and costs associated with their publications, pensions and jetons stand out in this category. The Bâtiments du Roi similarly subsumed royal observatories and royal botanical gardens, although none of these had jetons.[23] The Bâtiments du Roi superintended pensioned artists and royal manufacturing facilities such as the royal dye works of the Gobelins. The institution projected royal authority outward from the court and connected it to other government institutions and society at large. It issued its own jetons as part of the series of administrative jetons, but more than that, coordination with the court and funding for the production of royal jetons and medals took place through the agency of the Bâtiments du Roi.

The jetons of the Bâtiments du Roi were few in number before the 1660s, but from that point over 200 different jetons were struck for the organization before its last in 1758.[24] As elsewhere, purses of these jetons were handed out within the Bâtiments du Roi as perks of office, distributed to other exchequeries, and given to notables at the Étrennes. The iconography of the jetons of the Bâtiments du Roi is to be remarked on, as these jetons often depicted royal buildings in existence or under construction, such as its jeton of 1684 showing the hydraulic machine at Marly that supplied water for the fountains and ponds at Versailles, the one of 1690 depicting the new Hôtel des Invalides, or the jeton dated 1756 illustrated here showing the rebuilding of the facade of the Louvre (Fig. 3.2).[25]

In the two decades 1881-1901 the art historian and cultural administrator Jules-Joseph Guiffrey (1840-1918) published five volumes of the financial records of the Bâtiments du Roi covering

FIGURE 3.2: Jeton of the Bâtiments du Roi dated 1756.

22. On the Bâtiments du Roi, see Mainguy (1999), pp. 229–36; Bély et al. (2003), pp. 140–42; Barbiche (1999), pp. 239–44; Marion (1923), p. 41; see also Blanchet (1930), pp. 246–47; Veillon (1994), pp. 37–39.
23. McClellan and Regourd (2011), pp. 84–90, 107–10.
24. F.2968–3182.
25. On the iconography of the jetons of the Bâtiments du Roi, see Veillon (1994), p. 28; Mazerolle (1888), pp. 473–74. On these particular jetons, see F.3038v, 3182, 6033.

the period 1664-1715.[26] Guiffrey's volumes constitute an amazing record in and of themselves, but for our purposes Guiffrey's work shows how the Bâtiments du Roi served as the conduit through which funding was channeled for the jetons, medals and purses of the great state academies. These included the Académie Française, the Académie Royale des Inscriptions et Belles-Lettres, the Académie Royale des Sciences, the Académie Royale de Peinture et de Sculpture, the Académie Royale de Marine, and the Académie de Rome. The administration of the Bâtiments du Roi coordinated with the court, the Académie des Inscriptions et Belles-Lettres, and the Monnaie des Médailles in order to commission, design, manufacture, and pay for the broad range of jetons and medals that were produced for these institutions. Orders to the treasurer of the Bâtiments du Roi regularly authorized such payments. To pick but one example among many, take the order dated March 23, 1677, for payment to the silversmith Baslin (or Ballin) the sum of 1,114 *livres* 13 *sols* for 1,100 silver jetons destined for the Académie Française.[27] The Bâtiments du Roi may have kept its own collection of jetons, at least to judge from the 1687 payment of 168 *livres* to a certain Louis Daubancourt, sheath-maker, for 12 trays to hold His Majesty's jetons and three days of work on the gem cases of the king's cabinet.[28]

Treasury and Finance

Although fluid in its organization, from the sixteenth century the Trésor Royal became another substantial bureaucracy and the predominant financial institution of the state, generally responsible for collecting receipts from all sources and providing funds for other branches of government.[29] The Trésor Royal loomed large in matters both of finances and jetons. Over half of the administrative jetons were issued for the Trésor Royal.[30] The iconography of the jetons of the Trésor Royal is noteworthy in that images of security and plentiful resources were dominant themes, as in the jeton of Noah's Ark riding the waves against the great Flood (Fig. 3.3).[31]

The Parties Casuelles was another arm of financial administration that demands notice.[32] The institution dated from 1522 and was the grand office in charge of venal offices. It kept records and collected the requisite fees on the sale, transmission, and holding legal title of offices. From 1604

FIGURE 3.3: Jeton of the Trésor Royal, 1742.

26. See Guiffrey (1881–1901); see also the two volumes of accounts for the Bâtiments du Roi he published for the sixteenth century, Guiffrey (1877–1880).
27. Guiffrey (1881–1901), vol. 1, col. 1010.
28. Guiffry (1881–1891), vol. 2, col. 1203 [1687]: "20 juillet : à luy [Louis Daubancourt], gainier, pour douze tablettes à mettre les jetons de S. M., y compris trois journées qu'il a travaillé aux tabletter des agathes du cabinet… [*sic*] 168#." N.B., the Bâtiments du Roi also subsumed the Cabinet du Roi and the position of royal historiographer.
29. Bély et al. (2003), pp. 1227–30; Fontenay (1847), pp. 85–86; Mainguy (1999), pp. 274–99; F.1832–2103.
30. Above and Veillon (1994), p. 31.
31. Regarding such images, see also Veillon (1994), pp. 31–32.
32. On the Parties Casuelles, see Bély et al. (2003), pp. 920–21; and for its jetons see Mainguy (1999), p. 73; Veillon (1994), pp. 32–33, and the 236 different jetons of the Parties Casuelles listed in Feuardent, F.2580–2816.

it was responsible for collecting the *Paulette*—an annual direct tax on offices instituted that year by Henri IV, to the tune of four *deniers* per *livre* or one percent of the value of each office. In 1665 the Parties Casuelles brought in over two million *livres*.[33] It struck its own jetons annually from the 1620s and constituted another long-running series. A variety of other bureaus, such as the *payeurs des rentes* who dealt with state loans and the paying of interest, came to complement these other offices of financial administration.[34] As a general rule, the images portrayed on these jetons suggested bounty and plenty.

Army, Navy, and the Colonies

At the command of the king and his officers the military and naval establishments of Old Regime France were huge, and they entailed an equally huge bureaucratic apparatus to manage them.[35] The accounting offices for the military and their associated jetons allow us to see at least the main outlines of the powerful military arm of the state. In the case of the army, officials kept two separate sets of books and struck two separate series of jetons, the Ordinaire des Guerres and the Extraordinaire des Guerres.[36] The Ordinaire des Guerres was the accounting and disbursement ledger for the most ancient military branches most closely associated with the person of the king: contingents of bodyguards, a personal Swiss contingent, light cavalry, the *mousquetaires*, and other regiments of Swiss and French guards. The Extraordinaire des Guerres handled the general staff of the army and all the other military units enlisted in the service of France and the king across the provinces and units stationed in the Caribbean. From the 1620s both the Ordinaire des Guerres and the Extraordinaire des Guerres issued jetons as part of the totality of administrative jetons paid for by government, and a separate department of the Artillerie soon emerged, and it, too, struck its own series of jetons.[37] The iconography associated with these jetons is consistently marshal.

Headed by a separate ministry (the Ministère de la Marine et des Colonies), the French navy (the Marine Royale), complemented the army and was an equally large bureaucratic affair. With corps of sailors and officers, including 3,500 marines, the best-built ships of the line, extensive port installations, associated medical and training schools, elaborate supply chains, not to mention the Académie de Marine at Brest, the aristocratic royal navy developed considerably after reforms instituted first by Richelieu and then by Colbert and Louis XIV in the 1660s, and it easily rivaled, if not surpassed the better known British Royal Navy throughout the eighteenth century.[38] The French navy had its own set of financial controllers and treasurers and issued its own jetons from 1593 and more regularly from 1643–1758.[39] The handsome jeton of 1712 heading the preface to the present work shows a characteristic, if somewhat pacific nautical scene; many others are more bellicose or include such images as a rudder, a compass or the North Star. The depiction

33. Fontenay (1854), pp. 172–76.
34. Blanchet (1930), p. 276.
35. For an overview, see Barbiche (1999), pp. 195–226.
36. On these points see Fontenay (1854), pp. 168–69; Blanchet (1930), pp. 237–48, 265–67; Veillon, (1994), pp. 33–35. Feuardent lists 226 different jetons for the Ordinaire des Guerres from 1634–1758 and 287 for the Extraordinare des Guerres from 1615–1778, F.390–616 and 617–904 respectively.
37. Feuardent lists 180 different jetons issued for the Artillerie from 1563–1756: F.953–1133. On the jetons of the Artillerie and associated military engineering units see also Blanchet (1930), pp. 267–68; Mainguy (1999), pp. 169–72. On the French army in the colonies, see Lesueur (2014).
38. On this background see Bély et al. (2003), pp. 31–33; McClellan and Regourd (2011), pp. 14, 51, 57–70 and further passim.
39. Fontenay (1847), p. 94; on the 224 jetons of the royal navy issued from 1593–1758 listed in Feuardent (F.1178–1402), see further Mainguy (1999), pp. 249–72, 312; Blanchet (1930), pp. 268–69; Veillon (1994), pp. 35–36; Guichon de Grandpont (1854), pp. 11–14.

of Louis XIV in the jeton in question is somewhat unusual, too, in that more often Marine jetons carried the bust of the *amiral de France*, a post usually held by the Vendôme branch of the royal family.[40] Efforts in 1736 to standardize the *devise* of Marine jetons came to naught, and the annual series continued with new strikings and new *petits monuments* for another two decades.[41]

And let's not forget the galleys. Penal galleys were a separate quasi-naval bureaucracy connected to the legal system and administered under the ministry of the navy, and this department, too, struck jetons regularly. Rowed by convicts, fancy French galleys would occasionally make sorties from Marseilles venturing to Genoa or Barcelona, but mostly the galleys stayed in port where their manpower was deployed in penal servitude at the *bagne* or prison workhouse.[42] High aristocrats who never went to sea, including the same Vendôme princes, superintended the galleys and were featured on the obverse of their jetons.[43] In 1716 the treasury of the galleys distributed 26 purses of jetons. This series ran from 1654–1748, at which point the galleys became absorbed within the bureaucracy of the royal navy and jetons no longer issued for that unit.[44]

As for the colonies, on the other side of the Atlantic Samuel de Champlain established the first French settlement overseas at Quebec in 1608. The Canadian colony always struggled and never rivaled later Caribbean colonies, but by the first decades of the eighteenth century it had solidified itself and functioned normally as a settlement colony and an overseas extension of France itself. Royally appointed governors-general and intendants saw to royal affairs and French interests in an expanding territory, but the colony was effectively administered by the French Compagnie des Indes. Various French trading companies came and went over the seventeenth and eighteenth centuries, but, following the fall of John Law and his bank in 1720, the Compagnie des Indes became a major player in long-distance trade on par with the British East India Company (also operating in Canada) and the Dutch *Vereenigde Oostindische Companie* (the VOC). With headquarters at the Hôtel Tubœuf in Paris and its own navy and port facilities at Lorient, the French Compagnie des Indes controlled the lucrative trade in Canadian beaver pelts and other commodities while it likewise expanded its reach into the Indian Ocean and the spice trade. Although not a royal or government institution per se, the Compagnie des Indes was nevertheless tied tightly to the state, essentially a state within a state. The king had shares, royal representatives sat de facto on the board, and the corporation was otherwise controlled by insiders. It takes us away slightly from the centers of royal power and direct administration, but it can be considered here alongside the royal navy.[45]

In the years 1751–1758 the Compagnie des Indes issued a lovely series of jetons for its colonies in North America.[46] Several are in copper or brass, and they may have been used on company counting tables in France or in Canada; they may have been distributed ceremonially in purses at the Étrennes. We know that French authorities handed out historical *medals* to native chiefs in Canada, and so some of these jetons may have made their way into indigenous hands as well.[47] These jetons have attracted considerable attention, notably from Canadian scholars and numis-

40. On this dynastic connection see the several works by Jules Chautard listed in the bibliography.
41. Blanchet (1930), p. 269.
42. On the galleys, see Zysberg (1987) and Bély et al. (2003), pp. 585–88.
43. Feuardent lists 147 jetons issued for the galleys from 1680 to 1748 (F.1403–1550); see further Guichon de Grandpont (1854), pp. 14–17; Blanchet (1930), pp. 269–70; Veillon (1994), pp. 36–37.
44. Guichon de Grandpont (1854), pp. 9–10; on these points see also Chautard (1881); Chautard (s.d.).
45. On these points see McClellan and Regourd (2011), pp. 54–56.
46. F.1558–1584; Wellington (2018); Betts (1894), pp. 172–78.
47. Wellington (2016), pp. 302–03, details the demand and regular supply of French medals to local native Americans allies; see also Zay (1889).

matic experts.[48] Individually and collectively, what with images of beavers and corn on some, Mercury flying across the ocean, or Jason and the Argonauts on others, the distinctive iconography of these jetons bespeaks a particular view of the French North American outposts. For present purposes the 1751 jeton (Fig. 3.4) will have to be representative. It shows a Native American with bow and quiver looking over his shoulder to a field of blossoming lilies and, signaling the exotic, an alligator and a palm tree in the distance on the left with the legend, SUB OMNI SIDERE CRESCUNT (They grow beneath every constellation).[49] Alas, the Seven Years' War (1756–1763) and the Peace of Paris (1763) ended Compagnie involvement in Canada and this series of jetons, as well as French government itself in North America.

FIGURE 3.4: Jeton: Colonies françaises de l'Amérique, 1751.

Clergy and Nobility

This chapter so far has focused on the royal court and the wider institutions of government. Jetons and practices surrounding jetons came to be part of royal government and the French state in astounding ways—think only in this instance of state control over the production of jetons (to be treated more fully later) or the ceremonial exchange of purses of *jetons d'étrennes* at court or among the various government treasuries. As much as the court and the bureaucracies of state constitute perhaps the major pole, the historically fascinating aspect of the story of Old Regime jetons is the extent to which they so thoroughly permeated contemporary society well beyond court and government. Jetons allow us to see a wider panorama and other centers of power and practice. That would begin, for contemporaries at least, with the formal division of Old Regime society into the classic three orders of society or the three estates: the clergy, the nobility, and commoners or the Third Estate. To be sure, a rather wide segment of the Third Estate made contact with jetons and participated in jeton culture—again, from the tavern keeper keeping tabs, to the manorial manager, to members of city councils, to guild members, to academicians, to these selfsame government bureaucrats and administrators, such as Colbert, for example, who were not all noble, at least to start out with. But there were no distinctly Third Estate jetons, and there is no evidence the author has come across to suggest that jetons played any role on the village level or in peasant life. The story is different for the other two estates, and, although not affecting the village priest or the poor noble on his scrap of land, jetons did become part of the larger worlds of the clergy and the nobility.

48. The Paris Mint issued a set of restrikes in the 1960s and '70s; see sale catalogs (1968) and (1976) by Monnaie de Paris in ANS library, Pamphlet file. Further on Franco-American jetons, see Frossard (1899); Pages (1931); Pédery-Hunt (1979); Blanchet (1930), pp. 270–71; Zay (1892).

49. F.158v and further below, List of Figures. Robert Wellington believes this jeton was intended for the French colony in Martinique, but the contemporary source that might suggest this conclusion probably mistook the "lam" in the exergue for "la Martinique" instead of "l'Amérique"; Wellington (2018). For other jetons of the Compagnie des Indes, see Weil (1988).

FIGURE 3.5: Jeton of the general assembly of French Clergy, 1775.

As for the clergy, church tokens (*méreaux*) continued to be struck and used in major urban parishes in the eighteenth century. Archbishops and other high-ranking clergy would occasionally have true jetons emblazoned with their arms and station struck for their personal use. But jetons are most prominent in the general assemblies (*assemblées générales*) of the French clergy.[50] These meetings were held regularly from 1561 with perhaps 100 such gatherings from 1586–1785, or one every couple of years.[51] The French Catholic Church was a formidable institution of its own of course, and nominally on the invitation of the king, prelates and clerical retinues ultimately from 14 provinces would assemble in Paris to review the affairs of the Gallican Church. The First Estate was legally free of any and all taxation, but these general assemblies voted the so-called *don gratuit* or "free gift" the church offered the king in lieu of taxes, and this was a major source of revenue for the state. By 1582 purses of jetons began to be given out as swag to attendees, and the tradition continued all the way through 1789. The Church struck 1,500 jetons for its convocation in 1582.[52] In 1608 it struck 3,800 in silver at a cost of 4,800 *livres*, in addition to paying for 38 velvet purses (two in scarlet garnished with silver thread, the rest in green velvet with silk).[53] The jeton struck for the meeting of 1755 was remarkable in that it was stamped on a small silver bar and not a circular coin, with the bust of Louis XV on one side and a similar inscription dated 1755 as in the figure above.[54]

The practice of having jetons fashioned for themselves and their families spread widely among the upper levels of the French nobility over the three centuries from 1500–1800.[55] Unlike the institutional jetons surveyed so far, this broad class of jeton production was more private and oriented toward personal use. These were jetons for noble individuals.[56] The arms and insignia that bedecked these jetons evidence a concern for genealogy and the grand traditions of the French nobility. Heraldry and family crests, ciphers, and emblems are prominent (Fig. 3.6). The French crown and other signs of due submission do appear on these personal jetons, but no royal busts adorn them. These jetons may have been used on counting tables, but for nobles and noble families they also functioned as a kind of visiting card, a little something that distinguished its owner

50. On these assemblies and their jetons see the wonderful study by Florange and Strowski (1927); see also Blanchet (1930), pp. 282–86.

51. Florange and Strowski (1927), pp. 87–88; Church assemblies took place nominally only every five years and only every ten years to vote the *don gratuit*; more pressing affairs dictated more frequent meetings.

52. Florange and Strowski (1927), p. 10.

53. Payment order quoted in Florange and Strowski (1927), p. 12.

54. F. 3971v; such an item showed up on eBay in 2005.

55. On personal/family jetons with crests and ciphers, see J. Florange (1902, 1907); Valin (2016), vol. 1, pp. 122–24. Blanchet (1930), pp. 409–71; Parsons (2014), p. 276; Fontenay (1854), pp. 308–30 for the Burgundian case.

56. Rossignol (1851), pp. 53–55, makes the distinction between personal/family jetons and official/institutionalized ones, as does La Tour (1899), p. xii, and Blanchet (1890), vol. 2, p. 403.

and his family from the common run.[57] At the turn of the twentieth century the French numismatist Jules Florange published his great two-volume compilation of jetons with arms and insignia, and his work signals just how deeply the personal jetons of the nobility diffused into the culture of the Second Estate. Similarly, Florange's work shows the depth of penetration of jetons into the provinces and noble families in the countryside and provincial urban centers.

FIGURE 3.6: Jeton of the La Burgade family, 1715.

Nobles needed permission to strike their own jetons, and cases that came before the royal monetary court (the Cour des Monnaies, seen earlier) show control and concern over the noble status of the petitioner. In 1577 one Nicolas Bernard, for example, received permission for jetons "that are rightfully his because of his station."[58] Earlier in 1570, as the Baron de Torcy, Jean de Blosset had no trouble in his request for jetons to be struck with his and his wife's respective coats of arms, but the court insisted that these jetons were "for their personal use only."[59] At about the same time one Abraham Martineau, *sieur de la Besne*, asked the Cour des Monnaies for permission to have personal jetons struck. The engraver refused to engrave the dies without court permission. Martineau was a noble of the sword (*noble d'épée*), a knight (*chevalier*), and only a captain of a company of the royal guards, but he boosted his noble status before the court in writing about his wife's noble background: she is "Isabel Justinien, issued from the illustrious family of the Justinians, sovereigns of the island of Chio in Greece" where they exercised sovereign justice and could mint money. The judges referred the matter to the royal attorney (the *procureur général du roi*) who gave permission. The court approved the striking of 300 jetons under its close supervision at the mint. Martineau then wrote to request that five of his jetons be struck in gold. But Martineau may have pushed his noble credentials too far, for a follow-up court order gave permission for only 300 in silver.[60]

Provincial Estates

Provincial estates in the Old Regime were institutions that united the three orders of a province in a legal, representative body with certain powers and authority, notably over taxes, in the province and before the king. Not every province had its own estates, but many did, including the more important outlying provinces, such as Burgundy, Brittany or Languedoc.[61] From the sixteenth

57. J. Florange (1902), p. ii; Fontenay (1847), pp. 25–30.
58. "...qui luy sont deuz à cause de sondict estat," quoted in Mazerolle (1902–1904), vol. 1, p. 556.
59. "...pour leur usage tant seullement," quoted in Mazerolle (1902–1904), vol. 1, p. 571.
60. "Isabel Justinien, issue de l'islustre famille des Justiniens, souverins de l'isle de Chio en Grèce"; this episode, per Mazerolle (1902–1904), vol. 1, pp. 580–81.
61. On the provinces, see Bély et al. (2003), pp. 103–06; Blanchet (1890), vol. 2, pp. 429–36. As a counter example, the Estates of Provence didn't meet between 1639 and 1787; however, two jetons were struck in 1698 and 1713 for the "trésoriers des États de Provence," presumably for accounting purposes; see Raimbault (1903), pp. 5, 22–23.

FIGURE 3.7: Jeton of the Estates of Languedoc, 1787.

century provincial estates regularly had jetons struck for themselves. The Estates of Burgundy, for example, issued jetons from 1575, the Estates of the Artois from 1579, Lille at least from 1598, Brittany from 1641, Cambrai at least from 1678, and so forth.[62] Initially these jetons were struck for use on counting tables and then as *épices* or part of the package of perks for office holders. But more than that, the distribution of purses of jetons at the time of assemblies became part and parcel of the institutional life of various estates, just as elsewhere in other institutions in contemporary society. Jetons issued by various provincial estates have been the subject of specialized studies by numismatists, and as a group the estates and their jetons must figure largely here.[63]

In the previous chapter we observed that by 1789 the Estates of Brittany handed out some 6,200 silver jetons in 62 purses biannually.[64] For its triennial meetings the Estates of Burgundy similarly issued 6,650 silver jetons, 200 in gold, and another 10,000 in copper.[65] Thanks to Émile Bonnet's 1899 study, the Estates of Languedoc provide another close-up example of the role and functioning of jetons within the context of a notable provincial estate.[66]

The formal body that united the three estates of Languedoc dates to 1418. By the 1600s the Estates met annually, and from 1736 that meeting was always in Montpellier. Although the numbers fluctuated slightly over time, the Languedoc estates generally consisted of 110 members: 22 clergy, 22 nobility, and 66 Third Estate members drawn from across the province. The Archbishop of Narbonne was the ex officio head of the estates.[67]

The first jeton of the Languedoc estates dates from 1634. The estates struck jetons annually and in silver from 1677. According to Bonnet, for the period 1651–1790 they issued 348 different jetons.[68] From the later seventeenth century the annual production of jetons hovered in the range of 3,000 to 4,000.[69] The estates spent roughly 10,000 *livres* a year on jetons, the number for 1787 being 13,154 *livres* for 3,700 jetons, one of which is pictured here (Fig. 3.7).

The Languedoc estates distributed at least 34 purses of jetons annually. Recipients included the archbishop-president, the leaders of each of the three estates, syndics from regions of the province, and individuals in the accounting offices of the estates. That figure does not count additional

62. Bonnet (1899) in Depeyrot (2007), p. 76; per p. 81 n., the Estates of the Artois didn't begin to strike silver jetons until 1748, however. See otherwise Feuardent, passim.
63. See Daniel (1980); Blanchet (1930), pp. 323–30; Fontenay (1854), pp. 268–308; Deyperot (1980) and (2007) and Armandel (1901) on the Languedoc estates.
64. See above, Ch. 2 and Fig. 2.2.
65. Feuardent (F.9771–9924) counts 153 jeton strikings by the Estates of Burgundy and related parties from 1575 through 1790.
66. Bonnet (1889) and as reprinted in Depeyrot (2007); see also Depeyrot (1980) and Vian (1949).
67. On these points see Bély et al. (2003), pp. 510–12; Bonnet (1899); Amardel (1901). Major clerics typically headed provincial estates.
68. Feuardent (F.10900–11014) lists over 100 of these.
69. A peak number of 4,100 for 1731; numbers for other years: 1775 (3,100); 1777 (3,300); 1787 (3,700); 1788 (2,800); see Bonnet (1899) in Depeyrot (2007), p. 129 n.; Depeyrot (1980), p. 9.

purses sometimes given out to royal representatives, the person who came up with the *devise* for the jeton that year, or people who might have rendered special service to the estates that year. Originally, jetons were struck for use within the accounting bureau (the *bureau des comptes*) of the estates consisting of fifteen annually-appointed members. In 1677 the estates created an additional office (the *bureau des recrues*) with 12 members to assist in rectifying accounts at the beginning and end of the year. Members of both of these two bureaus received a substantial package of benefits. The top two deputies in each office were paid 900 *livres* for their service; everyone else received 450 *livres* in cash, and they all received a purse of 100 jetons and six pounds of candles. (If two members of the Bureau were from the same town, they had to share.) Noteworthy for our purposes, toward the end of the eighteenth century, certain members of the estates who were entitled to jetons took money instead. They were paid the value of the silver (250 *livres* for 100 jetons) plus the cost of the purses themselves. The archbishop of Narbonne had his own octagonal jetons struck, subsidized by what he received from the estates in lieu of their ordinary jetons to which he was entitled.[70]

The Languedoc estates was a powerful institution, and every year it sent an official delegation of its three orders on a *voyage d'honneur* to present its *cahiers de doléances* to the king at Versailles. These were formalized affairs, and it would undoubtedly have been the case that jetons were sprinkled about along the way. While unclear in the case of the Estates of Languedoc, the archival record shows this happening during the same ritual visits made by the Estates of Burgundy.

FIGURE 3.8: Jeton of the Estates of Burgundy, 1785.[71]

The Estates of Burgundy met triennially, so only every three years did it undertake its *voyage d'honneur* and send a delegation to present the Burgundian *cahiers de doléances* to the king.[72] These visits were highly scripted. The trip taken in 1737 is entirely typical.[73] The royal governor-general extends an invitation for a delegation of the estates to travel to Paris. The delegation consists of a representative from each of the three orders of society (clergy, nobility, and Third Estate), the secretary of the estates, its treasurer, and its syndic for a total of six individuals. The delegation arrives in Paris in January and is received at the Hôtel de Condé by the governor-general, the duc d'Orléans. Elaborate negotiations ensue over the formal document presented to the king. As part of these discussions, the group retires to the duke's private office to settle on the design for the jeton of the estates to be handed out during the *voyage d'honneur*, and it compiles a list of who is to get jetons. Finally, a royal audience is arranged for Monday, June 10, 1737. Accompanied by the duke as governor-general, the representative from the clergy dons the long robe with a hood.

70. On these points, see Bonnet (1899) in Depeyrot (2007), variously pp. 76–91.
71. See also above Fig. In.5.
72. On these episodes see AN, K 683[1] [Bourgogne: Voyages d'honneur], #17, #23, #55, #67, #79, #116, #123, and #135, which give accounts of the delegations from the Estates of Burgundy and royal audiences from 1737 through 1779.
73. AN, K 683[1], #17: "Rélation de ce qui s'est passé au voyage d'honneur fait en 1737 pour présenter au Roy les Cahiers de la Province de Bourgogne, en conséquence des décrets des Etats tenus en la Ville de Dijon au Mois de May 1737."

The representative nobleman wears his "ordinary clothes" ("*dans son habit ordinaire*"); the Third Estate representative sports the short cloak; gloves and hats are specified. Two royal masters of ceremonies usher the group into the hall at Versailles reserved for receiving ambassadors. The king (Louis XV) is seated in a chair in front of the fireplace, his major ministers (Saint Florentin, Orry, Cardinal Fleury) by his side. Bodyguards and *gentilshommes de la chambre* stand nearby. The governor-general makes three profound reverences and introduces the delegation to the king. Everyone takes his assigned place and uncovers. The king takes off his hat, keeping it by his side. The Third Estate representative, the syndic, and the treasurer kneel before the king. The other two estates remain standing. The clerical representative, the Abbé de Periguy, makes a speech and passes the Burgundian *cahiers* to the king, who says just a word and immediately hands the document on to his ministers. Then, after further bows, they retire.

Rounds of visits ensued with stops at the apartments of the queen, the dauphin, and various other royal family with more palaver and purses of jetons given away. This traipsing through Versailles culminated with another meeting with the duc d'Orléans in his private apartments. Two of His Serene Highness' gentlemen met the party at the doors of the apartments and escorted them in. The duke received them standing and uncovered (i.e., without a hat) and without a chair behind him. The Abbé de Periguy made another speech to which His Serene Highness responded. Monsieur de Montigny, the general treasurer of the estates, then presented the duke with a purse embroidered with his arms containing one hundred gold jetons stamped with the arms of Burgundy and the *devise de la Trenialité*. His Highness offered dinner to the party, and his gentlemen then escorted them out.

The 1758 delegation, to pick another example, followed the same script.[74] They, too, arrived in January and, coordinating through the royal governor-general, the prince de Condé, they met with the king (again, Louis XV) on Sunday, April 30, 1758. This time the king took his hat off on seeing the delegation, then put it back on when they approached. There followed the "customary visits" (*visites accoutumées*) to the queen, the dauphin, the dauphine, other ranking court ladies with households, and various dukes and counts, the Comte d'Artois "being in the arms of his wet nurse."[75] The delegation from Burgundy again scattered its purses of jetons at each stop. It stayed the night in Versailles with further meetings the next day including one with the controller-general who hosted a dinner. After that, the visitors were treated to a waterworks display at Versailles and the Trianon palace, traveling the short distance from the main palace to the Trianon by gondola. They did not go on to the royal menagerie because it was too late in the day. They took a carriage back to Paris that evening. Further ceremonial encounters followed in Paris, including one with the Comte d'Artois, the future Charles X, who was then just a baby, so his governess replied to the speech by the Burgundians. Then, on August 28, 1758, the delegation met with the controller-general again to receive the king's response to the *cahiers de doléances* (called the *réponse des Cahiers*) of the Burgundian estates. The king was naturally pleased with all of his subjects in Burgundy. The delegation left the next day to return to Burgundy. The *voyage d'honneur* of 1767 followed the same scenario except that some meetings took place at Fontainebleau and that no waterworks were possible at Versailles because of a drought, but the king (still Louis XV) insisted (via intermediaries) that he did not mean to change anything in the traditional practice of receiving representatives of the Burgundian estates. There would be a next time, and jetons would again figure in the ceremonies.[76]

74. AN, K 683¹, #67: "Voyage d'Honneur 1758."
75. "...le Comte d'artois qui étoit sur les bras de sa nourrice...," AN, K 683¹, #67.
76. AN, K 683¹, #116, "Procès Verbal" 1767.

The City of Paris

In typical Old Regime fashion several overlapping centers of power exercised authority in and over the great city of Paris. One of these was the Paris Parlement, the sovereign law court of the region. Another was the *prévôté* de Paris, seated in the Châtelet fortress, which handled police and civil matters on the part of the king. These were institutions of medieval origin. Yet a third, with his spies and agents, the powerful position of chief of police (*lieutenant de police*) arose only in 1667 and reported directly to the king. These governing bodies notwithstanding, the nominal head of the merchants' guilds, the *prévôt des marchands* and a *bureau de la ville* (city council, or sometimes *conseil* or *corps de la ville*) constituted the chief administrative body that ran the city. With its origins in the fourteenth century, the *bureau de la ville* consisted of the *prévôt des marchands*, four aldermen (*échevins*), and 26 additional councilors, not to mention an outsized staff and others who worked for the city in one capacity or another. The *prévôt* was essentially the mayor of Paris, the aldermen and councilors the city council. Located in the Hôtel de Ville, this large and important organization was in charge of the water supply, maintaining the streets and port facilities, provisioning the city with grain, wine, wood, and charcoal, and collecting city taxes.[77] And, as might be expected, jetons figured largely in the minds and purses of city fathers.

FIGURE 3.9: Jeton. Ville de Paris.

Paris issued its own jetons at least by 1440. By 1461 the city began to have its jetons stamped in silver as well as baser metals. Over almost four centuries leading up to 1789 nearly 500 different jetons issued from city hall.[78] These jetons were struck for the city of Paris itself (*la Ville de Paris*) (Fig. 3.9), for individual *prévôts*, and for various aldermen (*échevins*).[79] The city of Paris issued jetons in huge numbers and at great expense: 17,750 jetons in 1647 (9,550 in silver; 8,200 brass) at a cost of 8,816 *livres*.[80] The number was 17,200 in 1661, 20,000 in 1679, and over 21,000 by 1685 (13,600 in silver; 7,500 in brass).[81] The numbers may have reached over 24,000 jetons annually.[82] A total of 136 individuals received silver jetons in 1661; 183 purses were handed out in 1685, 221 in 1692.[83] The numbers tail off in subsequent years—only 4,595 silver jetons handed out in 1739,

77. On city administration, see Bély et al. (2003), pp. 955–57; Veillon (1994), p. 39; Blanchet (1930), pp. 274, 277–80; Blanchet (1890), vol. 2, pp. 420–21.
78. Feuardent counts 470 such jetons, F.3352–3823; d'Affry de la Monnoye's earlier catalog (1878), pp. 1–206, lists 362. On the Hôtel de Ville and its jetons, see also Rouyer (1879), especially pp. 7–15; Barnard (1916), pp. 75–76. See also d'Affry de la Monnoye (1878), pp. 209–290 for a remarkable selection of administrative records of the *prévôté des marchands* and the Ville de Paris concerning ordering and paying for jetons.
79. Rouyer (1879), pp. 20–21 points out that, whereas the former were struck at the expense of the city, jetons of the *échevins* (F.3770–3822) were privately struck like the individualized jetons of the nobility at the personal expense of the *échevin* in question. Municipal funds otherwise paid for the jetons of the city of Paris, as d'Affry de la Monnoye (1878) notes, p. xx.
80. Order reprinted in d'Affry de la Monnoye (1878), p. 244.
81. AN, K 1019 [Orfèvres, jetons], #76 (1685), #73 (1661), #75 (1679).
82. Veillon (1994), p. 28.
83. Counting both silver and brass; AN, K 1019 [Orfèvres, jetons], #78 (1692).

and only 112 purses offered in 1769.[84] But still, we find 100 jetons in gold for the king and 8,954 others in silver for distribution in 1771.[85] In 1691 the city of Paris spent 15,537 *livres* for jetons and purses to give out at the Étrennes. Throughout the middle of the eighteenth century Paris spent around 12,000 *livres* annually on its jetons. That number rose to almost 30,000 *livres* in the second half of the century, but that number includes both jetons given out at the Étrennes and jetons used internally as *jetons de présence* at council meetings, about which we will say more momentarily.[86] The point now is that the incorporation of jetons within the Hôtel de Ville was extensive and parallels the historical development of jetons and their use that we have already observed. The abundant archival detail available for the city of Paris allows us to see this development in an even closer manner.

Naturally, then, we see jetons first used at city hall as tokens deployed on counting tables and then quickly becoming in the late Middle Ages perks of office and part of packages of *épices* for officeholders and those tending accounts.[87] Nascent city government undoubtedly used jetons already in the 1300s out of accounting necessity. By the 1440s a new *prévôt* or *échevin* received a leather purse with either 50 or 100 jetons, "for their use in the Hôtel de Ville."[88] The legal right of the *prévôt* and others to receive jetons as a token form of remuneration came in 1537, although it was consistently exercised only after 1581.[89] So, in 1544, for example, the Ville de Paris ordered a modest 400 jetons in silver and 1,000 in brass, which the *prévôt* distributed in packages of 50 silver jetons and 100 brass ones to the four *échevins* and three other officials in city hall, with the remaining two hundred brass jetons "left in the office of the said town for common use in its affairs."[90] The numbers of jetons struck by and for city officials and the social complexity of the uses to which jetons were put exploded dramatically thereafter, but this core function of jetons for accounting use and as a perk of office never ceased within the halls of city hall.

At the Étrennes persons affiliated with the Bureau de la Ville received their packages of *épices* ("spices") that included not only jetons but a small cornucopia of other items: real spices, letter openers, gloves, canes, reams of paper, quills, spiced wine, candy, jellies, torches, candles, and matches. In 1661, for example, strictly internal to the Bureau de la Ville and omitting jetons and presents handed out at the Étrennes, the numbers came to 7,700 silver jetons, 4,800 brass ones, 2,400 top quality quills, 5,550 second quality quills, 111 reams of fine paper, and 19½ reams of fine paper with gilded edges.[91] These packages went to a wide range of individuals within the Hôtel de Ville: not only to the *prévôt*, the *échevins*, and the 26 councilors, but also to the heads of the secretariat (*greffier*) and accounting offices (*receveur*), royal officers and agents (*procureur, avocat*), the keeper of the seals of the city, the internal notary and keeper of cadastral records, the 16 neighborhood officers (*quarteniers*), and retired *échevins*, who until 1731 annually received a purse of 50 silver jetons.[92] Walking canes and gloves of a quality appropriate to the station of their recipients

84. AN, KK 1056² [Fêtes].
85. See numbers given in AN, K 1019 [Orfèvres, jetons], #80 (1739), AN, KK 1056 and 1056² [Fêtes]. See also Rouyer (1879), p. 13.
86. See below and AN, KK 1056² [Fêtes], 1772.
87. Veillon (1994), p. 28; La Tour (1897), p. viii.
88. "...pour eux servir oudit hostel," quoted in Rouyer (1879), p. 7.
89. d'Affry de la Monnoye (1878), pp. xix, xxi.
90. "mys au Bureau de ladite Ville pour servir en commune pour les affaires d'icelle," quoted in Rouyer (1879), p. 10; A. Barthélemy (1875), p. 240, provides similar details.
91. Extracted from AN, K 1019 [Orfèvres, jetons], #73 (1661). See also related documents #72–#180 detailing the distribution of jetons and *épices* of the Ville de Paris from 1637–1789. On the distribution of jetons by the Ville de Paris, see also Rouyer (1879), p. 11.
92. AN, K 1019 [Orfèvres, jetons], #88 (1731).

went to the *prévôt*, the *échevins*, and half a dozen ranking city officials "for visits to the sewers and to the waters at Rongis, Arcueil, Belleville, and Saint Gervais."[93] One can only imagine such scenes. Canes were an annual expense of roughly 500 *livres* that went on year after year for decades.

These distributions happened at New Year's, and formal lists of recipients were kept.[94] Still internally, jetons and presents went to everyone at the Hôtel de Ville at the Étrennes. That would include what we might call an engineering staff: the chief engineer, the chief carpenter, inspectors and intendants of fountains, masters of public works (*maîtres des oeuvres de la Ville*), down to plumbers (*égoutiers*). A support staff of assistant secretaries, clerks, and trainees received jetons, as did ushers and doormen (*huissiers*), armed guards (*Suisses*), masters of ceremonies, and concierges, as well as the priest and a raft of other servants, bailiffs, porters, pages, coachmen, lackeys, and *garçons* of various sorts. Fifty different *categories* of people and roughly 100 individuals received something at the new year from the city. In addition, authorities set aside 600 jetons (for 1692 at least) for later use and distribution during the year.[95]

As time went on the full package of *épices* came to be paid simply in additional jetons or sometimes in cash, as, for example, in 1583 when jetons substituted for actual spices and spiced wine.[96] By 1697 Paris had shifted to cash payments for its Étrennes, and cash seems to have held for the early eighteenth century, but by the 1760s jetons were again distributed for the Étrennes.[97] Probably lesser in-house New Year's gifts were paid in cash and statutory recipients in the Bureau de la Ville continued to receive their annual purse of jetons, quills, and reams of paper. In a revealing case of 1672 regarding the value of jetons versus cash, the maître d'hôtel of the *prévôté* asked that the thirty *livres* given to him at the Étrennes be exchanged for 50 of the city's jetons, undoubtedly a greater honor.[98]

At the Étrennes the Ville de Paris rewarded outside individuals or more occasional employees who had otherwise assisted city government that year. Details of archers, gunners, town trumpeters, royal drummers, bassoonists, and oboists (*joueurs de hautbois du Roi*) served the city of Paris, and as standard practice their representatives annually received jetons in unadorned leather purses. The spice merchant used by the town, the paper merchant, the city's printer, and pastry chef regularly appear on the list. In the 1660s, nephew of its founder and a councilor at the Cour de Monnaies, the similarly named Théophraste Renaudot received a purse of 100 silver jetons for a subscription to the weekly, the *Gazette de France*. And then, the fireworks master got his purse of jetons as a compliment for the city's pyrotechnic displays.

The jetons of the city of Paris served perhaps their most important function at the Étrennes when the Ville de Paris distributed purses of its jetons to the court at Versailles.[99] (Versailles—or whatever the locale of the king at that moment—must have been a madhouse on New Year's Day, as delegations scurried about bestowing little purses of jetons and exchanging New Year's greetings.) The *prévôt* and *échevins* headed the delegation from Paris, and the first visit was always to

93. "...pour les visites des Egouts, des Eaux de Rongis, arcueil, Belleville, et Prez Saint Gervais," AN, K 1019, #181, Ville de Paris, "Cannes distribuées aux membres du Bureau."
94. The archival series AN, K 1019 [Orfèvres, jetons] ["Ville de Paris, Étrennes"], #72–#180 contains these lists intermittently from 1637 and annually from 1731–1789.
95. AN, K 1019 [Orfèvres, jetons], #78 (1692).
96. d'Affry de la Monnoye (1878), pp. xxiv–xxv. For cash payments, see, for example, AN, K 1019 [Orfèvres, jetons], #87 (1697) and #88 (1731).
97. See, for example, AN, K 1019 [Orfèvres, jetons], #87 and #88 for distribution lists for 1697 and 1731. AN, KK 1056[2] [Fêtes] notes the distribution of gold and silver jetons as Étrennes presents in the 1760s and 1770s.
98. AN, K 1019 [Orfèvres, jetons], #78.
99. The following per AN, K 1019, #72–86. See also Veillon (1994), p. 28.

the king to whom Paris offered a purse of 100 gold jetons valued between 6,500 and 6,900 *livres*.[100] The king's jetons came in a special purse embroidered with gold and the arms of the city of Paris. For presentation the "king's purse" was placed inside another purse of white leather lined with taffeta, hemmed with silver and graced with a silver tassel.[101]

In scenes doubtless reminiscent of the visits of the Burgundian estates described above, the delegation from Paris then called on other royal personages and members of the royal household. The *Mercure galant* reported on the passage of city officials on New Year's Day in 1680:

> On the first of the year...Paris officials [Messieurs de Ville] with the *prévôt des marchands* at their head went to Saint Germain and had the honor of greeting the King, the Queen, Monseigneur, Monsieur and Madame. They then went to the apartments of all the princes of the blood and then discharged the same duty [*sic*] towards the Comte de Vermandois, Mademoiselle de Blois, the Duc du Maine, the Comte du Véxin, Mademoiselle de Nantes and Mademoiselle de Tours. They merely complimented the king without giving him presents, but they gave a purse of 100 gold jetons to all the princes and oranges, liqueurs and candy to the princesses.[102]

The party wouldn't have handed anything to the king directly, of course, but the purse likely passed to the king via a gentleman, and Jules Rouyer is correct in doubting that the princes, too, received gold jetons.[103] But gold or silver, they came in velvet purses threaded with gold, silver and silk with the arms of the city of Paris embroidered on the bottom of the purse.[104] Oranges, boxes of candies, jars of jelly, and bottles of *hypocras* (wine seasoned with sugar, cinnamon and other spices) went to royal ladies at court, a telling sign of gender relations at that level. Oranges in wintertime bespeak luxury, and *hypocras* was a coveted delicacy. In 1679 thirty packages of such treats were offered to the queen and other royal ladies, each according to her rank in the hierarchy of the court.[105]

At the Étrennes the Ville de Paris distributed its purses widely outside of court, too, and these made their way to government bureaus and other centers of power. The particulars might vary somewhat depending on the year, the office in question, and the office holder—we are speaking about a 250-year period after all, but recipients usually included the king's council (Conseil du Roi, when it existed), ranking princes and noblemen with high administrative charges (such as the duc d'Orléans or the prince de Conti), the chancellor, ministers and secretaries of state (Fouquet, Colbert, Soubise, Bertin), the controller-general (when that position was at play), intendants of one sort or another, and other administrative departments within royal government (war, foreign affairs, the Chambre des Comptes). In addition, Paris jetons went to the archbishop of Paris and

---

100. See AN, KK 1056[2] [Fêtes] for references for gold jetons for the king and their costs for 1767–1772.

101. "...bourse blanche doublée de taffeta brodée en argent avec glans d'argent pour servir d'Etuye a la Bourse du Roi." AN, KK 1056[2] [Fêtes], 1769. The value of this purse was 102 *livres*.

102. "... le premier Jour de l'Année...Messieurs de Ville ayant M[r] le Prevost des Marchands à leur teste, se rendirent à S. Germain, & eurent l'honneur de salüer le Roy, la Reyne, Monseigneur, Monsieur & Madame. Ils allerent en suite chez tous les Princes du Sang, & s'acquitterent de mesme devoir envers M[r] le Comte de Vermandois, Mademoiselle de Blois, aujourd'huy Madame la Princesse de Conty, M[r]. le Duc du Maine, M[r] le Comte du Véxin, Mademoiselle de Nantes et Mademoiselle de Tours. Ils ne font qu'un Compliment au Roy, sans lui faire de Présens; mais ils donnent à tous les Princes une Bourse de cent Jettons d'or, & aux Princesses, des Oranges, des Liqueurs, & des Confitures," *Mercure galant*, 1680(1), pp. 45–47; also quoted in Rouyer (1879), pp. 12–13.

103. Rouyer (1879), p. 13 n.

104. At least according to the distribution of 1739: AN, K 1019, #80 (1739).

105. AN, K 1019, #75 (1679).

other ecclesiastical dignitaries and to the president of the Paris Parlement and other jurists and personnel serving there.[106] To keep the wheels well greased, the city of Paris gave its purses to second-level assistant secretaries, bailiffs and clerks in the royal household and elsewhere who were paid otherwise by these bodies but who performed services for the Ville.[107] For many years a certain Desgranges, the chief of staff of the secretary of state Pontchartrain received a special bonus of 200 silver jetons, "in consideration of the good offices he renders the Town."[108] The maître d'hotel of the Ville de Paris, for many years a certain Pierre Lefevre, delivered these presents around town. In 1665, for example, he was paid 30 *livres* "for delivering *hypocras*, jellies, candies, and silver jetons to all the houses found on the Étrennes list for this said year as he is accustomed."[109] Interest in this example is not only that the city paid its maître d'hôtel to deliver these presents, but that they were delivered to *maisons* or "houses" and not so much to individuals.

If that were not enough, jetons played a role in Paris city government that extended well beyond what was already the case internally and externally with the distribution of Paris jetons at the Étrennes. Most notably, jetons functioned as *jetons de présence* or tokens handed out for attendance at meetings of the Bureau de la Ville. We have already seen jetons serving as a reward for attendance in the king's council, the Conseil du Roi, in the sixteenth century. We will see more of the role of *jetons de présence* in the royal academies of the eighteenth century, but members of the Bureau de la Ville (the *prévôt des marchands*, the four *échevins*, and the 26 municipal councilors) each received a jeton for the ordinary meetings of the city council scheduled for Tuesdays and Fridays of each week.[110] Members received four jetons apiece for each "extraordinary" meeting or "extraordinary" activity such as inspections, attending masses, participating in civic ceremonies, and the like. In 1785 the Bureau de la Ville was reorganized with special departments and committees to handle particular aspects of city administration.[111] The result was fewer ordinary meetings of the Bureau as a whole (62 fewer in 1786–1787) and more extraordinary meetings and committee meetings outside of the regularly scheduled ones (50 extra in 1786–1787). Jetons allocated for ordinary meetings not held were redistributed for extraordinary meetings at the higher rate. In order not to confuse jetons given out for attendance with the jetons distributed for the Étrennes, in 1787 the first *échevin* proposed that the entitlement of four jetons per meeting be converted into cash payments of eighteen *livres* a meeting, not a bad sum again considering that the average skilled tradesman might earn five *livres* a day. Plainly, jetons were a means for supplementing the income of members of the Paris city council. The accounts auditor, for example, received an extra 100 *livres* a year for his services from 1708.[112] Reconciling the accounts of deceased *échevins* seems to have been particularly rewarding.[113] The sums varied, but they could be substantial: commissioners for such occasions received up to 300 jetons for their work, with the reporter (*rapporteur*) getting three purses each containing 100 silver jetons; the *prévôt* and *échevins* each received 75

106. AN, K 1019, #168 for the distribution of 1774; see also AN, K 1019, #83 (undated).
107. AN, K 1019, #83 (undated).
108. "...en consideration des bon offices qu'il rend a la Ville," AN, KK 457, fol. 361r; d'Affry de la Monnoye (1878), p. 277.
109. "...ypocras, confitures, dragées et jettons d'argent par toutes les maisons contenues aux memoires des estreinnes de lad. année ainsi qu'il a accoustumé...," payment order of 31 August 1665, reprinted in d'Affry de la Monnoye (1878), p. 256.
110. Rouyer (1879), p. 14.
111. On these points see document dated September 4, 1787, "fixation des Jettons pour assemblee extraordinaire... Conversion de ces Jettons en argent," AN, K 1019, #86 (1787).
112. d'Affry de la Monnoye (1878) p. 282.
113. AN, K 1019, #85, "Délibération au sujet des Jettons à revenir a Messieurs du Bureau pour differents comptes," dated July 31, 1787.

jetons for their attendance at the reconciliation of accounts; other members of the Bureau in attendance received sixteen jetons. These sums added up.

Perhaps to rein in these practices, in 1783 Louis XVI suppressed the use of jetons for the Étrennes for the Ville de Paris; jetons were henceforth to be used exclusively for attendance.[114] This move met with considerable resistance from within the Bureau de Ville. As the Bureau disingenuously reported:

> Jetons, which our immemorial practice was to remit to the chamberlains of his Majesty and of the Queen, have ceased to be distributed to them, and this has occasioned on their part remonstrations which seemed all the more justified to the Bureau because the numbers involved didn't exceed twenty-three jetons and because one cannot imagine that His Majesty's intention could have been to suppress them.[115]

The notion that the *jetons d'étrennes* went exclusively to the court chamberlains was a ruse, and at issue is more likely 23 *purses* of jetons, not 23 jetons. One way or another, the king relented, but insisted that these gratuities be converted to cash and considered as part of the appointments of the office holders in question. This brought a further remonstrance.

> But after the refusal of the majority of those involved to receive cash in lieu of jetons, the *prévôt des marchands* and *échevins* thought that one ought to have a greater regard for their sensitivities, that the jetons in question substitute for any and all honoraria for the pensioners of this class, and that in fact the extra costs of these jetons add up to only an additional 2,000 *livres.*

The sensitivities of recipients regarding cash versus more dignified jetons is to be noted again. The Bureau recognized that whatever steps the *prévôt* and *échevins* took in this regard would require royal confirmation, and they petitioned the king to reinstate its traditional jetons for the Étrennes. Louis gave in and agreed "to the re-establishment of jetons to be given annually to the chamberlains of His Majesty and the Queen and to the distribution of jetons whose conversion into cash had been decreed earlier." And so, in 1785, the city of Paris again distributed 2,869 silver jetons in 14 embroidered purses and a number of other "sacks."[116] Still, 1789 was not far away.

In the period surrounding the death of Louis XIV in 1715 and then the succession of Louis XV,

114. On this episode, see AN, K 1019, #81, "Estrait des Régistres du Conseil d'Etat du Roy."

115. The original for these quotations is as follows: "...les Jettons qu'on etoit dans l'usage immoriale de reméttre, à l'epoque de chaque renouvellement d'année aux huissiers de la Chambre de Sa Majesté et de la Reine, ont cessé de leur etre distribués, ce qui à occasionné de leur part des réprésentations qui ont paru d'autant plus favorables au Bureau, que l'objet de cette distribution particulier n'excedant pas le nombre de vingt trois jettons, on ne pouroit pas supposer que l'intention de Sa Majesté eut été de les supprimer. [Le décret royal du 31 octobre 1783] a autorisé le Bureau à rétablier les Jettons qui se delivroient annuellement à titre d'honoraires, en convertissant toutesfois, d'après l'evaluation qui en seroit faite, les dits Jettons en espéces dont le payement seroit fait dorénavant sous la qualification d'appointemens. Mais que d'après le réfus du plus grand Nombre de ceux Intéréssés à cette distribution, de récévoir de l'argent au lieu de Jettons, les dits Prévot des Marchands et Echevins ont pensé qu'on devoit avoir autant plus d'egard à leur délicatesses, que les Jettons dont il s'agit tiennent lieu de tous honoraires quelconques à l'egard des pensionnaires de cette classe, et que dans le fait la plus Value du prix des Jettons ne forme pas en totalité la difference de plus de 2000#, que quelque soient les motifs qui ont determiné les dits Prevôt des Marchands et Echevins à faire les changements ci dessus énoncés, il ne leur à pas échappé que ces changements devoient etre soumis à l'approbation de Sa Majesté. [Sa Majesté est d'accord avec]... le rétablissement des Jettons dont lusage avoit consacré la rémise annuelle aux huissiers de la chambre de Sa Majesté et de la Reine, et la distribution de ceux dont la conversion en argent avoit été ordonnée par l'article 17 de la délibération du Bureau de la Ville du trente un Octobre mil sept cent quatre vingt trois...," AN, K 1019, #81, "Estrait des Régistres du Conseil d'Etat du Roy."

116. "sacs," AN, K 1019, #81 (1785).

jeton activity in city hall seems to have diminished somewhat.[117] Otherwise, the city of Paris struck and distributed its jetons in these ways every year *except* for 1710. Why? The deep and long-lasting freeze the previous year proved a disaster for the wheat crop, and hunger and the threat of starvation were rampant. That winter the city of Paris was forced to devote extra efforts and extra funds towards succoring the poor, and so for 1710 "the King found it good that the city employ to that end the funds it would have used to have jetons struck."[118]

The use of jetons and medals struck by the city of Paris extended even further. Each new neighborhood watch officer (*quartenier*—there were sixteen) owed a purse of 150 jetons to the doyen of the Bureau and 100 for his confrères; a further distribution was required on Saint Laurent (August 10) with 72 jetons to the doyen and 60 to the others.[119] When any members of the city council themselves got married, they owed twenty jetons to their fellows in lieu of the customary celebratory dinner.[120] The Bureau de la Ville in turn offered jetons and medals at the weddings of sons and daughters of *échevins*. On a more ad hoc basis, in 1720 medals were struck and given to the Bureau and its members, "in recognition for their pains and troubles in the supervision and distribution of firewood in the year 1720."[121] On one occasion at least, the city of Paris had 200 jetons struck especially for the clergy, and in 1766 it awarded two silver medals to winners of a marksmanship contest.[122] Rescuing people who fell into the Seine River became a matter of civic concern in the later eighteenth century, and in 1772 the city offered its medal as a prize for a successful method to revive the drowned.[123]

These examples make clear the manifold ways in which jetons and practices surrounding jetons and medals ramified throughout city government, but the most extraordinary example concerns the 600 marriages paid for by the city of Paris in 1751 on the occasion of the birth of the Duke of Burgundy, the son of the dauphin and grandson of Louis XV.[124] The city of Paris spent 500 *livres* each for 600 poor girls of Paris and young men capable of earning a living. That sum included a dowry, wedding outfits for the bride and the groom, wedding dinners, and jetons and medals. The Monnaie des Médailles struck 800 jetons featuring the bust of Louis XV on one side and the Ville de Paris on the other with the inscription, "Birth of His Highness the Duke of Burgundy. Marriages. Paris. 1751."[125] Authorities handed out these jetons to the couples as marriage medals, and later the city ordered a separate set of medals to otherwise commemorate the event.[126] In total, the city laid out more than 300,000 *livres* for the occasion, a truly enormous sum. These weddings were all celebrated on the same day, 9 November 1751, and a gigantic celebration took place that evening on the Place de Grève in front of the Hôtel de Ville.[127] The city had

117. On this point see the *Mercure* entry, 1722 (3), p. 104: "Depuis le nouveau Regne on avoit été peu exact à frapper au commencement de l'année les Jettons accoutumez. Ils furent rétablis en 1720, à l'exception de ceux de l'Artillerie et de la Ville"; also quoted in d'Affry de la Monnoye (1878), p. 323.

118. "...le Roy a trouvé bon qu'elle employast à cet usage le fond qu'elle auroit employé à faire frapper des Jettons," *Mercure galant* 1710 (2), pp. 199-200; also quoted in d'Affry de la Monnoye (1878), p. 322. On this episode see also Rouyer (1879), p. 13; Veillon (1994), p. 28.

119. d'Affry de la Monnoye (1878), p. xxv. Here, as elsewhere, it is not clear whether jetons went simply to the *prévôt* and the *échevins* or also to all the twenty-six *conseillers* of the Bureau de la Ville.

120. d'Affry de la Monnoye (1878), p. xvi.

121. "à cause de leurs peines et vaccations pour la police et distribution des bois en l'année 1720," document reprinted in d'Affry de la Monnoye (1878), p. 289.

122. See order reprinted in d'Affry de la Monnoye (1878), p. 283, and AN, KK 1056[2] [Fêtes].

123. AI, AIBL archives, A:66, minutes of the meeting of January 14, 1772.

124. These details are taken from Barbeau (1900). The Duke of Burgundy in question died at the age of 10 in 1761; his younger brother, born in 1754, became Louis XVI.

125. "Naissance de Mgr. le duc de Bourgogne. Mariages. Paris. 1751;" F.9768–69, in silver and copper.

126. Barbeau (1900), p. 4; see related archival reference AN Q[1] 1099, fol. 1381.

127. Today's Place de l'Hôtel de Ville – Esplanade de la Libération.

the Place de Grève illuminated at a cost of over 31,000 *livres*, and it spent over 23,000 *livres* for a fireworks display. The assembled crowd scrambled to retrieve the coins thrown to it as largesse in celebration—6,000 *livres* worth.[128] This example and other, lesser ones of marriages for the poor sponsored by municipal authorities bring home the point that Old Regime France was decidedly a world removed from the one we know today.[129]

A number of related offices operated in Paris more or less alongside the Bureau de la Ville and struck and used jetons. The Châtelet or Prévôté de Paris incorporated several departments that issued jetons, and the Châtelet in general seems to have adopted jetons in a manner similar to the Bureau de la Ville. Other Paris-based agencies included those handling government bonds (*receveurs* and *payeurs des rentes*), the Bureau des Pauvres (established 1544, responsible for poor relief, and made up of the *prévôt* and a few parlementarians), and the office responsible for collecting a special royal tax, the *Tontines*, that went to the city of Paris.[130] Messengers, town criers, money-changers, and mounted police also had their jetons. Paris firefighters (*la Compagnie des eaux de la pompe à feu*) in 1788 requested their jetons from the Académie des Inscriptions. Then as now, Parisian notaries controlled the transfer of property, and they issued their own jetons with the gnomon as their insignia from 1680 through the reign of Louis XVI and beyond in the nineteenth century.[131] Notaries and their jetons pop up everywhere, and make plain that jeton culture within guilds was omnipresent across the kingdom.[132] Jetons became deeply ingrained in the over 100 other guilds that exercised their authority over economic and productive activities in Paris. The same holds true for the various royal academies and for the faculty of medicine, as we will see in subsequent chapters. Jetons completely infiltrated the contemporary world, certainly in Paris.

But two major contemporary institutions in Paris are missing from the list of those that made themselves known through their jetons: the University of Paris (the Sorbonne) and the Paris Parlement. The Sorbonne was founded in 1253, but over the centuries did not strike jetons. (Feuardent lists only three jetons struck in its name, all in the later eighteenth century.[133]) The messengers of the university had their own jetons, but not the university itself.[134] The bailiffs (*huissiers*) of the Paris Parlement had jetons, but not the sovereign court of the parlement. Ranking jurists in the parlement received jetons at the Étrennes, but for some reason the parlement and the university did not hand any out. In the workings of the parlement and the Sorbonne jetons had to be used on counting tables when accounts needed to be maintained, but otherwise these two institutional pillars of life in contemporary Paris seem to not to have involved themselves with jetons. The explanation remains elusive, but in this case finding nothing is decidedly finding something.

128. Barbeau (1900), p. 5.

129. Nothing on the same scale ever recurred, but Paris again paid for the marriages of twenty poor girls in 1773, and in the 1780s authorities established a 100,000-*livre* fund to underwrite six marriages of poor girls annually. Other towns did the same; Troyes, for example, paid for four marriages at town expense, and there were others, as Barbeau (1900) indicates, pp. 5–6; see also F.9521 for another foundation established 1683 for annual support of 10 marriages.

130. Rouyer (1879), p. 27; see F.3212–3351 and 3823–3924; also d'Affry de la Monnoye (1878), pp. 1–206 passim.

131. Lerouge and Lerouge (1977), passim and pp. 93–112 regarding the notaries' gnomon; on notaries and their jetons also R. Gillet (1897); Boudeau (1908). Note that the guild of notaries was *not* disestablished in the Revolution. The jeton of 1683 of the guild of notaries heads the index below.

132. Per Feuardent, note other towns where notaries struck jetons: Auxerre, 1748; Bordeaux, 1756; Chartres, Louis XVI; Lyon, 1715, 1750; Nantes, Louis XV, Louis XVI; Orleans, Louis XV, Louis XVI; Rouen, 1719; Soissons, Louis XVI; Tours, Louis XV (1771).

133. F.4429–4431.

134. F.4432–4447.

Outside of Paris, parlements and universities across the realm seem equally absent in the world of jetons. Personal jetons of notables sometimes listed the parlementary positions they held, but the field is essentially blank for the institutions themselves. First-presidents of the Normandy Parlement in Rouen, for example, occasionally had jetons struck, and the attorneys before this parlement struck a nice silver jeton for themselves in 1789, but otherwise there was no sign of this parlement and little sign of other parlements.[135] A handful of presidents of the Burgundy Parlement struck jetons for themselves, and twice it seems the parlement itself struck jetons; once in 1645 at a politically delicate moment in support of the young Louis XIV and again in 1673 in homage to the Sun King.[136] The Parlement at Metz seems somewhat the exception. Funded by fines and induction fees owed by incoming parlementarians, from its foundation in 1633 jetons were distributed each semester along with paper and candles. But often no distribution took place, and the right to jeton was soon converted into straightforward cash payments. In 1643, for example, for his "jetons" each Metz parlementarian received 60 *livres*, a sum that moved up to 80 *livres* in 1648, but no further sign of jetons themselves.[137] Similarly for universities, no jetons. Almost every major city in France at the time had a university, but none struck jetons. The one-off strikings for universities in Caen, Lyon, or Reims listed in Feuardent's volumes were plainly exceptions that prove the rule. The university at Reims, for example, founded in 1547, struck its one jeton in 1756.[138] The author has been unable to account for this surprising, but characteristic absence of parlements and universities from the world of jetons.

FIGURE 3.10: Jeton: Valenciennes town council, 1726.

Other Towns

Although the largest and most complicated city in France, Paris was far from the only one to have adopted jetons. Quite the contrary, over 100 Old Regime French cities and towns incorporated jetons into municipal activities (Fig. 3.10).[139] They did so on a lesser scale than Paris to be sure, but they did so in characteristic and now familiar ways. The essentially universal utilization of jetons by so many town councils and civic organizations indicates just how deeply jetons and rituals involving jetons penetrated local elites, urban space generally, and the body of the nation as a whole.[140]

135. See jetons F.6099–6112, and index listing, Feuardent, vol. 3, pp. 488–89. Further on parlements, see Bély et al. (2003), pp. 960–65.

136. F.10124, 10128. The former with the legend PER.ME.REGES.REGNANT (Through me kings hold dominion).

137. Robert (1853), pp. 79–82; Feuardent lists two of these jetons, both from the 1640s, F.7729–7730.

138. See F.6374, 10754, 7944–7946, and index listing in Feuardent, vol. 3, pp. 505–506. Further on universities, see Bély et al. (2003), pp. 1236–1239.

139. See following note and Blanchet (1890), vol. 2, pp. 430–36.

140. The numismatic and antiquarian literature regarding French cities and towns and their jetons is not insubstantial and provides a solid starting point for considering same. Much of the detail presented here is drawn from Blanchet (1890), vol. 2, pp. 430–36; Blanchet (1930), pp. 331–400; Fontenay (1854), pp. 184–245; Feuardent, passim; and related, more purely historical sources indicated further.

The city council of Lyon and its jetons is the best-studied case outside of Paris.[141] The Lyon body issued its first jetons in 1624. From 1652 through 1789 it did so biennially coincident with the election of a new *prévôt des marchands* and new *échevins* for Lyon. The number of jetons varied between 1,600 and 7,700 at each issue.[142] Jetons served as *jetons de présence* and were distributed otherwise, likely at the Étrennes, to other bodies and local notables, just as in Paris.[143] At various points, for example, Lyon consular jetons went to the local learned society, the Bureau de Santé (after the plague of 1720), and the local company of archers. In addition to the *prévôt des marchands*, the *échevins*, and the city council itself, the archbishop of Lyon, the local intendant, and the local chamber of commerce struck jetons, making plain, as in Paris that jetons were seeping widely into urban institutions and the ruling élite.[144]

The same pattern repeated itself over and over in urban centers throughout France for at least two centuries. In Dijon, the capital of Burgundy, the chamber of deputies (the *Élus*) had its jetons, those of its "viscount mayor" being an outstanding series.[145] The *échevins* in Metz struck jetons from 1594 to 1754 but at their own expense.[146] The bishop of Metz had his, of course.[147] In Tours and Nantes, the city paid for jetons, the first in Tours dating to 1569, the last in 1789.[148] The mayor and *échevins* in Tours offered gold jetons to Louis XII at the early date of 1498 on his solemn entry into the town; the city of Rouen did likewise in 1610 on the ascension of the eight-year-old Louis XIII after the assassination of his father, Henri IV.[149] The city of Nantes averaged 300 silver jetons and 400 bronze ones a year at an average cost of 1,200 *livres*, and the town issued at least 104 different jetons from 1574 through 1790.[150] In Nantes the installation of a new mayor involved a big feast, a "decent" sword and scabbard for the chosen one, and undoubtedly purses of jetons came into play in one fashion or another. Similar usages for jetons are found in Nancy and Mantes.[151] Marseilles struck 3,000–4,000 a year.[152] Orleans issued its jetons from 1572 through 1786.[153] The mayors and councilmen of Angers had theirs from 1638 through 1785.[154] The town of Le Havre is known for distributing its jetons widely at court in the 1780s.[155] Jetons are well-embedded in La Rochelle. The city of Bordeaux was late in the game, striking its first jeton only in 1733, but it had a sustained tradition after that, striking an average of 1,800 jetons a year for internal use and for the Étrennes.[156] The city's agent in Paris got his 100 silver jetons for services rendered, as did the local intendant and governor-general whose purse was delivered embroidered with his coat of arms.

141. On the consular jetons of Lyon, see especially Loach (1995) and the impressive body of work by Jean-Louis Tricou, notably for present purposes Tricou (1954), Tricou (1955); see also Poncet (1883).
142. Loach (1995), p. 27.
143. Tricou (1954).
144. Feuardent (F.10497–10682) lists 185 consular jetons alone for the period 1638–1789; he counts 300 different jetons issued by various authorities in Lyon including the archbishop, governors, Intendants, treasurers and receivers, the local academies, chamber of commerce, and a variety of guilds. The register of jetons struck at the Monnaie des Médailles for December 1717 lists 16 different jetons struck for parties in Lyon; CAEF, Ms. F° 80, vol. 2, for 1717.
145. The *vicomte mayeurs* of Dijon were elected, but served for life; see Amaton (1871); Fontenay (1854), pp. 341–46; Swann (2003), pp. 103–05; Feuardent (F.9957–10121) lists 164 of these jetons.
146. Robert (1853), pp. 18–20, 57; d'Affry de la Monnoye (1878), p. xx; F.7731–7751.
147. Robert (1890; F.7758–7771.
148. Dauban (1858) and (1859); d'Affry de la Monnoye (1878), p. xx; F.8299–8370.
149. A. Barthélemy (1875), p. 243; d'Affry de la Monnoye (1878), p. xi; Thaurin (1857).
150. Dauban (1858), p. 15; F.8819–8937.
151. Lacronique (1902a), p. 413; Loir (1859), pp. 25–43; F.7666–7723, 5984–6001.
152. Évrard de Fayolle (1902), p. 158; F.11422–11427.
153. F.8101–8152.
154. F.8454–8534 for a total of 80 different jetons.
155. Guichon de Grandpont (1862), p. 2; F.6405–6407.
156. Évrard de Fayolle (1902), pp. 149–58; F.9175–9184.

The city of Bordeaux underwrote annual prizes awarded to students at the local college and a local art school, two medals (later jetons) in gold and four in silver went to winners of the rhetoric prize in the college and one in gold and two in silver for the best art students. Ones for the college came with little rings attached so that they could be strung with a ribbon. These were never struck in great numbers compared to those for the Étrennes, but still the town of Bordeaux spent 850 *livres* a year for prizes for the painting academy alone.[157] The local chamber of commerce in Bordeaux, as elsewhere, became involved with jetons as well, signaling once again the penetration of jetons into the common culture of urban elites.[158] We know these little vignettes because sources allow us to see them, but they are unquestionably typical of patterns and behaviors that spread across the urban landscape in contemporary France. They did not persist after 1789.

This chapter has explored the ways in which jetons became integrated into the governing bodies of contemporary society: the king and court, royal government and administration, the military and trading companies, the nobility, clergy, provincial estates, and towns, including notably Paris. Absent only parlements and the universities, these were apex institutions of political power and social control, and jetons formed part of that world. The next chapters examine another domain and form of power, that of specialized expertise, to see how this capability plays out in the further extension of jetons in the academies and the guilds.

157. For these details see Évrard de Fayolle (1902), pp. 171–79.
158. See previous note. On chambers of commerce and their on-again, off-again status, see McClellan (1985), p. 38, and below Ch. 5.

FIGURE 4.1: Jeton of the Académie Française, 1717.

4.

Assiduity and the Academies

The eighteenth century was the grand era of the learned society. Growing out of informal associations of humanists that arose in Renaissance Italy, dozens of these new-style academies and societies spread across Europe and America in the later seventeenth and eighteenth centuries.[1] The most notable were national institutions, such as the Académie Française (1635), the Royal Society of London (1662), and the national academies of Prussia, Russia, and Sweden. Others were more local, provincial associations, such as in Philadelphia, which was well represented by Benjamin Franklin and the American Philosophical Society. All such learned societies were chartered institutions, and by and large through their memberships, publications, prize competitions, and other activities they provided the institutional backbone to the contemporary Republic of Letters and supplanted contemporary universities as centers of research and learning.[2]

The learned society movement affected no nation more than France. Daniel Roche in his magisterial study of French provincial academies counts 29 of these institutions dotting the urban landscape of Old Regime France.[3] Paris alone was the world center bedecked with half a dozen royally chartered learned academies, testimony to an unmatched record of royal patronage and munificence. We have already mentioned these organizations: The Académie Française, the Académie Royale des Inscriptions et Belles-Lettres, the Académie Royale des Sciences, the Académie Royale de Peinture et de Sculpture, the Société Royale de Médecine, the Académie Royale de Chirurgie, and the Académie Royale d'Architecture. In Paris and throughout the provinces the learned societies of Old Regime France form an identifiable subset of institutions in which jetons came to play a defining role. This circumstance allows us to examine both the learned societies involved and a distinctive chapter in the history of jetons.[4]

The Académie Française

Cardinal Richelieu founded the Académie Française in 1635 as the national academy for the French language, and it always ranked first among the royal academies.[5] Forty outstanding writers and poets—the 40 French "Immortals"—attained the laurels that crowned (and still crown) liter-

1. For background, see McClellan (1985).
2. On this point see previous note and McClellan (1993), (1999).
3. Roche (1978).
4. On the jetons of the academies, see Feuardent passim; Blanchet (1890), vol. 2, pp. 428–29; (1930), pp. 287–90.
5. The much earlier foundation by François I in 1530 of the Collège Royal, today's Collège de France, is to be noted.

ary talent and achievement in the French language. They were elected for life, and each Immortal's chair had its own lineage, wherein its new holder pronounced the eulogy of his (or now her) predecessor. The king, beginning with Louis XIII, was the formal Protector of the Académie Française. Through the end of the eighteenth century the Académie Française met twice a week in the Louvre to rule on the formalities of French grammar, vocabulary and usage, and it had the particular charge to prepare an official dictionary. The Académie Française did not adopt jetons for itself until 1673, almost 40 years after its origin, and the amusing story behind its first jeton says a lot about the academy and about how jetons came to be used as *jetons de présence* in all of the royal academies and societies.

By the 1670s Louis XIV's great minister, Colbert, became annoyed at the Académie Française because of the irregularity of its meetings and lack of progress on the dictionary project.[6] He laid down rules for dates and times for meetings, and to underscore the point he provided a pendulum clock for the academy by which to regulate itself. In addition and as an incentive for diligence, Colbert provided *jetons de présence* to academicians who attended. In this way, the practice of receiving a silver jeton as a lagniappe or a small gratuity for attendance became standard in the academies with the Académie Française leading the way.[7] Thus, in the academies *jetons de présence* came to serve the same function as did *méreaux* in church circles.[8]

The Académie Française received its jetons through the Bâtiments du Roi beginning in 1673.[9] As pictured in the figure that heads this chapter, these had a standard format featuring the bust and titles of the king on the one side and on the other a crown of laurels and the motto, À L'IMMORTALITÉ (To Immortality) along with the legend PROTECTEUR DE L'ACADEMIE FRANÇAISE (Protector of the Académie Française).[10] These jetons are usually not dated after Louis XIV. As matters came to be settled, each of the Immortals received two silver jetons for every meeting he attended, and as the king did not provide pensions for members of the Académie Française, some academicians attended merely to collect their jetons. The playwright Pierre Corneille, elected to Chair 14 in 1647, coined the term *jetonnier* for such avaricious gentlemen. The scientist and encyclopedist, Jean Le Rond d'Alembert, elected in 1754 to Chair 25, was apparently a shameless *jetonnier*.[11] The poet Antoine Furetière, expelled from the Académie in 1657 because he used its work to publish his own dictionary, had harsh words for his former colleagues:

> The *jetonniers* make a big deal
> Of having expunged the wordy fat
> From the work they have in press.
> I know that they can never
> With pepper, butter and grease
> Guarantee a single one of its pages.[12]

---

6. On this story see Jaumonet (1985), p. 19; Bauer (1894), p. 137. The dictionary project became famous as never ending; see epigraph by G. Vallier (1879) that heads the present work.
7. Jaumonet (1985), p. 19; Lerouge and Lerouge (1977), p. 30, note that the practice soon spread to the notaries in the Châtelet.
8. Blanchet (1890), vol. 2, p. 408, makes this point. Barnard (1916), pp. 80–81 plays down the significance of *jetons de présence* in the larger world of jetons.
9. See Guiffry (1881–1901), passim; AN, O[1] 2064, "Registre des ordres de Paiment, 1711–1736."
10. See F.4352–4382 and his note, Feuardent (1904), p. 347.
11. Schodt (1873), p. 26 n; d'Affry de la Monnoye (1878), p. xxvii; Mainguy (1999), p. 107; Robert Darnton, personal communication. Voltaire uses the term derogatorily in his 1740 *Lettre au prince royal de Prusse*.
12. "Les jetonniers font un grand cas / D'avoir bien purgé de mots gras / L'ouvrage qu'ils ont sous la presse; / Je sais qu'ils ne pourront jamais, / De poivre, de beurre et de graisse, / Garantir un de ses feuillets," quoted in Schodt (1873), p. 26.

The Académie Française continued about its business until 1793. A period of turmoil began with the outbreak of the French Revolution in 1789 that affected the Académie Française and all of the royal academies in the kingdom. That turmoil only increased as the revolution turned more and more radical. The Académie Française and its sister academies limped along as best they could under these changing political circumstances. But then, with the French Republic having been declared, the Jacobins in power, and the Terror underway, on August 8, 1793, the revolutionary Convention voted its famous decree: "All the academies and literary societies patented or endowed by the nation are eliminated."[13]

The Académie Royale des Inscriptions et Belles-Lettres

Growing out of the Académie Française, the Académie des Inscriptions et Belles-Lettres is central to any discussion of jetons and vice versa. It began in 1663 when Colbert selected four of the Immortals to meet in his library to consult on the orders of the king regarding inscriptions for medals, jetons, and public monuments.[14] This erudite group was known as the Petite Académie, and it included such luminaries as the great playwright, Jean Racine, and the poet, Nicolas Boileau. Expanded in 1691 under minister Pontchartain, in 1701 the Petite Académie received letters patent as a full-fledged royal academy with 40 standing members under the name Académie Royale des Inscriptions et Médailles.[15] Its name at this point indicates both its propagandistic mission and the centrality of numismatics in its activities. As the Abbé Bignon, its president, said at its first meeting: "The single purpose of the Academy is to decode ancient inscriptions, explicate medals, and to work to immortalize in bronze and marble the glorious reign of our august monarch and the memory of his illustrious ancestors."[16] Numismatics remained a central activity in the institution, and in fulfillment of this mission, for example, in 1702 the academy produced a handsome illustrated limited edition of 50 copies of the historical medals of Louis XIV.[17] As time went on, although royal propaganda remained at the heart of the institution, with its expanded membership its mission became enlarged to encompass philology (Greek, Latin, Hebrew, and Near Eastern languages) and history and antiquity more generally, and in 1716 the crown authorized changing its name to the Académie Royale des Inscriptions et Belles-Lettres. So this academy remained until 1793 when it, too, was shuttered.

The Academy of Inscriptions is of high importance for the story of jetons largely because it was responsible for proposing the designs of jetons for other units of royal government, i.e., the administrative jetons discussed in an earlier chapter. Other parties also called upon the Academy for designs for jetons, medals, and inscriptions. This work occupied the Academy internally often to a degree that taxed its capabilities to produce thoughtful and apt suggestions in a timely fashion. In performing these tasks, the Académie des Inscriptions interacted with the department of the Bâtiments du Roi and with the mint in Paris that actually produced the jetons and medals in

13. "Toutes les académies et sociétés littéraires, patentées ou dotées par la nation sont supprimées"; see Hahn (1971), pp. 7–8.

14. On the Petite Académie, see Wellington (2015a), pp. 39–43; Josset (1978), p. 337; AIBL, H&M, vol. 1 (1736), Histoire, pp. 2–4. An important source that came late to the author's attention is Charton (2011).

15. Veillon (2008), pp. 57–60. The formal letters patent of 1701 are reprinted in AIBL, H&M, vol. 1 (1736), Histoire, pp. 10–19.

16. "…son unique étude étoit de deschifrer d'anciennes inscriptions, d'expliquer des médailles et de travailler à immortaliser sur le bronze et sur le marbre le glorieux règne de notre auguste monarque et la mémoire de ses illustres ancestres," quoted in Sarmant (2003), p. 111; see also Wellington (2015a), p. 140; (2015b), p. 21.

17. *Médailles sur les principaux événements du règne de Louis le Grand avec des explications historiques par l'Académie des médailles* (Paris: Imprimerie Royale, 1702); a revised, expanded edition followed in 1723; see also Sarmant (2003), pp. 109, 113; de Turckheim-Pey (2004), pp. 22, 49.

question, the Monnaie des Médailles. So significant is this aspect of the work of the Académie des Inscriptions to the present account and the story of jetons in contemporary French culture that we must devote a separate chapter to investigating and reporting on it.[18] Naturally, the Académie des Inscriptions et Belles-Lettres had its own jetons, so for the present, we can limit our focus to the role that these alone played in the institution.

What we find is typical of the place and role of jetons within a major royal academy. The academy met on Tuesdays and Fridays from 3:00 to 5:00 pm in the same room in the Louvre as the Académie Française. Despite the "Belles-Lettres" of its title, its academicians assembled for the *study* of words and writings, not as producers or governors of a language like its senior sister academy. The letters patent of the academy specified that 40 silver jetons be set aside for each meeting.[19] There was an attendance sheet. The call was made at 3:30 and a line drawn.[20] Academicians who had signed in at that point were entitled to share the 40 jetons for that meeting. These were sometimes distributed on the spot; at other times a period of some months would elapse, with packages of jetons then given out to each according to the number of meetings attended divided into the total number of jetons available for that run of meetings.[21] The arithmetic of *jetons de présence* was not exactly straightforward.[22]

Distrubution of *jetons de présence* continued in the Académie des Inscriptions through 1793.[23] On one occasion at least, the meeting when jetons were to be distributed was announced, with some academicians then in attendance receiving their shares in advance.[24] Visiting Free and Foreign associates received jetons for the meetings they attended. On one occasion, even though the expected bourse had not been received from the Extraordinaire des Guerres, the academy voted that members in attendance that day would be the ones to divide the purse when it came.[25] Jetons allocated for a postponed meeting were distributed at the following meeting, one occasion being the meeting scheduled for April 21, 1789, postponed because of festivities connected with the convening of the Estates General.[26]

The Académie des Inscriptions et Belles-Lettres as well as the other academies displayed a great concern for jetons and rules governing jetons. Regarding the distribution of purses received by the academy at the Étrennes, for example, the question arose as to what to do for academicians who were otherwise in regular attendance but happened not to be there for that particular meeting; the body decided not to deliberate on the point and establish a rule, but to leave the matter to the generosity of the members in attendance for that particular meeting.[27] It was often the case that absent members were marked present for the purposes of receiving jetons on account of "the regular attendance and the work they do for the Academy."[28] At least once, someone who arrived

18. See below, chapter 6.
19. Article XLVI of the letters patent of 1701; AIBL, H&M, vol. 1 (1736), Histoire, p. 19; Article L of the letters patent of 1787 repeats this specification verbatim.
20. In 1792 the hour was changed to 3:45 for drawing the line; AI, AIBL archives, A:71, *plumatif* of the meeting of November 30, 1792. *Plumatifs* are draft minutes taken in real time by an academy's permanent secretary.
21. H&M 1(1736), Histoire, pp. 4–6. See also, AI, AIBL archives, A:65, minutes of the meetings of August 7, 1770, September 4, 1772, April 23, 1773, January 7, 1774.
22. E.g., AI, AIBL archives, A:65, minutes of the meetings of February 1, 1765 and December 5, 1766.
23. AI, AIBL archives, A:71, *plumatif* of the meeting of 22 March 1793; see related distributions, *plumatifs* of the meetings of May 27, July 12, August 18, 1791, and January 27, 1792.
24. AI, AIBL archives, A:71, *plumatif* of the meeting of May 18, 1790.
25. See AI, AIBL archives, A:66, minutes of the meeting of Friday, January 7, 1774; these jetons were received and distributed at the next meeting, Tuesday, January 11, 1774.
26. AI, AIBL archives, A:71, minutes of the meeting of April 21, 1789.
27. AI, AIBL archives, A:63, minutes of the meeting of January 11, 1757.
28. "...leur assiduité continuelle, & de leurs travaux pour l'Académie," AI, AIBL archives, A:66, minutes of the meeting of Tuesday, January 9, 1776.

late was allowed to receive a jeton as a courtesy for that meeting.[29] Heirs inherited jetons due academicians who had deceased that year.[30]

Academicians received jetons for attending funeral masses and other Catholic services.[31] However, no jetons were distributed for the 1766 funeral mass for the dauphin held in conjunction with the Académie des Sciences.[32] The same was true for the funeral mass for the queen in 1768 and for the special mass the academy paid for on the death of Louis XV.[33] It was, apparently, undignified to receive jetons in such circumstances, attendees nominally wanting to show that they were there out of loyalty and devotion to royalty and the crown.

For the Académie des Inscriptions et Belles-Lettres the Monnaie des Médailles struck 3,400 jetons annually at a cost of 8,000 *livres* paid for by the Bâtiments du Roi.[34] Even with different dies and different strikings, the jetons of the Académie des Inscriptions stayed with the same design across the eighteenth century.[35] Figure 6.1 below depicts a representative example with the royal bust and titles on one side. On the other, as the Académie itself put it: "With regard to the jeton, nothing seemed more apt than to represent a Muse holding a crown of laurels with a half-column and obelisk behind her and for its soul [*âme*] this motto from Horace, VETAT MORI [It preserves from death]."[36] In the exergue comes the Latin: REGIA INSCRIPT. ET HUMANI. LITT. ACADEMIA (Royal Academy of Inscriptions and Belles-Letters).[37]

Mention should be made of the longtime permanent secretary of the Académie des Inscriptions et Belles-Lettres, Claude Gros de Boze (1680–1753). A scholar from Lyon, Gros de Boze became secretary in 1706, stepping down only in 1742 after 36 years as the "soul" of the company.[38] Everything concerning jetons, medals, inscriptions, and so forth within the Academy fell to de Bose to organize, and that was a lot. De Bose was the official intendant of inscriptions for royal edifices, and from 1719 until his death in 1753 de Boze concurrently headed the royal Cabinet des Médailles, and so through his person he linked it to the Académie des Inscriptions.[39] Gros de Boze had other overlapping institutional connections through his election to the Académie Française (Chair 34) in 1715 and the Académie Royale de Peinture et de Sculpture in 1727 and even the Royal Society of London (FRS in 1749). He had further ties to his native Lyon and learned circles and jetons there. Gros de Boze was one of those human pivots around which the institutions of the Old Regime turned, putting a personal face on the Académie des Inscriptions et Belles-Lettres and the world of jetons.

29. See AI, AIBL archives, A:67, minutes of the meeting of January 9, 1778.
30. Mainguy (1999), pp. 170–71.
31. Blanchet (1930), p. 287.
32. AI, AIBL archives, A:65, minutes of January 17 and 21,1766.
33. AI, AIBL archives, A:66, minutes of the meeting of June 7, 1774; A:65, minutes for August 5, 1768.
34. Mainguy (1999), pp. 170–71.
35. Compare F.4383–4401.
36. "A l'égard du Jetton, rien ne parut plus convenable que d'y représenter une Muse tenant à la main une couronne de laurier, & ayant derriére elle des cippes & des obélisques ; & pour ame ce mot d'Horace, VETAT MORI," AIBL, H&M, vol. 1 (1736), Histoire, p. 24; Mainguy (1999), pp. 106–08.
37. An academy jeton dated 1704 with Feuardent bust #245 of Louis XIV (Feuardent, vol. 4, plate XII) has in the exergue: REGIA INSCRIPT. ET NUMISM. [*sic*] LITT. ACADEMIA, (Royal Academy of Inscriptions and Numismatics) showing again the role of numismatics in the institution; this jeton in the author's collection is not listed in the Feuardent catalog.
38. Per Sarmant (2003), p. 183. On Gros de Bose, see also Sarmant (1994), pp. 91–126; Blanchet (1930), pp. 224–27; Sarmant (2003), pp. 110, 112, 183–84.
39. On the Cabinet du Roi, its administration, and transformations, see studies by Sarmant (1994), (2003). Sarmant (2003), pp. 115–16, notes that relations between the Académie des Inscriptions and the Cabinet des Médailles were not always harmonious.

Of lesser note, but still relevant, a scandal associated with jetons, the "Scandale Félibien," broke out in the Académie des Inscriptions et Belles-Letters in 1716.[40] The architect Jean-François Félibien Des Avaux (1656?–1733) served as the Academy's treasurer from 1701 to 1716. In that time he fraudulently took 1,520 silver jetons and gave himself 1,600 more to which he was not entitled. Félibien seems to have remained in the academy, going on to become permanent secretary at the Académie Royale d'Architecture in 1718, but in these numbers for Félibien the value of jetons proved irresistible.

The Académie des Inscriptions et Belles-Lettres had a formal tie to its sister academy, the Académie des Sciences, in that each was required to regularly report on its work to the other, and jetons show up in this connection, too. For example, in August of 1777 the librarian of the Académie des Sciences, Jean François Clément Morand (1726–1784), went to the Académie des Inscriptions to read the science academy's report and present copies of its most recent volume of *Mémoires*. Morand was invited to sit as an ordinary academician and was given a jeton at the end of the session.[41] In 1788 the science academy sent a delegation to the Académie des Inscriptions et Belles-Letters to consult over an artifact; they were invited to stay and received their share of the jetons distributed at the end of the meeting.[42] Academicians from the Académie des Inscriptions et Belles-Lettres were in turn guests at meetings of the science academy and no doubt received a jeton along with the science academicians.[43]

The Académie des Inscriptions et Belles-Lettres would occasionally entertain noble and royal guests and famous visitors, and jetons played an integral part in these ceremonies.[44] Traveling under the name of the Comte de Falkenstein in 1777, for example, the emperor of the Holy Roman Empire, Joseph II of Austria (brother of Marie Antoinette) unexpectedly showed up at the Louvre as a meeting was in progress. The academy interrupted the meeting to go and greet the emperor at the stairs. Joseph was asked to preside as the meeting resumed, but he refused. The minutes note that the prince accepted with pleasure the jeton that was offered him and each of the four gentlemen accompanying him.[45] In 1782 the Grand Duke and Duchess of Russia, the future Emperor Paul I and the future Empress, Maria Feodorovna, (traveling as the Comte and Comtesse du Nord), "accepted the jeton that was presented to them. Jetons were distributed to everyone in their suite."[46] The academy entertained the king of Denmark (Christian VII) in 1768; Prince Henry of Prussia (traveling as the Comte d'Oels) and each of his entourage were given two jetons on their visit in 1784.[47] The previous year, 1783, Benjamin Franklin visited the academy, and in a reverse manner the academy accepted a medal offered by Franklin struck by the Congress of the new American nation on the occasion of its independence.[48]

And then, jetons came to into play in odd, one-off occasions within the Académie des Inscriptions et Belles-Lettres. For example, in 1785 one Cassieri of the Académie de Peinture et de Sculpture presented a bust of Colbert to the academy. In return for this gift, Cassieri received forty

40. Mainguy (1999), pp. 172–73.
41. See AI, AIBL archives, A:67, minutes of the meeting of August 12, 1777.
42. AI, AIBL archives, A:70, minutes of the meeting of June 13, 1788.
43. The academician appointed to report to the Académie des Sciences received 400 *livres* for this duty.
44. Hahn (1971), pp. 73–74, lists important heads of state visiting the Académie des Sciences. The same list would apply to other academies.
45. See AI, AIBL archives, A:67, minutes of the meeting of May 16, 1777.
46. "...ont accepté un Jeton qui leur a eté presenté. On en a distribué à toutes les personnes de Sa Suite," AI, AIBL archives, A:67, minutes of the meeting of June 7, 1782.
47. AI, AIBL archives, A:65, minutes of the meeting of December 3, 1768; A:68, minutes of the meeting of September 7, 1784.
48. Here called the *États unis*, the United States; AI, AIBL archives, A:68, minutes of the meeting of June 6, 1783.

jetons; apparently he was given all the jetons set aside for that meeting.[49] On another occasion in 1782 a certain M. de Grace presented a copy of his book on gardening, *Le Bon Jardinier*, to each academician at a meeting of the Académie des Inscriptions, and the academy voted unanimously to offer M. de Grace forty jetons for his gifts.[50] In another odd example from 1783 an outsider, the Abbé Mongez, natural history keeper at Ste. Geneviève church, asked to be admitted to read his paper on singing swans (*Sur les Cygnes Chantants*); he was granted permission and a jeton.[51] In the highly charged political context of 1793, perhaps sensing the impending closing of all the academies later in August of that year and out of fear, if not merely a nod to political correctness perhaps, at its meeting of March 22, 1793, the Académie des Inscriptions et Belles-Lettres voted to give all the jetons that it had accumulated over the last four meetings (not held because of political disturbances) worth 520 *livres* to an employee in the secretariat of the academy for over 50 years who was gravely ill.[52]

The Académie des Inscriptions et Belles-Lettres kept a close eye on its jetons and pensions and a wary eye out for its sister academy, the Académie Française, and because they were linked administratively they regularly squabbled over jetons. At various points between 1717 and 1726 the Académie des Inscriptions et Belles-Lettres did not get jetons, whereas the Académie Française did, provoking a rupture between the two institutions which only gradually healed.[53] Another episode in 1779 shows even more of this friction.[54] The custom was for the treasury to make payments in March and December to cover the expenses for jetons for the three academies without making a distinction as to how the funds should be divided up among them. In 1779 the minister authorized a March payment of 16,000 *livres*, but somehow only d'Alembert and the Académie Française received funds for its jetons and not the Académie des Inscriptions or the Académie des Sciences. D'Alembert claimed he didn't know anything about the matter, but on behalf of their institutions the secretary of the Académie des Inscriptions, the Abbé Louis Dupuy, and the treasurer of the Académie des Sciences, Mathieu Tillet, coordinated a further response to the situation. At first, they tried to check things out by going down to the Monnaie des Médailles on the Quai du Louvre and speaking with its director, Jules-François de Cotte. Nothing came from that course of action. Next, they petitioned the Controller-General of Finances, Jacques Necker, who replied via a subordinate that the March payment had already been made and he could not make another until December. Worried about their jetons, the two academies alerted the mint to proceed with work on the jetons as usual for the year. In the meantime they solicited a meeting with Necker, naming a delegation to go and discuss the matter. The plan was to request one payment a year, which, if made in June, could carry the academies through to the end of the year and the next funding cycle; if this wasn't acceptable, they stood ready to propose two payments a year, in June and January, which would suffice for their immediate expenses, the minutes of the academy of inscriptions noting that "we don't need the totality of our jetons at the beginning of the year. It suffices that we have the necessary quantity of them for distributions as they occur."[55] The specific outcome of this episode is lost to us, but the jetons of the three academies continued to be struck by the thousands for another decade.

49. AI, AIBL archives, A:68, minutes of the meeting of July 22, 1785.
50. AI, AIBL archives, A:67, minutes of the meeting of December 10, 1782.
51. AI, AIBL archives, A:68, minutes of the meeting of July 22, 1783; this was seemingly the only time an outsider was admitted to read a paper.
52. AI, AIBL archives, A:71, *plumatif* of the meeting of March 22, 1793. These were the last jetons distributed.
53. Mainguy (1999), p. 173.
54. On this episode, see AI, AIBL archives, A:67, minutes of the meetings of May 21 and June 1, 4, 8, and 28, 1779.
55. "...nous n'avons pas besoin en effet de la totalité de nos Jettons au Commencement de l'année, il suffit que nous en ayons la quantité necessaire pour la distribution à mesure qu'elle se fait," AI, AIBL archives, A:67, minutes of the meeting of March 8, 1779.

Another incident involving the Académie des Inscriptions in 1784 speaks to the same set of circumstances. That year the king doubled to three *livres* the value of jetons given to the Immortals at the Académie Française, and Académie des Inscriptions ardently wished the same favor.[56] The secretary was charged to draw up a memo to be presented to the minister by a formal delegation from the academy consisting of its officers and two other members. The document claimed historical equality with the Académie Française and complained that the Immortals receive the equivalent of 1,000 *livres* a year in jetons whereas their academicians only 240 *livres* in jetons. This petition faced the awkward reality that some academicians at the Académie des Inscriptions received formal pensions, whereas there were none in the Académie Française. The government already spent 20,000 *livres* annually on pensions in the inscriptions and belles-letters academy, where the 10 senior academicians (out of 40) each received the sum of 1,800 *livres* a year, jetons actually representing a bump of 40% beyond that.[57] The memo replied to this point by noting that academicians were generally elected at the age of 40 or 45 and usually only qualified for a pension after 24 or 25 years of service; only half lived long enough to qualify as among the 10 oldest in the academy; and those who did only enjoyed these pensions for a few short years before they themselves died. At the Académie Française everyone received 1,000 *livres* from the day of their election. Ordinary academicians at the Académie des Inscriptions et Belles-Lettres who received *jetons de présence* for attending meetings merited and could benefit from an increase in the weight of their jetons as much as any Immortal at the Académie Française. The minister then in charge of the Bâtiments du Roi, the Baron de Breteuil, gave in and acceded to their request.[58]

This hiccup may have triggered the reworking of the whole budget of the Académie des Inscriptions et Belles-Lettres two years later in 1786.[59] Regular payments for jetons were still a problem apparently, and the academy moved to request additional funds for pensions and jetons. To discuss the matter, the president of the academy, Clément-Charles-François de L'Averdy, the Marquis de Gambais (1723–1793), met privately with the chief financial officer of the kingdom, the Controller-General of Finances, Charles-Alexandre, vicomte de Calonne (1734–1802). L'Averdy himself was a former controller-general under Louis XV, and he and Calonne had been school chums. There was not going to be a problem here. Calonne agreed that jetons needed to be paid, that they were needed to supplement minimal pensions beyond what might provide for basic necessities, and academicians shouldn't have to wait so long to get a pension, given that many did not live long enough to get one.[60] The minister was sensitive to the modest recompense in jetons for attendance, that no funds were provided for the maintenance of the library or for a servant or custodian to attend to the academy's quarters or put books back on the shelves when academicians were too old and decrepit to do so themselves. (Imagine this scene.) He accepted the proposal that associate members should get a small pension to carry them over until they qualified for the big pension. On the basis of these discussions a memorandum was drawn up in the academy, although the memo made it seem that the Académie des Inscriptions et Belles-Lettres was itself initiating everything. Calonne then made a recommendation to the king for additional funding for the Académie des Inscriptions et Belles-Lettres, albeit at a somewhat lower level than the academy proposed.

56. See AI, AIBL archives, A:68, minutes of the meeting of January 25 and February 1, 1785, the latter with the formal memo prepared for the minister.
57. Mainguy (1999), pp. 170–71.
58. AI, AIBL archives, A:68, minutes of the meeting of February 15, 1785.
59. On this episode, see the details presented here in AI, AIBL archives, A:68, minutes and materials copied into the minutes of the meetings of May 23 and June 13, 1786, including Calonne letter dated, "De Versailles le 15 Mai 1786." See also royal order and other materials related to this episode reprinted in AIBL, H&M, vol. 47 (1809), pp. 3–16.
60. The record refers to these pensions as "*alimentaire*."

A letter from the controller-general dated Versailles, May 15, 1786, gives the details of the new arrangement. The annual budget for the academy was raised from 28,000 to 40,200 *livres*, a 44% increase. The ten senior academicians now received 2,000 in pensions straight out; the oldest five associates could be nominated for pensions of 800 *livres*. The emoluments for the permanent secretary remained set at 1,000 *livres*, but the office received an additional 3,000 *livres* for expenses. A custodian (*huissier*) was appointed at 600 *livres*. The royal presses were ordered to give one copy of each book it printed to the secretary of the academy who was implored to make this a gift to the academy. More to the point here, funds for jetons were raised from 6,400 to 10,000 *livres* a year. In this instance, clearly, jetons had become part of a salary package for academicians at the Académie des Inscriptions et Belles-Lettres. The document was marked "*bon*" by the king, Louis XVI. A year later, as reported on March 30, 1787, the controller-general allocated an additional 2,000 *livres* to be distributed to eight free associates of the academy, a step that "had previously been forgotten."[61]

As a result of this refunding effort and the augmentation in the number and weight of its jetons to match the heavier new jeton of the Académie Française, the king authorized new, larger dies to be cut, and the Académie des Inscriptions appointed commissioners to supervise the design of its new jeton.[62] On Friday, August 4, 1786, the secretary of the Academy reported on the commission's proposal, which was then adopted by the academy as a whole. The result? The new jeton was bigger, but the design was for all intents and purposes the same as it had always been: a muse with a crown of laurels and an unfurled scroll, a column, and an obelisk in the background and in this case, an overturned urn out of which spill coins and medals. The same VETAT MORI (It preserves from death) adorns the reverse. The same REG. INSCR. ET HUMAN. LITT. ACADEMIA in the exergue. Depicted below (Fig. 6.1), it was the same jeton struck since the days of Louis XIV and the early years of the century.

The Académie Royale des Sciences

Founded in 1666 by Louis XIV following a suggestion by Colbert and receiving formal letters patent in 1699, the Académie Royale des Sciences was a full-fledged academy funded by the French state. With a resident membership of about 40, 20 of whom were pensioned, the state underwrote other of its costs including its publications and numerous scientific expeditions. With an exceptional body of men of science making up its membership, publishing the single most important scientific series (its *Histoire et Mémoires* series), and sponsoring an unmatched series of scientific prize contests at the cutting edge of contemporary science, the Académie des Sciences in Paris was indisputably the world's leading scientific institution of its day.[63] It, too, was headquartered in the Louvre palace where its twice weekly assemblies took place, and in typical fashion it, too, had and used its own jetons provided by the state through the Bâtiments du Roi.

With thousands struck every year the academy's jetons had a characteristic design that maintained itself for almost a century. The undated academy jeton is typical (Fig. 4.2). With, in this case, Louis XV on one side, on the reverse it features a seated Minerva, the Roman goddess of wisdom—in other academy jetons with her owl—surrounded by the emblems of science and in particular of geometry, mechanics, chemistry, natural history, botany, geodesy, and astronomy, and especially to be noted in the background is the Observatoire Royal built in Paris in 1671, for

61. "...l'objet qui avoit eté oublié l'an passé," AI, AIBL archives, A:70, letter of L'Averdy with the *bon* of Calonne copied into AIBL minutes of the meeting of March 30, 1787.
62. AI, AIBL archives, A:68, minutes of the meetings of July 25 and August 4, 1786.
63. Hahn (1971); Gillispie (1980); McClellan (1985). On the Royal Society of London and its *Philosophical Transactions* as a comparison, see also McClellan (2001), p. 7.

FIGURE 4.2: Undated Louis XV jeton of the Académie Royale des Sciences.

which medals were struck but no special jetons. The *devise* reads INVENIT ET PERFICIT (It discovers and perfects) with the Latin at the bottom indicating that this is a jeton of the Académie Royale des Sciences.

We needn't tarry over the Académie des Sciences and its jetons, as jetons functioned within the Académie des Sciences essentially as they did in the Académie des Inscriptions et Belles-Lettres. Same 40 jetons set aside, same sign-in sheet, but the Académie des Sciences tallied and distributed *jetons de présence* strictly quarterly. Taking 1760 as a representative year, the academy set aside 800 jetons each quarter and over the course of the year distributed 3,089 as *jetons de présence*. For three funerals it handed out an additional 46 jetons to academicians who attended, and so on.[64]

Other Parisian Royal Societies

Other royal academies graced the great city of Paris.[65] Next on the list and in seniority after the Académie Française would be the Académie Royale de Peinture et de Sculpture established in 1648 under the authority of a young Louis XIV for royal patronage of artists and the fine arts, and in this connection we should not lose sight of the general importance of traditions of patronage for the fine arts and social role of artists and their creations in contemporary European society and culture. All in all, for a century and a half from its creation in 1648 to its suppression in 1793 the Académie Royale de Peinture et de Sculpture was the dominant art institution in the world.[66] In the seventeenth century with the official court painter Charles Le Brun (1619–1690) as its director and backed with an iron hand by chief ministers Mazarin and then Colbert, the academy of painting and sculpture and the associated art school strictly controlled official French art—and there was no other—dictating acceptable style and deciding the careers of artists. A rigid hierarchy reigned, and artists could climb the greasy pole through various levels gaining ever more prestigious commissions and academy pensions which could reach 6,000 *livres*. But membership in the Académie de Peinture et de Sculpture was not restricted in numbers and included women artists, such as Marie-Suzanne Giroust-Roslin (1734–1772) and Élisabeth Louise Vigée Le Brun (1755–1842).[67] Painters, sculptors and fine art engravers were accredited through a presentation piece to the Academy. Although its financial fortunes varied, most costs were underwritten by the crown

64. See, for example, PA, Service des Archives, "Distribution des jetons" that starts off the *plumatif* archival folder for 1760. Tallies for such jeton distribution are preserved in folders in the academy's archival *plumatifs* for virtually every year of the eighteenth century in the present-day archives of the Académie des Sciences of the Institut de France.
65. On the urban geography of learned institutions in contemporary Paris, see notably Belhoste (2011); McClellan (forthcoming).
66. This picture of the academy has been gleaned from various websites, notably the unsigned http://academieroyalepe.canalblog.com/; last consulted September 23, 2018.
67. The Academy elected sixteen women over its 145 year history, although one, Marguerite Haverman, was soon struck from the list for failing to present her reception piece; see Fidière (1885), passim and p. 27.

or other noble patrons or student fees. From 1667 academicians and professors and their students exhibited their work publicly in an academy-controlled official royal art exhibition, what became the famous French art salon held in the Louvre palace annually from 1737. From the outset under its aegis, the Académie Royale de Peinture et de Sculpture trained the next generation of artists, although the formal art school came into being only in 1749, academicians serving as professors, one of the ranks in academy membership. The school had a formal four-year program of study, and from 1684 it awarded medals for its top art students, first for painting and sculpture with engraving and architecture added in 1720. Also, annually from 1663 the academy awarded the prestigeous Grand Prix de Rome. This munificent royal patronage provided three- to five-year fellowships (and from 1666 residential and studio space) for further art study and training in Rome.

The Académie Royale de Peinture et de Sculpture, too, incorporated jetons into its institutional identity and practice.[68] It received its jetons quarterly through the accounts of the Bâtiments du Roi.[69] They served as *jetons de présence* and doubtless were handed out on other occasions. The novelty concerning jetons within the Académie Royale de Peinture et de Sculpture was that they were awarded as prizes to art students in various categories in contests organized first quarterly and then monthly within the school to spur work and competition.[70] Winners enjoyed other privileges, such a being entitled to enter the building through a special entrance way. The 1764 jeton of the Académie Royale de Peinture et de Sculpture heads the list of figures below, graced with a bust of Louis XV and his denomination as ARTIUM PARENS (Father of the Arts).[71]

Founded in 1671, the Académie Royale d'Architecture was another Parisian royal academy associated with the arts. It was the home for royal architects, and it taught architecture to rising stars. It had its own jetons, such as the Louis XV jeton pictured at the head of our list of abbreviations.[72] We can surmise that jetons went to academician-architects when they met as a group and as prizes to architecture students and otherwise within the institution. The notable point about the style of the jeton of the Académie d'Architecture is the utter simplicity of its reverse. There is no art here, just three lines on a blank field: ACADEMIE ROYALE D'ARCHITECTURE. None of the fine art and artisanship seen in so many seventeenth- and eighteenth-century jetons is to be found; the jetons of the Académie d'Architecture were simple tokens testifying to simpler sensibilities that took hold as the eighteenth century progressed.

In turn, the Société Royale de Médecine (1778) was actually a royal *academy* like the others encountered so far.[73] Its area was medicine. It was not part of the great efflorescence of Louis Quatorze royal academies of the seventeenth century, and it emerged comparatively late—only in 1776/1778—due to the longstanding opposition of the Paris Faculté de Médecine to any competing form of organization for medicine. The Société Royale de Médecine arose out of the bureaucracy, court physicians and medical correspondence circles, but it quickly morphed into an ordinary academy, this one attached to the ministry of foreign affairs. It received letters patent in 1778 and an academy's internal organization; funded to the tune of 20,000 *livres*, it held its meetings

68. See F.4418–4425.
69. Note the 1685 payment of 1,800 *livres* by the Bâtiments du Roi for 1,600 jetons for the Académie Royale de Peinture et de Sculpture; Guiffry (1881–1901), vol. 2, col. 780: [1685] entry for April 22.
70. See F.4424–4425 and other entries in Guiffry (1881–1901) regarding these prizes, vol. 2, col. 441 [1690]; col. 305 [1689]; vol. 3, col. 537 [1684], cols. 583–84 [1691].
71. Suppressed in 1793, the school was revived in 1797 and persists today as the École Nationale Supérieure des Beaux-Arts on the rue Bonaparte in the sixth arrondissement in Paris.
72. Feuardent notes only two jetons of the Académie Royale d'Architecture of these, both under Louis XVI, F.4426–4427.
73. A full study of this great institution still needs to be written; otherwise, see a not insubstantial literature, including Gillispie (1980), pp. 194–203; Garnière (1964); McClellan (1985), pp. 3, 8, 221–22; McClellan and Regourd (2011), pp. 125–32; Kluyskens (1884), pp. 8–10.

in the Louvre, published volumes of memoirs, offered prizes, and otherwise engaged the outside world regarding matters medical, meteorological, and epidemiological.[74] A good part of that engagement was with its sister institution, the Académie Royale des Sciences, and in this sense the Société Royale de Médecine was the medical arm of the science academy and the whole concatenation of royal French academies operating in Paris at the end of the eighteenth century. It was a progressive and successful organization, all told, and in the 1770s and 1780s the Monnaie des Médailles struck thousands of jetons for the society to distribute as *jetons de présence* and otherwise as we have seen in other academies. They are of a bland style, as shown, like the ones from the Académie d'Architecture.

One little tidbit allows us to say one thing more about the Société Royale de Médecine and about how jetons were used in the academies. Every year for the years 1776–1789 the royal medical society published a thick volume of *Histoire et Mémoires* like those of the Académie des Sciences. D'Affry de la Monnoye (1878) reports that the Société Royale de Médecine presented its volume of *Mémoires* to the king as soon as they appeared, and when the delegation did so, along with the volume, they offered the king a gold jeton as the founder and protector of the institution (Fig. 4.3).[75] If we didn't already know better from other evidence, this offering would seem extravagant and hardly believable. But we know that this gold jeton existed.[76] We know that the Académie Royale des Sciences and the Académie des Inscriptions et Belles-Lettres, on invitation from the minister, also sent delegations to Versailles to present their volumes to the king when they appeared. An almost comical report of 1774 details a delegation from the Académie des Inscriptions et Belles-Lettres traipsing around Versailles looking for notables to whom it could offer its volumes and presumably its jetons.[77] The case of the Société Royale de Médecine confirms that these ceremonial academy presentations centrally involved an offering of jetons and made the occasions akin to the exchange of jetons at court at the Étrennes or the royal reception of the delegations from the estates.

FIGURE 4.3: Jeton of the Société Royale de Médecine.

Unlike the doctors, the surgeons of Paris welcomed the formation of their own learned society in part because in so doing they further distinguished themselves from barbers and wigmakers.[78] The guild of surgeons or the community of Saint Côme, after the patron saint of surgery, harked back to the fourteenth century, but in the sixteenth century the group was recognized as an aca-

74. On the society's funding, see Ganière (1964), p. 30.
75. d'Affry de la Monnoye (1878), p. xxv.
76. F.4694.
77. For a closeup look at the distribution of academic volumes to the king and court by the Académie des Sciences and the Académie des Inscriptions et Belles-Lettres, see McClellan (2003), pp. 88, 93.
78. See especially Gelfand (1980) for these details; also Lacronique (1902b), (1902b); Ganière (1964), p. 38; Julien (1973).

demic unit with surgeons who wore the long robe and whose theses were delivered in Latin. They ran the university-level college of surgery (the École de Chirurgie). Although turf-wars broke out with other barber-surgeons and with the physicians of the Faculté de Médecine, by the eighteenth century professional boundaries had been largely settled. With complementary cadres of royal surgeons at court—particularly the first royal surgeon—the stage was set for French surgery to be completely organized by the government under the control of the crown and the surgical establishment in Paris and Versailles. An academy of surgery was the next step. On the initiative of the first royal surgeon and royal confidant of Louis XV, François Gigot de La Peyronie (1678–1747), the Académie Royale de Chirurgie emerged in 1731, more than four decades before the doctors had their academy. Official letters patent followed in 1748. With an M.D. degree from Reims and as Free Associate in the academy of sciences, La Peyronie remained president of the surgery academy and in control of French surgery from 1731 to 1747, and his successors did so down to 1793. The organization of official surgery melded the royal academy of surgery, the royal school of surgery, and control over surgical practice through the guild of master surgeons, and jetons played a role in each division of this surgical trinity.

FIGURE 4.4: Jeton of the Académie Royale de Chirurgie, 1751.

Although all of the 200–300 licensed surgeons in Paris were nominally members, the Académie Royale de Chirurgie was organized along the lines of a typical academy under the presidency of the first royal surgeon. It met in the surgery amphitheater erected in 1694 for the community of surgeons and the school of surgery at the church of Saint Côme on the rue des Cordeliers near the Latin Quarter. La Peyronie's will funded the jetons for the surgeons. Figure 4.4 features Minerva and a genie with arms and surgical instruments; another type shows Apollo and Hygie, the goddess of health.[79] Forty jetons were available for each meeting of the surgery academy as *jetons de présence*, each jeton worth one *livre* and five *sols*.[80] Curiously, the rule was that the academy secretary received half of the jetons otherwise destined for academicians not in attendance with the other half distributed to adjunct members in rank order of seniority.[81] La Peyronie's bequest in addition funded an annual prize medal worth 500 *livres*. The academy published its own *Mémoires* that were undoubtedly offered to the king along with a gold jeton.

A governing board ran the surgery school, the École de Chirurgie, and the associated surgical hospice that consisted of six and then from 1783 12 beds. The first royal surgeon and his lieutenant, the director, vice-director and secretary of the academy, the senior surgeon in the school, and three adjunct members constituted this board. They met on the first of every month, and each

79. On these and related surgical jetons, see F.4700–4762.
80. Ganière (1964), p. 42.
81. Lacronique (1902b), pp. 42–43.

member received his *jeton de présence* for attending. In 1783 the board added a lawyer from the Paris Parlement, a royal notary, and a royal prosecutor, and with this development board members received three jetons for each meeting they attended.[82]

Jetons show up in a major way in the community of Saint Côme or the guild of master surgeons of Paris. This group of about 30 was a formal guild, and we will see more about guilds in the next chapter. In particular here, jetons came to play a part in the examination and certification of new surgeons, and the expenses involved for such certification were high. For duties (*droits*) owed in 1699, for example, candidates paid the first royal surgeon or his deputy six *livres* to convoke the examiners. The first royal surgeon, his lieutenant, four *prévôts*, the receiver and secretary at Saint Côme received eight jetons apiece and two pairs of gloves, one simple and one fancy. Indicating the surgeons' formal subservience, the dean of the Faculté de Médecine and two other doctors received three *livres* and two pairs of gloves, one plain and one fancy. For duties owed to the Faculté de Médecine itself candidates paid three *livres*, 12 *sols* and six *deniers*. Each of the other master surgeons of the guild in turn received two jetons worth two *livres*, 12 *sols* and two pairs of gloves worth three *livres*.[83] This kind of initiation is characteristic of guilds and enriched the holders of power, and it is telling for our purposes that we find jetons here.

Jules Chautard (1896) notes an odd jeton of the Paris surgeons of 1635.[84] He suggests that surgeons may have given it to patients to give to a barber for a prescribed bleeding. This function for jetons seems unusual, and no other evidence in the literature about jetons suggests anything like it. Still, Chautard's suggestion is not beyond the realm of possibility, and for the story of jetons it would be nice if it were true.

The Faculty of Medicine (Paris)

The medical faculty, the Faculté de Médecine, of the University of Paris was a powerful institution and a redoubtable center of organized medicine in Europe across the entire span of time under consideration here. Dating from the thirteenth century, the Faculté was responsible for teaching future doctors and granting degrees, and at least according to its own lights, it represented official medicine and the best of medical thinking and experience. It had authority over licensing doctors, surgeons, pharmacists, and midwives, and it inspected apothecary shops and the spice trade.[85] The Faculté de Médecine embodied a powerful cadre of medical doctors at the pinnacle of organized medicine in an institution—the university—that remained outside of direct royal, central government, or town control. The Faculté de Médecine fiercely defended its interests, its authority, and its control over organized medicine, as we have seen in the case of the late emergence of the Société Royale de Médecine or the subservience of the surgeons to the Faculté de Médecine and the doctors reigning there, literally doctors regent. The Faculté had its own jetons, and jetons became deeply ingrained in the workings of the Faculté in ways that say a lot about both the Faculté and jetons. The Faculté de Médecine looms large in the story of jetons and speaks volumes about the state of the academies and organized medicine in Paris prior to the Revolution.

The first jetons specifically struck for the Faculté de Médecine date from 1398 and were mundane objects struck or even cast in tin. These early Faculté jetons were probably used as tokens on counting tables, but their origins are more to be found in their function as *méreaux* and more like

82. Lacronique (1902b), pp. 49–50.
83. Guild rules quoted in Lacronique (1902b), pp. 39–40.
84. Chautard (1896), pp. 88–89.
85. On this point see Desavis (1933), p. 15. On the Faculté de Médecine see also Blanchard (1901), pp. 161–72; Corlieu (1887), p. 5.

church tokens dispensed to physicians for attending university convocations, masses, and funerals for deceased colleagues.[86] Like mass tokens, they may have been redeemable for cash with the bursar of the Faculté.[87] The jetons/*méreaux* of the Faculté de Médecine continued to be ordered and produced at intervals for two hundred fifty years through 1638, at which point the institution began to issue a new kind of true jeton, its decanal jeton.

FIGURE 4.5: Decanal Jeton of the Paris Faculté de Médecine, 1666.

The Faculté de Médecine was one of the three faculties of the university, along with the faculties of theology and of law, and was led by its own dean. The dean was a top administrator in the university along with the rector and the deans of the other two faculties. Members of the medical faculty elected their dean every two years, and it was possible to be elected for a second two-year term. Regularly from 1638 through 1792, then, the Faculté de Médecine issued new jetons for each new dean, the dean so exercising a right to have jetons struck in his name (Fig. 4.5). From 1652 these were graced with the bust of the dean. Struck in total were approximately 90 different jetons of 65 deans of the medical faculty in an essentially uninterrupted series of Faculté jetons spanning a further 155 years.[88] Each new dean was required to deposit a silver and bronze jeton with the librarian of the Faculté, and in this way a more or less complete record of this series has survived.[89]

The decanal jetons of the Faculté de Médecine have attracted considerable attention from numismatists and collectors, and the resulting literature is surprisingly large.[90] Perhaps not surprisingly, medical doctors stand out as collectors and interested parties when it comes to these jetons.[91] Already in the nineteenth century physicians were recognized as a distinct group of collectors.[92] This interest stems largely from antiquarian impulses, interest in the history of medicine, and pride as a medical professional. The results have been historiographically primitive.[93] Analogous to efforts seen for numismatists and collectors to compose a *histoire métallique* or history told through medals and jetons, a notable strand in this literature has been for doctors and those

86. On this point, see especially Rouyer (1892), pp. 13–15; see further below, List of Figures and Tables.

87. Corlieu (1887), p. 6; Blanchard (1901), pp. 85, 87, 91; Ch. Florange (1933), pp. v–vi.

88. Josset (1978), p. 340; Desavis (1933), p. 15. Feuardent (1904–1915) lists 199 variants (F.4492–4691) of jetons of the Faculté de Médecine for the period 1638–1789.

89. Desavis (1933), p. 24.

90. See Kluyskens (1884); Corlieu (1887); Fournié (1907); Desavis (1933); Ch. Florange (1933) and his bibliography, pp. 1–2; Freeman (1946); Freeman (1964); see also Rouyer (1892), pp. 13–15, 19–30; Blanchet (1890), vol. 2, p. 429; Blanchet (1930), pp. 290–94; Blanchard (1901), passim and pp. 267–68.

91. A number of jetons of the Académie de Chirurgie in the author's collection have a bit of solder on them, indicating that they were once part of a single display.

92. See Renauldin (1851), itself a remarkable piece of numismatic erudition.

93. See, for example, the M.D. thesis by Madeleine Desavis (1933), a work heavily relied on here, but which is more a simple catalog of the deans of the Faculté de Médecine than an analytical history. She intended her work as a visitors guide to the medical history exhibitions at the Paris medical faculty and for "collectors and amateurs of medical numismatics"; Desavis (1933), pp. 169–71.

interested in the history of medicine to use medical medals and jetons to write a *histoire métallique médicale*.[94] One way or another, the result is that much is known about the decanal jetons of the Paris Faculté de Médecine and a detailed report is possible.

Although the roles that jetons played within the Faculté expanded considerably with its new jetons, as we will see next, the case reveals that the Faculté de Médecine linked itself unusually closely to the Catholic Church and to church masses where its jetons continued to serve steadfastly as *méreaux*. The Faculté held its own mass every Saturday, except during vacations, and a special mass in the university chapel every year on the eighteenth of October, the name day of Saint Luke, the patron saint of physicians. Similarly, for funeral masses for deceased doctors-regent and other special services, the identity and activities of the Faculté centered on the altar. The Faculté strictly enforced regulations regarding attendance at mass, and in one case in 1572 a candidate was not allowed to take his exams because he hadn't been assiduous in attending church.[95] At mass each doctor-regent received a jeton. The dean and the senior member of the company were given two. The attending priest got one or two jetons, as did the four ushers at the service, each jeton worth one and three-quarters *livres*.[96] Raphaël Blanchard in his study of 1901 reports that these jetons could be redeemed for cash.[97] In this way a limited number of decanal jetons circulated and notably *recirculated* among the doctors of the Faculté as did *méreaux* in religious communities.

But masses were just the beginning. For funerals, the Faculté stood graveside at burials of doctors-regent, but if these were far away everyone who went got an additional jeton, the dean two. The Faculté sent parties of three doctors to present condolences to the families of its deceased colleagues, each representative receiving a silver jeton for this duty. At public ceremonies, such as those of 1781 on the birth of the dauphin, attendees received a jeton, but in these cases Faculté jetons begin to function more as *jetons de présence* than as *méreaux* associated with a religious community.

The Faculté de Médecine along with the faculties of law and theology constituted the university under the authority of the rector, and the medical faculty marched in formal university processions led by the rector. Thanks to Robert Darnton, we can readily imagine these promenades around the neighborhood of the Sorbonne.[98] For this, each doctor-regent present and those serving as ushers and beadles received a jeton, the dean again getting two.[99] The Faculté distributed 29 jetons for parades in 1777 and that many again in 1778. And then, six representatives of the medical faculty attended general university convocations four times a year to listen to the rector's quarterly report, and each received a jeton for this duty. The Faculté de Médecine was the only faculty to have and use jetons, and to repeat a point made earlier, the case shows up the general absence of jetons in contemporary universities and university culture.[100]

The Faculté de Médecine itself held a general faculty meeting at the first of every month to discuss reigning illnesses and other matters medical. Although any doctor-regent could attend, only the dean and the twelve senior doctors-regent received jetons for attending these monthly meet-

94. See notably the catalog complied by Freeman (1964) and the catalog of the collection of Dr. Géza Faludi (1929); for a "histoire métallique" of the Faculté de Médecine in 1792, see Nocq (1910). See also Corlieu (1887), p. 5; Blanchard (1901), p. 91; and Dillemann (1997), p. 5, regarding jetons and the history of pharmacy.

95. Blanchard (1901), p. 88.

96. Desavis (1933), pp. 16–19, 23; Blanchet (1890), vol. 2, p. 429; Corlieu (1887), pp. 6–7.

97. Blanchard (1901), p. 88. It is not clear to the author that such redemption was actually the case, however.

98. Darnton (1984).

99. See Desavis (1933), pp. 19–20, for these details; see also Ch. Florange (1933), pp. vi–x; Blanchet (1930), p. 291. Blanchard (1901), pp. 89–90 summarizes the distribution of decanal jetons in 1782.

100. Messengers associated with the university had their jetons, but not professors or other faculties; compare F.4429–4431 and F.4432–4447.

ings, known as the *Prima mensis*. The dean and the most senior doctor-regent received two jetons and the other doctors a single coin handed out at the end of the meeting, and these had clearly become *jetons de présence*.[101] The dean was furthermore responsible for the financial accounts of the Faculté. Silver and bronze jetons went to those assisting him in keeping accounts, as well as for reconciling the accounts at the change of deanship.[102] In addition, after mass every Saturday six doctors-regent were supposed to make themselves available to the poor for free medical consultations. They were to receive a jeton for this community service, but apparently the doctors were not diligent in performing this duty.[103] And then, Faculté doctors were given jetons for participating in examinations for the conferral of the degree of doctor of medicine. Seven jetons went to each examiner and 14 to the dean, with a total of 30 distributed at one exam in 1778, for example.[104] The Faculté sponsored medical prize contests similar to prize contests of the academies. The Cuvillier Prize of 1772, for example, for the best paper on epidemics was won by a certain Lebrun, who received a purse of 100 decanal jetons for his submission.[105] Thirty jetons went to people who served on these prize committees and 60 to the dean, a comparatively large number presumably because real work was involved.[106] The Faculté, like the Paris city council, handed out jetons to individuals who performed special services, and it's not for nothing that lawyers defending the interests of the Faculté were regular recipients of purses of 50 or 100 jetons.[107] The dean also handed out decanal jetons to friends and dignitaries.[108] From November, 1777 to November, 1778 the dean of the faculty of medicine distributed 1,510 jetons worth 2,620 *livres* 15 *sols*.[109] In total, as of the 1770s and 1780s the dean of the Faculté de Médecine supervised the annual distribution of 1,200–1,500 jetons.[110] The Faculté de Médecine was a landmark institution in contemporary Paris, and its jetons entirely emblematic of its character and station.

Provincial Academies

Altogether, its extraordinary concentration of royal academies made the city of Paris prior to the Revolution the world center of expertise, organized knowledge, and state patronage of expertise in the service of governance. No other metropolitan center in Europe or elsewhere in the world could compare. Plainly, the city of Paris dominated the rest of contemporary France much as it does today, but the social and institutional movement to found learned societies was hardly limited to the great metropolis, and in no nation did learned societies implant themselves more deeply than in the provinces of Old Regime France.[111] Outside of Paris and across the provinces urban elites collaborated to found learned societies. Every provincial city of any note had one, typically an *Académie royale des sciences, arts, et belles-lettres*. The entire spectrum of French cities and towns

101. Desavis (1933), p. 20.
102. Desavis (1933), pp. 22–23.
103. Desavis (1933), p. 21; Blanchard (1901), pp. 89–90, per the accounts of the dean in 1782; and again Blanchet (1930), p. 291, and Ch. Florange (1933), pp. vi–x.
104. Desavis (1933), p. 21.
105. At other times one or two gold jetons of the Faculté went to prize winners; Ch. Florange (1933), p. ix; Desavis (1933), p. 22.
106. Blanchard (1901), pp. 89–90; Desavis (1933), p. 22.
107. Desavis (1933), pp. 23–24.
108. Desavis (1933), p. 48; Blanchard (1901), p. 96 n.
109. Blanchet (1930), p. 291.
110. See numbers given in Desavis (1933), p. 24.
111. Roche (1978); McClellan (1985), pp. 89–99. Note that the phenomenon was international, indeed, global, as learned societies sprang up from Philadelphia, across Europe to Russia and as far as the Indies, East and West. England, Spain, Italy, Holland, German-speaking lands, and Scandinavia had their share; on the academies as capitals of the Republic of Letters, see McClellan (1993) and (1999).

became graced with academies of various grades. Major provincial academies in the metropolises of Bordeaux, Lyon, Montpellier, Toulouse, and Rouen were founded comparatively early. By mid-eighteenth century the academies in Paris were backed up by a network of other societies in towns occupying the second level of cultural and economic importance in France after Paris. Twenty-nine such institutions dotted the urban landscape of the provinces in the eighteenth century.[112] The king gave them letters patent and royal recognition, but with the exception of the Académie Royale de Marine at Brest, these organizations were not royal institutions so much as ones tied to local and regional élites and centers of power and government. Here, too, we find jetons woven into the identities and practices of provincial learned societies. Not all provincial academies struck jetons, far from it. Most were too small, weak, and financially precarious to incorporate jetons into their practices, but a few—at least ten—did, and with these jetons we can see further into life in provincial urban centers and the sociability of provincial academies.

The Académie Royale des Sciences, Belles-Lettres et Arts of Lyon can serve as an example.[113] The Lyon academy began in 1700 as a private group, essentially a club of gentlemen interested in science and literature. With the backing of the local archbishop in 1724 it gained recognition as the Société Royale des Sciences et Belles-Lettres, and in 1752 it was elevated to the status of a royal academy. From 1735 the town council voted an annual subsidy for 300 jetons, and these were duly minted and distributed as *jetons de présence* at meetings and otherwise through 1789. A corresponding Académie des Beaux-Arts came into being in Lyon in 1713, and from 1744 the town likewise provided it 140 jetons annually. In 1758 these two organizations merged to form the above-titled academy that lasted until 1793. With the fusion of these two academies, a new jeton was designed and new dies prepared, but apparently the Monnaie des Médailles back in Paris for years kept using the old dies, drawing regular protests from the academicians in Lyon.[114]

The jetons of the Lyon academy are to be situated in the larger urban context of jetons elsewhere in the city of Lyon. The Feuardent catalog, for example, lists 300 different jetons and variants struck for town councilors, archbishops, governors, intendants, other administrative offices, and guilds in Lyon in addition to those commissioned and distributed by the local academy.[115] The governor-general of the province and protector of the academy, the Duc de Villeroy, and his family struck their own jetons, and these, too complemented those of the local learned society and others in the town.[116] To a greater or lesser degree, Lyon typifies this entangled social context involving jetons, provincial academies, and local and regional élites and governing bodies. Indicative of this entanglement provincial academies usually met in the local city hall, and on one occasion when new jetons had just arrived from Paris, the Lyon academy invited in the Lyon town council that was meeting simultaneously and offered them jetons by way of thanks (Fig. 4.6).[117]

The Académie Royale des Sciences, Belles-Lettres et Arts of Rouen (1744–1793) provides another revealing example of a provincial academy and its jetons. Rouen was a vibrant city on the Seine in the north of France, connecting Paris to the English Channel port of Le Havre. Led until his death by its secretary, the first royal surgeon in Rouen, Claude-Nicolas Le Cat (1700–1768), the Rouen academy was an active academy with twenty ordinary members that met regularly

112. McClellan (1985), appendix I.
113. Morin-Pons (1900); McClellan (1985), pp. 98, 271.
114. Morin-Pons (1900), p. 24. At some point the proper jeton (F.10749) did issue forth with the correct designation (ACAD.LUGDUN.SCIENT.LITTER.ET.ARTIUM).
115. F.10494–10795. See again Loach (1995).
116. E.g., F.10827–10828.
117. Morin-Pons (1900), p. 19. The reported largesse ("à prendre tel nombre de jetons qu'ils souhaiteroient") is to be doubted somewhat.

FIGURE 4.6: Jeton of the Académie Royale des Sciences, Belles-Lettres et Arts of Lyon.

through the second half of the eighteenth century.[118] Although it didn't publish memoirs, it had a library, botanical garden, and regional contacts with other learned societies in France and overseas, and it offered public courses on medicine and hydrography. It struck and used jetons at least from the later 1740s.[119] Nevertheless, attendance and participation at meetings was a problem that bedeviled the Rouen academy. In its first years anyone who read a new paper of at least 15 minutes whether for publication or simply deposit at the academy's secretariat received two silver jetons after a second reading. Forty-four jetons went out this way in 1749 and 30 in 1750 before the practice was suspended in 1751.[120] The academy then instituted what amounted to dues as a means of encouraging attendance and paying for jetons.[121] Each academician was dunned 60 *livres*, but each received a silver jeton for each meeting, 30 handed out to those in attendance at 4:30 precisely.[122] Thus, if you paid your 60 *livres*, you would get the value back through the jetons you would receive at meetings.[123] The practice was sporadic, and repeatedly every year or so dues became the subject of further discussion and deliberation in the Rouen academy. Finally, in the 1780s the body decided that academicians could not be required to pay dues; voluntary contributions to underwrite jetons were welcome and expected in a certain sense, but anyone attending a meeting would receive his jeton.[124] In this context it's not surprising to find that benefactors came forward to endow jetons, one in 1780 to create a fund for 36 jetons, another in 1786 to produce 90 *livres* a year for 40 jetons for attendance at mass on the morning of the annual public meeting of the academy. Both propositions were naturally accepted by the Rouen academy, but what came of them is unknown.[125] The case of the Rouen academy yields up the interesting example of jetons being used to fund a medal for the academy, when on July 26, 1746 M. de la Bourdonnaye

118. McClellan (1985), p. 277; McClellan and Regourd (2011), pp. 444–45.

119. F.6113–6122, and BM-Rouen, Ms. B4/1: "Registre journal des Assemblées et Deliberations de l'Academie des Sciences, Belles-Lettres, et Arts, De Rouen etablie en 1744": Minutes of the meeting of August 16, 1752: "L'Academie a choisi pour type de jettons le temple de la gloire avec trois portes, avec la devise TRIA LUMINA PANDIT. Mr. Descamps a été prié de faire le dessein, que sera examiné par les officiers et communiqué a M.r de Luxembourg [le protecteur de l'Académie]."

120. See BM-Rouen, Ms. B4/1: Rouen academy minutes of the meetings of August 27, 1748, January 21, 1749, August 19, 1749, and August 24, 1751.

121. BM-Rouen, Ms. B4/1: Rouen academy minutes of the meeting of August 16, 1752.

122. BM-Rouen, Ms. B4/1: Rouen academy minutes of the meeting of August 16, 1752, and p. 68, "Suite de l'assembleé du 23 août 1753."

123. BM-Rouen, Ms. B4/2, Rouen academy minutes of the meeting of March 23, 1768; see also BM-Rouen, Ms. B5/2, Procès-verbaux, classe des sciences, 1752–1765, pp. 1–2.

124. Thirty-six of these were set aside for 36 meetings a year. BM-Rouen, Ms. B4/2, Rouen academy minutes of the meetings of January 12, 1780, January 19, 1780, and August 1,1780; the record indicates previous discussion of these matters in in 1748, 1751, 1752, 1753, 1756, 1767, 1768, and 1774.

125. BM-Rouen, Ms. B4/2, Rouen academy minutes of the meetings of February 23, 1780 and August 30, 1786.

FIGURE 4.7: Jeton of the Académie Royale des Sciences, Belles-Lettres et Arts of Rouen.

made a gift of 140 silver jetons for that purpose (Fig. 4.7).[126] Finally, the Rouen academy had its rituals like other academies, and these included guests, such as one M. Enslau who presented his newly invented siren, much talked about in the public press. The academy felt obligated to invite M. Enslau and his associate to remain through the meeting and grudgingly offered each a jeton.[127] One can easily imagine similar scenes playing themselves out at meeting after meeting in other academies elsewhere throughout the provinces.

Duke Stanislas of Lorraine, ex-king of Poland and father-in-law of Louis XV, underwrote the Société des Sciences et Belles-Lettres (1750–1793, a.k.a. the Académie de Stanislas) at Nancy and its jetons.[128] Over at Metz the duc de Belle-Isle, Maréchal de France and governor-general of the province, endowed the Société Royale des Sciences et des Arts to the tune of 60,000 *livres* in 1760, and its jetons, graced with the bust of its founder, served as *jetons de présence* and otherwise for the Metz academy.[129] The local bishop underwrote the jetons of the academy in Montauban, and the academies in Amiens, Toulouse, and Bordeaux each had their jetons. With a single issue, jetons seem underrepresented in the otherwise vibrant provincial academy in Bordeaux, the Académie Royale des Sciences et Belles-Lettres (1712–1793).[130] Missing from the record of provincial academies and jetons are the Dijon academy (the Académie des Sciences, Arts et Belles-Lettres, 1725–1793) and even more remarkably the Société Royale des Sciences of Montpellier (1706–1793). The latter was a prominent and active royal institution formally linked to the science academy in Paris. Furthermore, Montpellier was the major university in the south of France and home to a medical faculty that vied only with its counterpart in Paris. Geography and distance from Paris may have played a part here, as well as the different arithmetic techniques of the quill versus the counting table, but where we don't find jetons in contemporary provincial academies reveals the notable contours of where we do find them.

The Académie Royale de Marine or Academy of Naval Sciences at the port of Brest is exceptional in the general run of French provincial academies in the eighteenth century in that it was not a

126. The academy graciously accepted the gift and proceeded to deliberate on the proposed medal; BM-Rouen, Ms. B4/1, Rouen academy minutes of the meetings of July 26, 1746 and August 2, 1746, where "On delibera sur la devise destinée pour la medaille de l'academie."

127. BM-Rouen, Ms. B4/2: fols. 412–412v, Rouen academy minutes of the meeting of November 23, 1785.

128. The French crown took up funding after Stanislas' death in 1766. See F.7581–7583 and image above, Fig. In.3. Note also the jeton Stanislas had struck for his *cabinet du roi*, F.7585–7587.

129. Robert (1853), pp. 83–84; F.7756–7757; McClellan (1985), p. 273.

130. On these jetons see F.6536, F.9204, F.1034–11035, F.11069–11070. On the Bordeaux academy, see McClellan (1985), pp. 97–98, 265.

regional or local body, but attached directly to central royal government and the royal navy (the Marine royale) in particular.[131] At the entrance to the English Channel the naval port at Brest was one of the royal navy's three major ports in addition to Rochefort and Toulon. In addition to extensive dock, harbor and refitting operations, Brest was the home of a large general staff, a naval training school, a hospital and medical school, and a facility to manufacture nautical instruments. Enlightened naval officers—virtually all of them noblemen—began private assemblies in 1749. With the support of the minister of the navy and the powerful inspector general of the navy, Duhamel de Monceau, the Académie de Marine emerged with a formal charter from the crown in 1752 followed by letters patent in 1769. The Académie de Marine was organized as a typical academy, meeting weekly in a room graced with two portraits of the king, and particularly after the Seven Years' War it was well funded at 6,000 *livres* a year. But its members were all officers or engineers associated with the navy, and the Académie de Marine was formally attached to the ministry of the navy and the colonies. Another distinctive feature of the marine academy at Brest was the formal affiliation effected in 1771 with the science academy in Paris. The Brest academy worked closely with the ministry and the Académie Royale des Sciences on a range of research regarding matters nautical, functioning something like the naval arm of the science academy. It didn't sponsor prizes and published only one volume of memoirs, but the Marine Academy at Brest otherwise was a dynamic institution that stood out among French academies in the provinces.

FIGURE 4.8: Jeton of the Académie Royale de Marine, 1778.

The Marine Academy struck its jetons first in 1769 and regularly thereafter (Fig. 4.8).[132] Its first featured a bust of Louis XV on one side and a ship under sail on the other with the legend PER HANC PROSUNT OMNIBUS ARTES (By such proceed all the arts). In 1777 the academy petitioned the minister for new dies and 1,000 new jetons showing a bust of the new king, Louis XVI. The academy complained that the depiction of the ship on its jeton of 1769 was a nautical impossibility, and it sent its own design to Paris drawn by an official navy painter. It also requested a change in the legend to BELLE AC PACE NUSQUAAM IN GLORIA (In war and peace nowhere but in glory). The minister rejected the latter request, sticking with the original and official legend of the academy, but working through its agent in Paris and with the ministry and the mint in Paris, the Monnaie des Médailles, the Académie de Marine got its new jeton in 1778. It thought the king looked a little old and asked the mint to touch up the portrait. It was still not happy with the depiction of the ship, and it insisted that any future costs for jetons would be for jetons alone, not for any new dies, the old ones to be kept at the Monnaie des Médailles. The academy felt good enough about the new jeton to accede to the request of the engraver, Benjamin Duvivier, for two jetons as a tip

131. On the Académie de Marine, see McClellan and Regourd (2011), pp. 110–24.

132. On these points, see Blondeau letter dated "27 Juin 1777," and Sartine letter dated "Versailles le 12 Juillet 1777," SHM-Vincennes, Ms. 093; Académie de Marine (2002), p. 14; F.1551–1555, 9203.

in addition to his fees. The Brest academy paid for these jetons through a draft on a bank in Paris, and it received them back in Brest through an intermediary, possibly in several shipments, possibly for fear of them being lost or stolen.[133] In 1780 the Brest academy ordered another 500 jetons from Paris, again asking the engraver to redo the ship which the navy men at Brest still found inadequate.[134]

A word needs to be added about the agricultural societies that arose alongside the academies and learned societies of the provinces.[135] With antecedents sponsored by the Estates of Brittany in 1757, in 1761 the national government launched a drive to install local improvement societies across the kingdom. The result was 20 agricultural societies in 20 provincial towns and the *généralité* of Paris. Their charge was the improvement of agriculture. They were variably successful—the Paris group got off to a great start with illustrious members and an initial publication, but its vitality declined sharply after that. Other agricultural societies had an on again, off again existence, too. A few, such as the agricultural society of Lyon, struck and used jetons, but the details remain murky.[136] Certainly, compared to the many chambers of commerce that dotted ports and provinces in the day, about which more later, jeton use was rare among the agricultural societies. In the 1780s, however, facing drought and famine, authorities revived the Paris society, which in 1788 became an official Société Royale d'Agriculture. It was well-funded and an active institution alongside its sister academies in Paris, having publications, prizes, correspondence networks, and an illustrious set of members with powerfully overlapping institutional connections. The Société Royale d'Agriculture had and used its own jetons. The jeton or medal (Fig. 4.9) commemorates a 1785 act of royal largesse, the gift of a herd of cows to fend off precarity.[137] In this way the Société Royale d'Agriculture fit into and strengthened the larger complex of royal academies in Paris on the eve of the French Revolution.

All that being the case, the most remarkable French provincial academy of the eighteenth century was not in the provinces at all, but overseas in the Caribbean in the French colony of Saint Domingue on the western end of the island of Hispaniola, today's Haiti. In the 1780s Saint Domingue was the richest and most productive colony in the world, bringing in more annually than the whole of the depleted Spanish empire.[138] Saint Domingue was the jewel in the crown of French and, indeed, European overseas expansion. The colony originated as a pirate enclave in the 1660s, followed soon by other settlers and the beginnings of government. The treaty of Ryswick (1697) made official French possession of the colony which grew steadily and rapidly after the Seven Years' War (1757–1763). Contemporary Saint Domingue was the most intensely slave society ever to have existed, and soon with the French Revolution it was the locus of history's largest

133. On these points see Academy correspondence with the navy artist, Nicolas Ozanne in Paris, SHM-Vincennes, Ms. 092, letters dated "27 Juin 1777," "8.e Aoust 1777," "9 Janvier 1778," "23 fevrier 1778," and "le 8 Mai 1778."
134. Academy letter to Duvivier dated "15.e X.bre 1780," SHM-Vincennes, Ms. 092.
135. On these organizations see especially Justin (1935); Roche (1978), vol. 1, pp. 280–84; Gillispie (1980), pp. 368–87; McClellan and Regourd (2011), pp. 134–42; Regourd (1998).
136. F.10776 and F.10776[a].
137. Feuardent (1904–1915), lists the jetons of the royal agricultural society, F.4899–4908.
138. For these details, see McClellan (1992/2010), Part I.

FIGURE 4.9: Medal of the Société Royale d'Agriculture, 1785.[139]

and most successful slave revolt that gave rise to the Republic of Haiti (1804), the second independent republic in the Western hemisphere. At the time in question 500,000 black slaves toiled on sugar, indigo, and cotton plantations spread out in a remote heart of darkness on fertile plains below the mountains of Haiti where coffee plantations thrived.[140] The colony harbored a mere 32,000 white people and 28,000 people of mixed race, both free and slave, everyone crammed into an area the size of the tiny state of Delaware.[141] Authorities kept a tight lid on the cauldron that was old Saint Domingue through the most atrocious violence. The colony was governed from Versailles through the Ministère de la Marine et des Colonies in the manner of any French province with a governor-general and intendant on the spot and in this case the royal navy providing necessary support. Although smaller towns existed, two major ones dominated the colony. Cap François in the north (today's Cap Haïtien) had a population in 1789 of 18,500, making it about the same size as Boston in North America or Dijon in France. Port-au-Prince in the center had a population of 9,400. As they grew, particularly Cap François, so came the institutions and infrastructure of contemporary French civilization: parish churches, government offices and officers, military installations and parade grounds, law courts, hospitals, prisons and poorhouses, printing presses and a newspaper, bookstores, a theater, coffeehouses, dance halls, bars, prostitution, public fountains, and paved streets. A small segment of colonial society was attuned to currents back in France and in Europe, and so in August of 1784 an elite group of royal physicians, surgeons, lawyers, merchants and landowners came together and formed a nascent learned society in Cap François they modestly called the Cercle des Philadelphes.

The group that started out as the Cercle des Philadelphes in 1784 enjoyed quick success becoming the Société Royale des Sciences et des Arts du Cap François in 1789.[142] The founders, notably its secretary, the first royal doctor in the colony, Charles Arthaud (1748–1793?), made the proper overtures to the governor-general of Saint Domingue, César-Henri, the comte de la Luzerne (1737–1799), who approved a set of formal statutes right away. Approaches through the ministry of the navy resulted in provisional royal recognition and 3,000 *livres* in funding in 1786. The scientific and utilitarian orientation of the Cercle des Philadelphes meshed well with government policies for the development of the colony and accounts for this fast-tracked treatment. With a core membership of about 30 in Cap François, through 1792 the body elected a total membership of 163 from elsewhere in the colony, the Caribbean, and back in France, including the great

139. Feuardent labels this item (F.4905) a jeton, but it seems more a medal; see further below, List of Figures.
140. In dramatic contrast, only 700,000 slaves inhabited the whole of the United States in 1790.
141. This area is only slightly larger than the Île de France region in France itself.
142. The present account is drawn from a not undeveloped literature surrounding the Cercle des Philadelphes; see McClellan (2000); McClellan (1992/2010), especially chapter 11; McClellan and Regourd (2011), pp. 461–75.

Benjamin Franklin who was genuinely pleased to be elected. The Cercle meet twice a month in rooms on the rue Vaudreuil in Cap François that housed a cherished bust of the king. In a short period the Cercle des Philadelphes published an agricultural survey and works on mineral waters, tetanus, preserving paper, and epizootic diseases. It inaugurated a botanical garden where it raised cactus for experimental cochineal production. It held annual public meetings, well-attended by the cream of Cap François society. The Cercle launched a vigorous series of prize questions, and it began offering public courses on botany and medicine. The Cercle des Philadelphes was an unusually vigorous, yet still typical French provincial academy, only this one happened to be located in the tropics.

FIGURE 4.10: Jeton of the Cercle des Philadelphes, 1788.

Sometime in late 1787 or early 1788 the Cercle des Philadelphes took some of its subvention and sent off to France to have jetons struck at the Monnaie des Médailles. Figure 4.10 depicts the jeton, engraved by the king's royal engraver, Simon. On the front of the coin, in positively grotesque relief (he was then 34 years old) is Louis XVI, king of France and of Navarre, and the ominous yet quintessential date of 1788. On the back, the insignia and legends of the Cercle des Philadelphes: a trite beehive and hovering bees beneath the rays of a meridian sun encompassed with the motto, EXERCET SUB SOLE LABOR (Work is done under the sun). In the exergue in two lines the inscription In French: CERCLE DES PHILADELPHES /ETABLI AU CAP 1784 (the Cercle des Philadelphes, founded at Cap François, 1784). These jetons were struck in silver and in gold. Each associate of the Cercle received a silver jeton, and the Cercle sent coins to the colony's administrators and their wives.[143] The gold ones undoubtedly greased the machinery, and one probably went to the king whose bust was featured on the obverse of this piece.

But the story of the Cercle des Philadelphes and its jetons does not end there.[144] In late 1788 the Cercle sent representatives to France to petition for letters patent. On December 11, 1788, a distinguished delegation met at Versailles with the minister of the navy and the colonies, the selfsame comte de la Luzerne who had moved on from Saint Domingue to head this key ministry. The delegation consisted of Jean Barré de Saint-Venant, a rich colonist and the current president of the Cercle, Moreau de Saint-Méry, the colonial jurist sponsored by the government to compile a work on colonial law, and the comte Du Puget d'Orval, a military man, national associate of the Cercle des Philadelphes, and tutor to the dauphin. At this meeting La Luzerne gave his approval for formal letters patent. The government raised funding for the Cercle to 10,000 *livres* and authorized further jetons, but as La Luzerne noted in the margin of the draft, only if struck in France. The final version of the letters patent cleared La Luzerne's desk on April 30, 1789. The papers were

143. On the jeton of the Cercle des Philadelphes, see McClellan (1992/2010), pp. 242–43 and notes, p. 350; Zay (1892), p. 234; Blanchet (1930), p. 270. This jeton is not listed in Feuardent.

144. These developments are described more fully in McClellan (1992/2010), pp. 247–56, and McClellan and Regourd (2011), pp. 469–72.

shown to the king for his approval on May 13, 1789, and parchment letters patent prepared. Louis XVI formally signed the document on May 17, 1789, creating the Société Royale des Sciences et des Arts du Cap François. At that very moment elsewhere on the grounds at Versailles, the Estates General of France—called for the first time since 1614—had been meeting for a fortnight and was deadlocked over the verification of deputies. The declaration of a National Assembly and the storming of the Bastille were just weeks away.

The granting of letters patent was a capital achievement for the group in Saint Domingue, but success crowned their fortunes in another way in 1789: a formal union with the great Académie Royale des Sciences in Paris. Earlier, on Wednesday, February 25, 1789, Barré de Saint-Venant, joined again by Moreau de Saint-Méry and, this time, by Barré de Saint-Leu, a naval officer and French national associate of the Cercle, addressed the royal science academy in its meeting room in the Louvre palace.[145] The initiative for the meeting might have come from the state minister La Luzerne himself, elected an honorary member of the Academy of Sciences in September of 1788. La Luzerne was vice president of the Academy for 1789, and he went on to succeed as president of the Academy of Sciences in 1790. It certainly flattered La Luzerne's career to see an association between the colonial science society of which he was soon to become protector and the mother of all such institutions in Paris in which he served as a ranking member and officer.

In his speech Barré de Saint-Venant introduced the Paris academicians to the lush Caribbean paradise whence he came, and as much as possible kept his focus on points of scientific interest. He elaborated the research program of the Cercle des Philadelphes in medicine, botany, agronomy, chemistry, and meteorology, emphasizing the useful at every turn. Only at the end of his speech did Barre make clear that, in coming before the Academy of Sciences, the Cercle des Philadelphes had something more in mind than simple institutional goodwill or receiving the blessings of a senior society, or even establishing an ordinary institutional "correspondence" that often connected such organizations. Barré de Saint-Venant proposed something more formal and specific: an actual union of the two institutions. Barré concluded his presentation—and this is the point to be noted—by asking the Academy, regardless of its decision on affiliation, to accept the Cercle's jeton, and he ventured to request one of the Academy's jeton in return. "May this exchange and mutual gift put a seal on the union we desire."[146] After a committee report the Paris Academy approved an unprecedented formal association with the Société Royale du Cap François at its meeting on April 1, 1789.[147] The oldest and newest of France's science academies thus forged a remarkable new association. Here, symbolically, the exchange of jetons was more than token.

The new royal society in colonial Saint Domingue already had sustained contacts with the Société Royale de Médecine in Paris over tropical medicine. The Saint Domingue group forged formal ties to the academies in Rouen and in Bordeaux, and the Cercle's jetons may have played a role in fashioning these associations. In the end the Cercle des Philadelphes/Société Royale became a notable cog in the bureaucratic machine that ran the nontrivial overseas endeavors of the French in the eighteenth century.[148]

145. See above and the 14-page "Discours prononcé à l'Académie royalle des Sciences, le mercredy 25 Feb. 1789 par Mr. Barré de Ste Venant Président du Cercle des Philadelphes du Cap François," PA-PS (February 25, 1789); copied into PA-PV 108 (1789) ("Du Mercredi 25 Fevrier 1789").

146. See previous note and McClellan (1992/2010), p. 254; McClellan and Regourd (2011), p. 471.

147. See above and report on this association, letter signed Le Vavasseur and dated "Cap Le 1.er août [1789]," BM-Rouen, Ms. C25.

148. On this "colonial machine," see McClellan and Regourd (2011).

One last vignette links the Cercle des Philadelphes/Société Royale with numismatics and Old Regime academies. Riding the crest of the wave of its success in the spring of 1789, the new Société Royale of Cap François in the far away Antilles made a request of the Académie des Inscriptions et Belles-Lettres in Paris for suggestions for "a medal to consecrate the memory of its foundation" in 1784.[149] Even as modernity was about take a radical turn, the spirit animating this request in 1789 was the time-honored one that motivated Louis XIV a century earlier and emperors and rulers even earlier in antiquity: to solidify a place for themselves in a future past through a record stamped in enduring metal.[150] The commissioners of the Académie des Inscriptions came up with two proposals for this medal. One, in French, the other in Latin. Both had the usual royal bust and legends in French and Latin respectively on the obverse. On the reverse the French version had Minerva seated beneath a covering strewn with fleurs-de-lys. Around her, the attributes of the sciences and arts: "...a sphere, compass, instruments of physics, chemistry, ~~astronomy~~ [*sic*], etc., and beside her sugar canes, coffee trees, etc." The legend in French on this side: "For the progress of the sciences and the arts" and in the exergue, "The Royal Society established at Cap François in 1784."[151] On the reverse the Latin version had Minerva buckled in armor standing on a promontory surrounded by the attributes of the arts and sciences, holding in her right hand a crown of laurels and leaning with her left on a sheaf of canes, with a shield with three fleurs-de-lys at her feet. The inscription read MINERVA TRANSMARINA (Minerva overseas), and in the exergue came SOCIETAS REGIA SCIENTIARUM ET ARTIUM APUD PROMONTORIUM GALLICUM INSTITUTA ANNO 1784 (Royal Society of Sciences and Arts established at Cap François in 1784).[152] Both these versions put the foundation of the vigorous, new royal society—and not the originally frail Cercle des Philadelphes—back to 1784. The commissioners reported these suggestions to the Académie des Inscriptions at its meeting on Tuesday, July 3, 1789, and they received unanimous approval from the assembled group. Nothing came of this initiative, however, as other, more dramatic matters erupted eleven days later with the storming of the Bastille on the fourteenth of July, 1789.

Jetons infiltrated French society and culture in manifold ways, but this chapter concerning assiduity and the academies shows both a distinct set of institutions in Paris and the provinces and the ways jetons became characteristically associated with them. Still, the story goes on, and a last big body of organizations that incorporated jetons and practices surrounding jetons into their activities remains to be considered: The guilds.

149. "...une Médaille pour consacrer La Mémoire de son etablisement," AI, AIBL archives, A:71, minutes of the meeting of June 19, 1789. On this episode, see also AI, AIBL archives, A:71, *plumatifs* and formal minutes of the meetings of June 26 and July 3, 1789.

150. This is great insight one finds in Wellington (2015a).

151. "R. Minerve assise sous une voile semée de fleurs de lis et attachée à deux arbres du pays : autour d'elle sont les attributs des sciences et des arts, sphère, compas, boussole, instrumens de Physique, de Chimie, ~~astronomie~~ [*sic*] &c. à côté des cannes à sucre, des caffiers, &c. Légende : Pour le progrès des sciences et des arts. Exergue : Société Royale etablie au Cap François en 1784," AI, AIBL archives, A:71, minutes of the meeting of July 3, 1789. The ~~astronomie~~ was struck presumably because its instruments fitted in the category of "Physique."

152. "R. Minerve debout sur un promontoire, entourée des attributs des arts et des sciences, L'Egide sur la poitrine, tenant de la main droit une couronne de Laurier, et appuyée de la gauche sur un faisseau de cannes à sucre ; à ses pieds son bouclier sur le quel sont trois fleurs de lis. Légende : MINERVA TRANSMARINA. Exergue : SOCIETAS REGIA SCIENTIARUM ET ARTIUM APUD PROMONTORIUM GALLICUM INSTITUTA ANNO 1784," AI, AIBL archives, A:71, minutes of the meeting of July 3, 1789.

FIGURE 5.1: Jeton of the corduroy-makers' guild (*Corroyeurs du Corduroi*), 1755.

5.

Guilds and Yet Others

Almost unfathomable in our day is the fact that guilds—trade corporations—from the Middle Ages onward through the French Revolution monopolized trade, manufacture, and a large portion of the labor force in France and elsewhere in Europe. It's an exaggeration, but not far from the truth, to say that there was no free enterprise. Although the rudiments were there and exceptions are to be made, Old Regime France was decidedly not a modern capitalist society. It was likewise not industrialized. The primary economic activity of agricultural production took place on farms and vineyards around rural villages still associated with a manor and local lord. While not excluding a small measure of private exchange, at fairs, for example, or traveling salesmen, all manner of feudal obligations and village traditions governed the economy on this level, and no jetons are to be found there except sometimes on manorial counting tables. Craftspeople and specialized artisans, such as the local cobbler or mason, are present in the countryside, but no evidence suggests that these local craft or artisanal activities, much less long-distance trade, were organized into guilds in the countryside. In towns and cities, on the other hand, the situation was different. There, economic productivity was completely controlled by the guilds. These were incorporated bodies of trades and crafts that monopolized and policed both merchant trading and the many artisanal specialties that formed themselves into guilds from the late Middle Ages onward.[1] In this way the guilds not only were key elements in the economic life of France before the Revolution, they structured society and the urban scene in particular. Although they originated in the medieval era and persisted for hundreds of years, the guild system reached its apogee in the eighteenth century.[2] Guilds occupied a notable and distinctive place in contemporary society, and jetons became an integral part of guild life. Jetons in turn have much to tell us about the world of the guilds.

A Feudal Mode of Production

In 1268 an early Paris city administrator (*prévôt*) named Étienne Boileau (1200–1270) compiled his treatise on the contemporary guilds in Paris and their regulation, *Le Livre des Métiers*, which is, as Wikipedia puts it, "a faithful mirror reflecting the smallest details of the industrial and com-

1. An extensive literature backs up any consideration of medieval and early modern guilds. See especially Martin Saint-Léon (1941), whose editor, p. viii, rightly calls this work "la base de toute étude approfindie sur les corporations et le corporatisme en France." See also the substantial Coornaert (1941); Lespinasse (1886-1897); Lespinasse and Bonnardot (1879). For more recent, specialized studies see Gallinato (1992); Moulin (1995).
2. Coornaert (1941), p. 15: in the later eighteenth century "le système corporatif, sur le tard, atteint son apogée."

mercial life of Paris in the 13th century."[3] In the 1260s Boileau already could enumerate 100 recognized guilds in Paris. As the totality of guild and corporate structure developed, the Parisian guilds came to be headed by the so-called Six Trades (the *Six Corps*) who early on grouped themselves together above all others and whose head, the *prévôt des marchands,* ultimately came to serve as mayor of the city of Paris. The six trades and associated monopolies were for the weavers and drapers (*drapiers*); apothecaries and spice merchants (*épiciers*); jewelers and retail merchants (*merciers*); tanners and leather merchants (*pelletiers*); milliners and furriers (*bonnetiers*), and the silversmiths/goldsmiths (*orfèvres*).[4] Later, the wine merchants tried in vain to join the *Six Corps*, but had to content themselves with being entitled to march directly behind them in civic parades.

Guilds had a legal existence and a civic personality. They could own property, make contracts, take on debt, and conduct business. They had guild halls. They organized a body of guild members. They took in money and spent money, and they possessed the further attributes of legal corporations in the Old Regime: a patron saint, an affiliated parish church, a treasury and archives, a banner, a coat of arms, a costume, a seal, and jetons.[5] As distinct as a guild or trade corporation was, it is to be distinguished from the associated *confrérie* or brotherhood of guild members who worshiped and organized themselves at a parish church. Although related and having the same members, guilds and *confréries* were and are often confused. A guild controlled a trade, while a *confrérie* constituted a religious fraternity associated with the trade. *Confréries* were supposed to have their own organization separate from the related guild.

The number and variety of guilds grew quickly and wildly. The third of the *Six Corps*, for example, the *merciers*, split into 20 specialized trades each with its own guild structure. Paris in 1766 hosted 113 active guilds, and records for 1726 show 35,000 licensed masters practicing their trades there: 4,000 in the *Six Corps* and the rest in other guilds, including 380 master distillers and lemonade-makers and 260 master leatherworkers.[6] Eighty-five different guilds could be found in Nantes in the seventeenth century; Bordeaux had 115 trades united in 85 guilds; the same number could be found in Rouen. The atomization of economic activity in the trades is astounding.[7] There were tinsmiths, mirror makers, fishmongers, dyers, coopers, and the corduroy finishers and merchants (Fig. 5.1). The guild of purse makers fabricated purses for jetons. In Rouen in 1776 14 different guilds dealt with food, another eight with drink, eight for clothing and hair, eight for shoes, four for woodworking, seven in construction, 28 different trades in total for iron, steel, and metalwork....[8] In Montpellier there were three different communities of innkeepers and six different guilds devoted to baking bread.[9]

All guilds had a roughly similar organization. Usually guild members elected four sworn officers (*jurés*) who served rotating two-year terms. One of the *jurés* generally took special responsibility for the financial accounts of the guild. Together the sworn officers called and led ordinary and extraordinary meetings of the guild and handled elections and the succession of accounts. They had oversight concerning the contracts of apprentices. Along with other masters and guild

3. https://en.wikipedia.org/wiki/Étienne_Boileau; accessed September 29,2018; on this landmark work see further Lespinasse and Bonnardot (1879) and praise for Boileau's work by Préau (1887), p. 12.
4. Martin Saint-Léon (1941), p. 113; Marion (1923), p. 152; Mousiner (1990–1992), vol. 1, pp. 362–68. Coornaert (1941), p. 139 adds money-changers to what is already a complex set of trades forming the *Six Corps*. A distinction to be borne in mind when considering the guilds concerns artisans versus merchants, that is, between a craft or art and strictly merchant or commercial activity.
5. Regarding these seals in particular, Lousse notes that "celui des corporations de métiers porte un ou deux outils particulièrement caractéristiques de l'activité professionnelle"; Lousse (1943), p. 215; see also p. 211.
6. Martin Saint-Léon (1941), pp. 497, 501.
7. Belaubre (1965) gives a short history of the dyers guild, its conflict with drapers, and a description of its jetons.
8. See listing, Moulin (1995), pp. 4–36.
9. Coornaert (1941), p. 250.

officers they examined candidates to become masters. They enforced rules regarding trade practices, and the guild carried out mandated visits to the workshops and businesses operated by guild members, upon whom they could impose fines and other penalties for infractions. Guild officers were the outward face of the guild regarding tax and other civic matters, such as the convocation of watchmen and the local militia.[10]

The status and financial strength of guilds varied. The richest owned their own guildhalls and sometimes other property from which they collected rents.[11] A large part of guild income came from initiation fees paid by apprentices and by candidates for admission as masters; other income derived from an annual fee for the associated *confrérie*, charges for shop visits, fines collected, other dues or charges self-imposed by guild masters, donations, and as bequeathed through wills. A guild had to maintain itself, and guild expenses included rent (if necessary), upkeep of the guildhall and guild property, and a bailiff if the guild could afford one. Outlays went for firewood, candles, paper, printing, transportation of the guild strongbox, and tips. Expenses for guild dinners and celebrations took a tidy sum, and so did those for participation in public functions such as parades. A guild had legal and notarial expenses to meet. And, a guild on its own or through its associated *confrérie* spent substantially on the upkeep of a chapel and on masses, funerals, and charity for sick masters and their widows. A guild budget could easily top several thousand *livres* a year.

Towns taxed guilds both on the products they produced (e.g., meat, drink, fish) and for the services provided by towns, including the registration and seal of ordinances, renting space at markets, public charges for the sheriff, and the like.[12] Guilds regularly bought back from towns the obligation to serve as watchmen or in the militia. A guild handled the collection of the royal head tax (the *capitation*, instituted in 1695) and what was basically an income tax (the *vingtième*, instituted in 1749), but it did so as a vehicle for individual masters and not for the guild itself. In order to exclude outsiders and maintain control of their own affairs, guilds bought back licenses to practice a trade offered from time to time by the king as a measure to enhance royal revenue. More patriotically, in times of war guilds participated with their towns in contributing to the French war effort. Apparently, in 1759, for example, the *Six Corps* offered up an incredible 514,000 *livres* in patriotic support, and it more than matched that sum again in 1762 with a promise of 700,000 *livres* for a 72-cannon ship that was built in Lorient and christened the *Six Corps*.[13]

But many guilds were weak and their financial strength often minimal. In these cases the *confrérie* played a stronger role, and clerics in the parish church sometimes had charge of keeping guild accounts. Occasionally, it seems, guild officers, either deliberately or out of ignorance, played fast and loose with the accounts of the trade's *confrérie* and those of the guild per se. Guild and *confrérie* administrators stubbornly refused to make their accounts public.[14]

From the time of Louis XIV the guilds were constantly at odds with the government, as central authority tried to impose state control over the economy in general and over the guilds in particular.[15] At issue were government audit and supervision of guild finances and especially the desire to eliminate debt on loans taken out by guilds. A law of 1714 required guilds to submit their accounts

10. On these points see Martin Saint-Léon (1941), pp. 111–12, 386, 388; Lousse (1943), 222–23.
11. On guild finances, see especially Gallinato (1992), pp. 170–76; also Martin Saint-Léon (1941), pp. 113, 202; Coornaert (1941), pp. 224–25.
12. Gallinato (1992), p. 202.
13. Martin Saint-Léon (1941), p. 465.
14. Coornaert (1941), p. 225; Martin Saint-Léon (1941), pp. 113, 493.
15. On these points see Martin Saint-Léon (1941), pp. 464, 492–933, 514–16. Colbert's creation of inspectors of manufacturing has to be understood as part of the background to these disputes. There were 34 *inspecteurs des manufactures* in 1715; see Coornaert (1941), p. 144. For further background consult Minard (1998).

to police and prosecutorial bodies; another of 1740 called on guilds to give an account of their debts and sought a wholesale examination of the books of the guilds. An edict of 1749 particular to makers of musical instruments forbade them from taking on additional debt without permission, and it required them to keep government-audited records (with receipts) and to maintain their accounts separate from their *confrérie.*[16] Guilds resisted these moves at every step. The situation remained tense through February of 1776, when the then-minister Anne-Robert-Jacques Turgot (1727–1781) suppressed the guilds and opened up trade to free enterprise. The outcry was so great that Turgot's ministry fell and the guilds were revived by May of that year, though reduced in number and still nominally subject to government oversight.[17] Article 24 of the new regulations specified that guilds could not demand dinners, jetons, or presents from artisans seeking admission, but these rules, too, were widely ignored.[18] On the night of August 4, 1789, the revolutionary Constituent Assembly famously and enthusiastically suppressed all feudal privileges, although the guilds were not technically eliminated until a budgetary bill in 1791. As the historian of the guilds Martin Saint-Léon put it, "A few lines inserted in a financing law sufficed to abolish an institution that for nearly seven centuries had been the foundation for the organization of national labor."[19]

Jetons in Guilds

Jetons formed an integral element of the organization and identity of the guilds.[20] Jetons were used almost universally in guilds, and that use only increased under Louis XIV, Louis XV, and Louis XVI. Arthur Forgeais lists 72 different Parisian guilds for which jetons were struck; one counts 86 in Adrien Blanchet.[21] Feuardent has categories for 114 trades in Paris alone that had and used jetons at some point, and he catalogs 672 different jetons and variants struck for Paris guilds over the course of the seventeenth and eighteenth centuries.[22] The *Six Corps* shows up for sure, as does a wonderful variety of other guilds with jetons. Butchers, brewers, clockmakers, shoemakers, stonemasons, locksmiths, tailors, tinsmiths, mirror makers, upholsterers, taverners, haulers, caterers, pastry chefs, joiners, acrobats, and the rod fishermen of the Seine among others all had their own jetons sometime during their histories. The pharmacists formally separated themselves from the spice merchants in 1777 coincident with the creation of the Collège de Pharmacie, and the group immediately struck 5,200 silver jetons "that should henceforth be the right of masters in different guild functions."[23] (They exhausted the supply by 1781 and ordered more at a cost of 4,000 *livres.*) The distillers were responsible for the especially charming guild jeton pictured in Figure 5.2.

16. Martin Saint-Léon (1941), pp. 458, 495.
17. Martin Saint-Léon (1941), p. 533.
18. Quoted in Martin Saint-Léon (1941), p. 588.
19. "Quelques lignes insérées dans une loi de finances suffirent pour abolir une institution qui, depuis près de sept siècles, avait été le fondement de l'organisaiton du travail national," Martin Saint-Léon (1941), p. 617; Jews were allowed in guilds only after the reforms and reestablishment of the guilds in 1776.
20. On jetons in guilds, see Lespinasse (1897); Tricou (1957); Blanchet (1930), pp. 295–303; Fontenay (1854), pp. 179–83; Forgeais (1874); Gombert (1974) adds a little; and again Lespinasse (1886–1897). On guilds in the United Provinces, see Roovers (1957). Per Erlanger (1959), p. 378, the formal abolition of Dutch guilds was in 1798, but they did not disappear there until entirely abolished in 1818.
21. Forgeais (1874), pp. 315–16; Forgeais gives a catalog, but doesn't unpack the function of jetons within guilds; see also Blanchet (1890), vol. 2, pp. 423–28.
22. Feuardent (1904), pp. 375–428, F.4764–4778.
23. "...les Prévôts furent authorisés à faire graver un coin à l'effet de faire frapper les jettons qui doivent désormais former les droits de présence des Maîtres dans les différents Actes," minutes of the general assembly of November 20, 1778, quoted in Dillemann and Roux-Fouillet (1976), p. 186; see also Dillemann (1997), pp. 10–11. The origins of French pharmacy in the spice trade and its migration toward medical officialdom, seen clearly in related jetons, are subjects to be underlined. With the Collège de Pharmacie, the pharmacists defined themselves, and they made their imprint on the city of Paris along with the surgeons and other expert institutions, all of which had their jetons, too.

FIGURE 5.2: Jeton of the distillers' guild.

Prior to the later 1500s, guilds used parish *méreaux* internally in guild activities. Redeemable at the guildhall, local *méreaux* served as *jetons de présence* for guild members at guild meetings and as examiners. They also served as alms given out by the guild, but most of a corporation's *méreaux* went for masses and related activities, indicating the historical roots of guilds in their ties to local parishes, to the local *méreaux* economy, and to *confréries*.[24] At a certain point, however, guilds began striking their own jetons. The Paris spice merchants and apothecaries' guild, for example, established in 1484 issued its first jeton in 1628.[25] The Rouen apothecaries, chandlers, and spice traders originated in 1508, but introduced jetons only in 1723. Similarly, dating from 1596 the pharmacists at Lyon had their first jeton in 1734. The earliest jeton for the notaries dates from 1681.[26]

One function jetons served in guilds was to keep internal accounts, and doubtless casting counters were to be found in many guildhalls. Guild jetons used for accounting were in brass or bronze, the silver ones reserved otherwise. A formal reckoning with jetons took place at the changing of the guard of sworn officers (*jurés*).

Another place jetons played an active part in the life of the guilds was at the examinations of candidates for admission as masters in the guild. We encountered this use of jetons above in the cases of the Avocats du Conseil du Roi and the Paris surgeons' guild.[27] Jetons and the use of jetons in these examination and reception ceremonies receive only rare mention in the formal rules or charters governing guilds.[28] But here and there sources show these examinations in action. For example, aspirants to the Rouen guild of apothecaries, spice and wax merchants owed two jetons to the chief examiner and two to each of his three associates at each session of the exam, and this process could carry over several sessions. The medical doctor present—note again the supervision and the authority of the physicians—received his, and successful candidates then owed four jetons to every other master and member of the apothecaries' guild in Rouen.[29] The 1669 reform of the Paris guild of painters and sculptors instituted by the *prévôt des marchands* shows jetons to be at the heart of these certification and initiation activities. Thus, prospective entrants owed four silver jetons (specified to be worth 15 *sols* apiece) to each examining guild officer, two jetons to retired members of the guild called in as examiners (but only up to 12), two jetons to the guild secretary, and one jeton to each of the youngest full members of the guild up to six in number. The edict notably stated that these jetons were to be offered in lieu of the customary banquet that new masters were otherwise expected to provide.[30]

24. Blanchet (1917–1918), pp. 176–77; Froeschlé-Chopard (1990).
25. Dillemann (1997), pp. 10–11; Dillemann, et al. (1996).
26. On guild tokens and their early origins and spread across the urban landscape of the Netherlands, see Roovers (1957); Erlanger (1959).
27. See above Ch. 3 and 4.
28. Lespinasse (1897), (1886–1897). There are only two incidental mentions of jetons in the magisterial work by Martin Saint-Léon.
29. This could amount to 150 silver jetons; Dillemann (1997), p. 26.
30. Lespinasse (1886–1897), vol. 2, pp. 211–13.

As revealing as these examples already are, the case of the guild of master surgeons in Bordeaux shows us even more. The organization of surgeons in Bordeaux mirrored the situation in Paris with a longstanding formal guild of Bordeaux surgeons, a Collège de Chirurgie (1753) and a "Société Académique de Chirurgie" (1762). This triumvirate began striking jetons in 1753 coincident with the establishment of the Collège de Chirurgie.[31] Legal authority to strike jetons ultimately came from the local parlement, and from 1763 through 1786 the Bordeaux surgeons struck 5–8 gold jetons and 50–200 silver ones annually.[32] As for examinations, guild rules of 1755 specified that aspirant masters pay 600 *livres* to the commonweal and give to each master surgeon present at his taking of the guild oath a pair of white gloves and a silver jeton worth 30 *sols*.[33] Then complaints came in that the gloves were of poor quality and the jetons too thin. Inflation apparently diminished the quality of a jeton of 30 *sols*, and candidates weren't living up to their obligations in terms of those gloves; on June 9, 1775, the guild rewrote the rules.[34] Gloves were henceforth to be deposited for inspection at the secretariat of the college three days prior to the formal reception; they were to be inspected by the same surgeons who inspected and approved bandages. The jetons in question were to be deposited at the same time. The text goes on to note:

> With regard to the jetons, the price of thirty *sols* is disproportionate to the cost of the die and cannot be set lower than forty *sols*, and the Community wishes that henceforth its jetons not be valued less. As this increase of ten *sols* per jeton could be onerous to aspirants, the Community on this account consents to what it has already done and that is to relieve aspirants of the obligation to visit each master after each of the exams in the course of his reception, which would entail an even larger transportation expense in carriages. This relief is even more advantageous to aspirants who are henceforth obligated to pay visits only to those masters who examined them.[35]

Jetons may have been distributed to other attendees at the reception of new masters, as the Rouen apothecaries apparently did, but one can only facetiously admire the consideration the Bordeaux surgeons displayed in this instance towards their prospective peers and colleagues.[36] This custom of visits paid by candidates has a long history in France that continues today, and we see in the eighteenth century that jetons were part of that picture, too.

Finally in this connection, jetons served as *jetons de présence* to boost attendance at guild meetings. This was certainly the case with the jetons of the apothecaries of Lyon and Rouen, and, incidentally, when these communities ran short they used the jetons of the Paris apothecaries.[37] For the chintzy Bordeaux surgeons again, there was a sign-in sheet and a line drawn, and the officers and twelve senior surgeons at the *collège* each received a silver jeton, but only after the

31. Évrard de Fayolle (1901); see also Chautard (1896), p. 83, for surgical schools elsewhere in the kingdom.
32. Évrard de Fayolle (1901), pp. 21–23.
33. Évrard de Fayolle (1901), pp. 25–26.
34. "Plainte de la Communauté au sujet des Jetons offerts aux Maîtres par les Aspirants," quoted in Évrard de Fayolle (1901), pp. 22–23.
35. "...et qu'à l'égard des jettons, dont le prix est de trente sols est disproportionné au quarré qui n'en peut contenir que de quarante sols, La Communauté désire que dorénavant ils ne puissent être d'un moindre prix ; et comme ce surcroit de dix sols par chaque jetton pourroit être onéreux aux aspirant, la Communauté consent en cette considération, comme elle l'a déjà fait, de dispenser les aspirants des visites qu'ils étoient obligés de faire à chaque Maître, après chacun des examens pendant le cours de la réception, ce qui les constituoit dans une dépense bien plus considérable en voiture ; ce dédomagement étant plus avantageux pour les aspirans qui ne seront tenus dorénavant que d'aller remercier ceux qui les auront interrogés," quoted in Évrard de Fayolle (1901), pp. 22–23.
36. On jetons at receptions in Rouen, see Rouyer (1892), p. 28.
37. Dillemann (1997), pp. 26, 31.

fifth session they attended in a row.[38] The new letters patent of 1781 for the guild of seamstresses and patternmakers specified that each officer collected two silver jetons worth a substantial forty *sols* each at every official meeting; other guild members in attendance received one jeton of the same value, but individuals leaving the meeting before it was over or refusing to sign the minutes would be deprived of jetons which would then be shared among those remaining in attendance and signing the minutes.[39] The community of weavers in the northern manufacturing center of Elbeuf provides a good example of guild tokens as *jetons de présence*. In 1758 the cloth-makers of Elbeuf decided, "to establish the custom followed everywhere else of handing out jetons."[40] They approached the Comtesse de Briône of the house of Lorraine, who were lords of Elbeuf, and she gave permission to use the cross of Lorraine on the guild's jetons; she also dictated the legend TALI FULCIMINE CRESCET (With such support shall it grow).[41] The guild ordered an initial 500 jetons in 1765, and along the way the masters assessed themselves dues, which, as we saw in the case of the Rouen academy, they then received back as *jetons de présence*. Just the same, guild officers worried about calling too many meetings and handing out too many jetons, so they ruled that only senior guild officials could call meetings. Uniquely it seems, jetons were handed out at the start of the meetings of the cloth-makers of Elbeuf, "a precaution we have thought prudent to take to have our brothers come to meetings on time" (Fig. 5.3).[42]

FIGURE 5.3: Jeton of the weavers of Elbeuf.

The Guilds of Engravers, Mint Workers, and Purse Makers

Three guilds draw particular attention: the guild of engravers, the guild of mint workers, and the purse makers' guild. The reason for this attention is that these guilds are to be associated with the manufacture of jetons and jeton purses. Their stories tell us more about the guilds in general, more about jetons in general, and more about the place and role of jetons in guilds.

The skilled artisans and artists who created the dies for jetons and medals came from the ranks of the engravers. Founded in 1623 and with letters patent in 1631, the community of engravers grew out of the guild of silversmiths, and like them engravers were subject to the authority and inspection of the Cour des Monnaies.[43] The guild was limited to 20 masters in the Paris region, and masters could open only a single shop and train a single apprentice. Two slightly different

38. Évrard de Fayolle (1901), pp. 21–23.
39. Lespinasse (1886–1897), vol. 2, p. 240. Note that this guild consisted primarily of women.
40. "... propose d'établir l'usage suivi partout ailleurs, de délivrer des jettons...," minutes of the guild meeting of March 11, 1758, given in Druet (1892), p. 5.
41. F.6400–6402.
42. "...précaution que nous avons cru devoir prendre pour engager nos confrères à y venir de meilleur heure," quoted in Druet (1892).
43. On these points, see Blanchet (1920a); Fontenay (1854), p. 46; d'Affry de la Monnoye (1878), p. xiii; Joly (1970), p. 371.

trades combined in the guild: ordinary engravers and *tailleurs* ("tailors"/"cutters") who worked with steel, and only *tailleurs* could cut dies for medals and jetons.[44] As was not uncommon in guilds, widows took over from their deceased husbands, as with the widow of the engraver Cousin in 1563.[45] In another case in 1552 an engraver with failing eyesight was given permission by the Cour des Monnaies to take on an assistant, who ultimately took over as a *tailleur* in the Paris mint, but only after providing a certificate attesting to his good conduct, upright morals and Catholic faith, and his swearing the customary oath.[46] Natalis Rondot in his 1904 study using original documents identifies 829 French engravers, many connected to silversmithing, from the fourteenth through the eighteenth centuries, although the absolute numbers decline over the later stretch of that long period.[47] Adrien Blanchet has uncovered addresses and descriptions of contemporary Paris engravers, such as a certain "Nicol, engraver living at Cloistre-St-Nicolas-du-Louvre, engraves seals, stamps, glass, and everything concerning the art." Similarly, a certain Charles Mavelot advertised himself around 1690 as, "Mavelot, ordinary engraver to Her Royal Highness Mademoiselle, living at the new courtyard of the Palais, at the bottom of the grand staircase by the arms of Madame the Dauphine, engraves medals, jetons, seals, stamps, and all sorts of works."[48] Engravers—and only engravers—were allowed to sell ordinary jetons (*jetons banaux*) for everyday use on counting tables and as gaming pieces, but they were forbidden dealing in silver jetons, and only *tailleurs* commissioned by or directly employed by the Monnaie des Médailles could have anything to do with engraving dies or manufacturing jetons. The engravers' guild itself struck its first jeton in 1718.[49]

The more ancient *Compagnie* of mint workers dates from 1354.[50] By the eighteenth century seventeen different royal minting workshops across the kingdom churned out coinage.[51] One in particular, the Monnaie des Médailles in Paris, we already know to be special in that it alone minted jetons and medals, but the guild structure was the same in all of these workshops. The actual production of coins and jetons involved several steps, and as it turned out the guild of mint workers incorporated three specialist subgroups: *monnoyeurs* or the men who operated the coin press (and before that the hammering), *ajusteurs* or the men who weighed and prepared the planchets, and *tailleresses* or the women who did the handwork on the blanks and planchets before striking.[52] This circumstance shows a marked division of labor and gender division of labor within the guild and the production process.[53] A remarkable contemporary archival docu-

44. Fontenay (1854), p. 47.
45. Mazerolle (1902–1904), vol. 1, p. 550.
46. Mazerolle (1902–1904), vol. 1, pp. 517, 554–55.
47. Rondot (1904), pp. 210–13. Mazerolle (1902–1904), vol. 1, pp. 501–82, gives a list of 43 Parisian engravers from 1488 to 1636. See also La Tour (1897), p. xi.
48. "Nicol, graveur demeurant Cloistre-St-Nicolas-du-Louvre, grave sceaux, cachets, vaisselle et tout ce qui concerne l'art…Mavelot, graveur ordinaire de S.A.R. Mademoiselle, demeurant Court neuve du Palais, au bas du grand Escalier, aux Armes de Madame la Dauphine, Grave Medailles, jettons, Sceaux, Cachets et toutes autres sortes d'ouurages [*sic*]," Blanchet (1920a), pp. 205, 203.
49. Blanchet (1920a), p. 202; see also CAEF, Ms. F^{o} 80, vol. 2, for 1718. Feuardent (F.5124–5125) notes an engravers' jeton of 1757.
50. Fontenay (1854), p. 54. On the mint workers' guild and its privileges, see also Blanchet (1890), vol. 1, pp. 8–11; vol. 2, p. 402; Mazerolle (1907a), p. 153.
51. Darnis (1996/1999), p. 16, and above Ch. 1.
52. Darnis (1996/1999), p. 17. Delcamp and Jambu (2018), pp. 67–68, note the existence of female *monnonayeuses*. Per Delcamp (2015a), after the introduction in the seventeenth century of the *coupoir*, the machine to punch out raw flans, the *ajusteurs* took over from *tailleresses* most of the handwork of finishing planchets, relegating *tailleresses* to a marginalized status within mints.
53. Delcamp (2015a), pp. 508–11, shows that the mint was a tense and hostile workplace for women workers, who labored in a separate room.

ment gives a count of the total number of these three specialists for the whole kingdom: 300 *monnoyeurs,* 261 *ajusteurs,* and 159 *tailleresses.*[54] Twenty-two *monnoyeurs,* 23 *ajusteurs,* and 30 *tailleresses* constituted the guild in Paris.

The Compagnie des Monnoyeurs was exceptional as a guild in that its members were effectively minor government employees who had to be accredited before the Cour des Monnaies.[55] Another distinctive, but not altogether exceptional feature of this guild is that positions were handed down in families. One needed to qualify, but one could only join the guild if one's parents were already members of the guild in one way or another. In essence, the jobs became hereditary, making the mint workers almost more a caste than a guild. Eldest sons succeeded automatically as *monnoyeurs*; second born sons did so as *ajusteurs*. Daughters could be received as *tailleresses*, and, if so, they would transmit to their male offspring (but not to female progeny) the right to be admitted to the corporation.[56]

As minor royal agents, mint workers enjoyed an enviable range of privileges and were people of some stature on the contemporary scene.[57] Although there were many local variants over the centuries, mint workers were essentially exempt from any and all taxes, duties, or public charges. They were free from the salt tax (the *gabelle*), the main royal tax (the *taille*), the head tax (the *capitation*) and any and all seigneurial assessments. They could not be forced to billet soldiers. They were free from the corvée and any duties to stand as watchmen or perform militia service. They were not subject to tolls on roads or across bridges. *Monnoyeurs* and *ajusteurs* could carry swords. Except in cases of murder or rape, in addition to other legal privileges, they escaped the jurisdiction of ordinary law courts and were subject only to the authority of the *prévôté* associated with the mint. Mint workers received five *sols* a day every day whether the mint was open or closed as their right (*droit de bouche à cour*) to be fed at the king's expense, and they had health insurance at the rate of two *sols* per day if ill or incapacitated. They sometimes received free burials and their daughters a dowry. These and other rights and privileges of the mint workers and their guild were reconfirmed as late as 1782. It's no wonder that other workers tried to pass themselves off as *monnoyeurs* in order to avoid taxes and to enjoy these privileges. True *monnoyeurs* carried with them proof of their identity in the form of small medals, an odd bit of numismatics showing up in this history.[58]

Mints and mint workers had their own jetons, and we are not surprised to find that these played a role in the reception of new *monnoyeurs, ajusteurs,* and *tailleresses.*[59] *Monnoyeurs* paid entry fees from the very beginning of the guild in 1354, and the custom continued through Louis XVI.[60] *Méreaux* may have been used at an early point, but soon the statutory obligation was for a certain number of silver jetons weighing two *gros* (or about a quarter of an once) each to be dis-

54. CAEF, Ms. A-1, undated eighteenth-century document, "Etat du Nombre de Monoyeurs, Ajusteurs, et Tailleresses des Monoyes du Royaume."

55. Delcamp (2015a), p. 504; Bély et al. (2003); Mousnier (1990–1992), vol. 2, p. 277.

56. Mazerolle (1907a), p. 153; Delcamp (2015a), p. 511.

57. See previous notes and CAEF, Ms. A-1 and Ms. 8, folders which contain several documents spelling out these privileges; see also Delcamp and Jambu (2018), pp. 67, 71; Delcamp (2015b), p. 220; Jambu (2013), pp. 187–88; Darnis (1996/1999), pp. 17–18; Moulin (1995), p. 24. Delcamp (2015a), p. 505, observes that extra expenses were involved if candidates skipped a generation, i.e., were grandsons or granddaughters of male mint workers.

58. F. 2130, F.11419; Mazerolle (1907a), p. 154.

59. F.2225–2236 for the jetons of the Paris guild covering from Louis XIII through 1767. See F.6278–6295 for jetons of the Rouen mint and the *monnoyneurs* and *ajusteurs* there, and other mint jetons listed in Feuardent per index listing (1914), pp. 480-481. On the reception of mint workers in Bordeaux, see study by Delcamp and Jambu (2018), passim.

60. Delcamp (2015b), p. 221, notes the rise of these fees for a *monnayeur* from 500 *livres* in 1691 to 1,500 *livres* in 1757. See also Delcamp and Jambu (2018), pp. 67–68; Barnard (1916), ppp. 77–78; Fontenay (1854), p. 54.

tributed upon entry into the guild. This use of jetons in the guild of mint workers would seem no different from the use of jetons in other guilds we have already seen, and there's little interest in pursuing the example except for one reason. That is, building up through the 1680s the obligations of new members of the community had become so baroque and complex and these obligations were so ignored that, on the initiative of the administrator of the guild (the *procureur-syndic*) who was having trouble collecting his due, on November 25, 1690, the Cour des Monnaies published an edict specifying in great detail the exact obligations of anyone seeking admission to the guild of Paris mint workers. In these specifications we see that jetons could not be done without, but we see as well that jetons formed just one part of elaborate social rituals involved in the examination and reception of new guild members. We have seen other things of this sort, too, but the documentation in this case is so extraordinary in unveiling this distinctive social practice that the details cannot be omitted.[61]

Thus, on admission *monnoyeurs* and *ajusteurs* faced an astonishing range of charges, obligations, and gifts to provide. The specifics break down as follows. Initiates had to pay or provide:

- A mass at the chapel at the mint.
- Twelve bouquets of flowers for the Cour des Monnaies, one especially big and beautiful for the *premier président*. In addition, new guild members owed the secretariat of the Cour des Monnaies two silver jetons, two candles, and a box of aniseed.
- Fifty-four *livres* in cash to the *prévôté* of the Paris mint, half for the *prévôt* himself and half to be distributed within the *prévôté*.
- To the master of the Paris mint: four *livres* cash, two silver jetons, two candles, and a box of aniseed for his right to furnish the material used in examinations.
- Six silver jetons each weighing two *gros* distributed to each of the officers of the guild. Formerly the requirement was four jetons, the extra two jetons now substituting for candles and boxes of anise owed heretofore.
- To existing *monnoyeurs* and *ajusteurs*: two silver jetons of the same weight, but the oldest 12 get four jetons each, in addition to two candles at 12-to-the-*livre* weight.
- To *tailleresses*, one candle.
- To the men who worked the forge (*recuiteurs*) and those whose muscles powered the coin press (*ricochons*), one candle.
- To the guards: 40 *sols* (i.e., two *livres*), two silver jetons of two *gros*, two candles, and a box of anise.
- To the chief engraver: four *livres* and two silver jetons.
- To the guardian: one silver jeton, one candle.
- To the assayer: two silver jetons, two candles.
- To the priest for his prayers and for accompanying the candidate: four *livres*.
- To the clerk keeping the minutes: five *livres*.
- Twelve *livres* in charity to priests and parishes of the candidate's choice.
- *Monnoyeurs* owed six bottles of wine, two plates of meat, and two dozen breads for those at the reception ceremony, more if necessary, depending on attendance. *Ajusteurs* got off

61. CAEF, Ms 8° 3: Ordonances des Monnoyes depuis 1680 Iusqu'en 1699, pp. 331–38: "Arrest de la cour des Monnoyes, du vingt-cinquiéme Novembre 1690, Portant homologation de l'estat en forme de Reglement, des Droits que les Ouvriers, Monnoyeurs, Tailleresses, Recuiteurs & Ricochons de la Monnoye de Paris, du Serment de France, doivent payer lors de leurs Accüeillement, Espreuves & Receptions en ladite Monnoye." Also CAEF, Ms 8° 3, p. 303: "Frais d'Espreve de chacun Ouvrier ou Monnoyeur." See also Bouclier (1924), p. 151. For related presentations of the costs and formalities of admission, see Jambu (2013), pp. 188–89.

with only four bottles of wine and one meat plate of a turkey and two capons or a turkey and three pullets and only one dozen breads.[62]

We needn't linger over similar obligations for *tailleresses*, except to note the obligation to their sisters of two silver jetons of 2-1/2 *gros* in lieu of the previously customary dinner.[63] New members of the guild swore to obey all laws, not to use metals other than provided by the master of the mint, and to report their fellows if they transgressed.[64] In a word, these regulations allow us to see life as lived inside a guild. They reveal a bygone world and a long-gone means of production.

Finally, we can say a word about the guild of purse makers.[65] Already by 1292 there were 45 master purse makers in Paris, and by 1447 we find a reference to purses specifically for jetons.[66] By the eighteenth century the trade had shrunk in size and importance, and purse makers and their guild counted little in the overall work force.[67] Louis XIV confirmed letters patent for the guild in 1659. It elected three *jurés* a year and conducted four workshop visits a year. Apprentices paid 20 *livres* and trained for four years, and the rules allowed only one apprentice per atelier. Journeymen needed three years of *compagnonnage* under their belts before standing for admission as a master, and they had to demonstrate their capacities by making "a jeton purse in velvet well and properly lined with a pouch well and properly made to close with a lock or a spring or a clasp."[68] Many of these *boursiers* were in fact master *boursières*, and purse-making seems another trade that involved women to a significant extent.[69] Widows of masters maintained their little shops, as did the master purse maker Catherine Lalmande, for example, or Catherine Vanier who succeeded her husband Pierre Alexandre Lefebvre as a purse maker sometime before 1689, but who seems quickly to have remarried for in 1690 the city of Paris placed an order for purses with "Robert Taillebois and Catherine Vaniere, his wife and formerly the widow of Pierre Alexandre Lefebvre, purse makers and merchants to the city." [70]

Purse-making required real skill in working with leather and fabric, in needlework and embroidery, and in other aspects of the craft. Purse makers obtained their raw materials from

62. See previous note. Delcamp and Jambu (2018) reveal the exceptional case of the guild mint workers in Bordeaux who for their reception were required to provide significant amounts of coffee and sugar, products coming into Bordeaux from the colonies.

63. The recent study by Marion Delcamp provides new details regarding the reception of workers in the Monnaie de Paris and of *tailleresses* in particular; in 1691 *tailleresses* owed 126 jetons worth a considerable 1,982 *livres*; in 1756 that sum had risen to an even more considerable 3,202 *livres*, but by 1767 fees had dropped to 318 *livres*; see Delcamp (2015a), pp. 503–04, 506.

64. Préau (1890a), pp. 127–28: CAEF Ms. A-1, 3a.

65. Lespinasse (1886–1897), vol. 3, pp. 401–20 provides these details; see also Valin (2016), vol. 1, pp. 214–16. The guild of *Boursiers, Tassetiers, Culottiers* grouped purse makers with other craftsmen who made and sold gaiters, chaps, and leather britches; they could sell gloves, but they could not make them, that a right reserved for the guild of glovers and perfumers.

66. A. Barthélemy (1875), p. 236 for the latter.

67. Lespinasse (1886–1897), vol. 3, p. 403. Still, the cost of a mastership in 1776 was 400 *livres*.

68. "... une bourse à jettons de velours bien et duement estoffée, avec une gibecière bien et duement faite fermante à serrure ou ressort, ou un brayer brisé," per letters patent given in Lespinasse (1886–1897), vol. 3, p. 420 n; see further pp. 401–20.

69. d'Affry de la Monnoye (1878), p. 287.

70. "A Robert Taillebois et Catherine Vaniere, sa femme, auparavant veuve de Pierre Alexandre Lefebvre, vivant M.[d] boursier fournissant la Ville....," quoted in d'Affry de la Monnoye (1878), p. 273; see also pp. 287 and 289 for other female purse makers.

other tradesmen, such as from velvet merchants or from the *fécandiers*, whose job it was to embellish fabric with silk, silver or gold threads.[71]

Jeton purses became evermore luxurious and expensive as time went on and as one moves up the social and institutional hierarchy. Simple black leather purses of the fifteenth century gave way to purses of white kid leather, then black leather lined with white leather, then white leather embroidered with green silk; then came green velvet purses, green satin ones, ones in green velvet lined with green satin, and other combinations, not infrequently interwoven with threads of silver or gold. Rarer and evidently of greater dignity were jeton purses of red velvet, similarly decorated with silver or gold.[72] Because of their luxury these purses required other purses—often of black velvet—in which they could be put for safekeeping. Claude Rossignol uncovered the 1688 case of a certain Henry Tresneau, "merchant purse maker of Paris, living near the clock of the Palais at the sign of the Bourse-Royale" who received an order from the Estates of Burgundy for 166 purses for their entire consignment of jetons that year.[73] The merchant purse maker promised to deliver within a fortnight, "nine purses of Genoese blue velvet, lined with pink Genoese satin, sown with lilies, and on the bottom of two of them will be embroidered the arms of their Serene Highnesses, Monseigneur the Prince and Monseigneur the Duke. The other seven will carry the arms of the province of Burgundy with drawstrings and tassels tufted with gold for a price of twenty-four *livres* apiece."[74] He further agreed to furnish at additional expense another purse in red velvet with the arms of the province, 36 more in half-crimson, half-green velvet lined with Genoese red satin with silk drawstrings and tassels with gold, and 120 plain leather ones with silk ties.

Such orders were not only large, they were expensive. In 1769 a master purse maker named Pontenier, purse maker to the city of Paris, received a large commission from the city for purses for its jetons for that year.[75] The order consisted of one jeton purse for the king in velvet knitted with gold with the arms of the king worth 96 *livres* along with its companion purse costing six *livres* made of white leather lined with taffeta, stitched with silver, and with a silver tassel to hold the official royal purse (Fig. 5.4).[76] Pontenier was to craft 51 additional purses in velvet and gold embroidered with the arms of the city at 24 *livres* apiece and five dozen plain leather ones for a total order worth 1,361 *livres* just for purses and the box in which they were delivered.[77] The purse makers' guild may have been marginal to the overall corporate/guild structure of the urban economy of Old Regime France, but privileged clients paid handsomely for the elegant and well-crafted products of their labor.

71. J. Guiffrey (1877-1880), vol. 2, p. 235 provides some sense of this division of labor in the payment to "Noel Michel, pour veloux vert à faire quatre bourses LXV[s], et à Jehanne, matelière broderesse, pour la façon et broderie des dites bourses, L sous"; see also reference in Feuardent, vol. 2, p. 434; F.10792.

72. Barnard (1916), pp. 83–84; Josset (1978), p. 337.

73. "M. Henry Tresneau, marchand boursier de la ville de Paris, demeurant proche l'horloge du Palais, à l'enseigne de la Bourse-Royale," Rossignol (1851), p. 52.

74. "...neuf bourses de velours bleu de Gênes, doublées de satin de Gênes, couleur de rose, parsemées de fleurs-de-lis d'or, au fond desquelles seront élevées en broderie, à deux bourses, les armes de leurs Altesses sérénissimes, Mgr le Prince, et Mgr le Duc ; et aux sept autres, les armes de la province de Bourgogne avec des cordons et houppes d'or, et tout pour le prix de 24 liv. pour chaque bourse," contemporary contract dated 17 December 1688, quoted in Rossignol (1851), pp. 52-53; see p. 164 for a similar order from the Estates of Burgundy in 1674.

75. See manuscript references and these details, AN, KK 1056[2] [Fêtes].

76. This purse ended up containing the 100 gold jetons the city of Paris offered Louis XV that year.The purse pictured in Fig.5.4 is a recent acquisition of the Département des Monnaies, Médailles et Antiques of the Bibliothèque Nationale de France, and the author thanks Ms. Inès Villela-Petit for drawing it to his attention. At one point the author encountered a similar royal purse held at the numismatic house O.G.N., and Thierry Sarmant lets it be known that the Carnavalet museum in Paris also possesses one.

77. The box cost three *livres*. The total for the order of 1771 came to 1,326 *livres*; see AN, KK 1056[2] [Fêtes], 1769 and 1771 payment orders.

FIGURE 5.4: A royal jeton purse.

Chambers of Commerce and the Coming of Capitalism

What we have seen of the guilds underscores that Old Regime France was not a modern capitalist or industrialized society. The guilds were the antithesis of free enterprise. By the same token, an independent economy and free merchant trading took root in Europe by the High Middle Ages, and the institutions of capitalism continued to develop in the world of early modern and Old Regime France. As far as this story is concerned, developments center on the dozen or so chambers of commerce that arose in cities across France in the early years of the eighteenth century.[78] The initiative came from the top and as part of central government efforts to get its arms around the economy. It principally involved a national-level Bureau de Commerce and the institution by Colbert of general inspectors of manufacturing and commerce.[79] In addition, a law of 1701 authorized the establishment of local chambers of commerce to organize and solicit advice from local merchants.[80] The chamber of commerce of Lyon dates from 1702, Rouen from 1703, Bordeaux from 1705, and others followed. Already in the creation of these institutions a certain tension existed between notions of free trade on the one hand and mercantile policies and government control on the other. Traditional centrifugal forces, as in Brittany again—but hardly as a voice of free enterprise—resisted the implantation of chambers of commerce by central authority. In the end what the chambers of commerce and other institutions of nascent capitalism show us is not the triumphant march of capitalism across the eighteenth century, but rather that business and proponents of free enterprise were suspicious of outside control yet very much on the outside seeking a way in and the favor of the more powerful governors of society and the economy.

Led by elected and variously titled directors, deputies, syndics, etc., eighteenth-century French

78. On contemporary chambers of commerce, see Bély et al. (2003), pp. 224–26, 319–20; Roche (1978), vol. 1, pp. 55, 280–84; Michel (1934), pp. 10–12; McClellan (1985), p. 38. Gillispie (1980), pp. 388–90ff., provides further background.

79. Several such bureaus preceded the one that operated continuously from 1722 to the Revolution; for detailed background see again Minard (1998), pp. 25–26ff.

80. The chamber of commerce of Marseille antedated this law in being formally established in 1650, and more than any other chamber it operated as an arm of the city and the crown in the control of trade with the Levant.

chambers of commerce were similar in organization. The chamber of commerce of Lyon can serve as an example, and it shows that "commerce" is tied to existing guild and trade activity, not so much to any self-conscious notion of laissez-faire capitalism, free markets, the profit motive, or business interests more as they might be thought of today.[81] The 10 directors served essentially *de jure*, and they included the local *prévôt des marchands*, a current-alderman, a former merchant-alderman, someone from the weavers' guild, two bankers or silk merchants, a spice merchant, a linen draper (*drapier*), a merchant from the thread or silver trades, and a merchant/manufacturer of silk cloth.[82] The chamber's secretary was always the town's secretary, and the chamber met at the Hôtel de Ville. In Rouen meetings were weekly. Annual general assemblies attracted numbers upwards of sixty to seventy-five attendees to elect the leadership of the chamber and, as important, merchants to serve as royal inspectors of manufacturing (*inspecteurs des manufactures*).

Unlike the contemporary agriculture societies which were weak and certainly lackluster when it came to jetons, the chambers of commerce waxed enthusiastic over them. Jetons were one of their major preoccupations. The chambers of commerce seem very status conscious and saw jetons and medals as institutional status symbols and a way to promote themselves and their interests. The first thing it did in Rouen, for example, the chamber of commerce spent 3,325 *livres* for 1,803 jetons.[83] The chamber soon ran out of jetons, and it made a new order with a private engraver in Paris for that many more, just to be reminded that only the Monnaie des Médailles could strike jetons. Because of the extra expense the chamber reduced the order to 500, but was soon commanding 2,000 jetons a year and a number of gold medals. On its jeton (Fig. 5.5) the king is featured as the protector of commerce (COMMERCII PROTECTOR). The chamber was concerned with the look of its jetons and negotiated carefully over every detail of cost and design, in one instance rejecting a design because it was dissatisfied with the size of the head of a snake in the image. Same story in Bordeaux, where upon its foundation in 1705 the chamber of commerce ordered up 1,200 jetons that it received in 1706. Working with the powers that be in Paris through its representative there, the Bordeaux chamber showed itself also to be punctilious about its jetons.

FIGURE 5.5: Jeton of the Chambre de Commerce of Rouen, 1712.

Naturally, jetons served as *jetons de présence* in assemblies. In the chambers of commerce of Lyon and Rouen, for example, each director received two silver jetons at the end of each meeting and a weighty gold medal at the end of his two-year term.[84] The same held true in Bordeaux where chamber directors collected up to 140 silver jetons a year, a gold medal for each year of service, *and*

81. Charvet (1878); Morin-Pons (1900), p. 50.
82. Lyon was the center of a vibrant silk industry, which accounts for this emphasis.
83. See Wallon (1897), pp. 10-52 for these details.
84. Morin-Pons (1900), p. 50; Charvet (1878).

another gold medal on their retirement from the chamber; the Bordeaux chamber handed out 10 gold medals a year to its leaders, including one to its agent in Paris.[85] But members had to attend meetings to get jetons, as became clear in the cases of two Bordeaux syndics and an inspector who wanted jetons and medals but who never came to meetings or performed their functions. They had friends in high places who pressured the institution, but the Bordeaux chamber ruled that to give the requested jetons and medals it needed explicit instructions from the controller-general or the intendant, and those instructions never came. In Rouen, the chamber gave 14 jetons to the estate of a dead syndic for the seven meetings he was not able to attend because he was ill and then died. An agent absent on official business in Paris was given jetons for meetings he couldn't attend. In December–January, 1730–1731, representatives of the Rouen chamber appeared before the controller-general to answer complaints lodged against the chamber, but they were tactful enough not to award themselves jetons for these meetings.

As *jetons de présence*, the jetons of the chambers of commerce of Lyon and Rouen and probably elsewhere served as *méreaux* that could be redeemed for cash. The Lyon chamber bought back jetons at their nominal value of two *livres*, five *sols* for redistribution.[86] In Rouen jetons could be handed in for cash (1 *livre*, 15 *sols* in 1706), so many of that chamber's jetons were thus recycled. In some instances the Rouen chamber's treasurer would buy back jetons and medals at prices higher than it cost to order them from Paris in order to make up for shortages. In effect, a secondary market arose for silver jetons and gold medals of the chambers of commerce. In 1709, for example, in Rouen a gold medal fetched 150 *livres*; 100 silver jetons went for 175 *livres*, and prices went up after John Law's bankruptcy in 1720 and the resulting scarcity of precious metals.[87]

As with other institutions, eighteenth-century French chambers of commerce distributed medals and purses of jetons widely to curry favor with other institutions and highly-placed individuals. Regarding the Bordeaux chamber again, town councilors each received a purse of 100 jetons from the chamber.[88] In 1708 at least the Archbishop of Bordeaux received a silver medal, and other jetons and medals were doled out in Paris, including a purse of jetons to its agent there. From 1706 to 1715 the chamber annually gave out an average of eight gold medals and 1,000 silver jetons; a long hiatus ensued from 1715 to 1750 because of lack of funds, but from 1750 through 1791 the Bordeaux chamber went back to distributing 1,000 silver jetons a year and eight gold medals.[89] For a while, for economic reasons, the Chambre de Commerce of Bordeaux used the dies of the Ville de Bordeaux for one face, but then, in 1783 the chamber, "understood that a great Company should have its own dies and have fashionable jetons struck," and so it stopped borrowing dies from the town and ordered more expensive octagonal jetons which it distributed in blue velvet purses embroidered in gold.[90] Henceforth, to pay for the additional expense of octagonal jetons, directors received only one purse every two years and only one gold medal. Later financial support from the province and from the king himself, "as a mark of his satisfaction for their services" improved the financial circumstances of the Bordeaux chamber.[91] In Lyon the chamber of commerce annually handed out 10 purses at a cost of approximately 2,000 *livres* a year.[92] In

85. On the Bordeaux chamber of commerce and its jetons, see Évrard de Fayolle (1900), here p. 55.
86. Charvet (1878), p. 23.
87. For these details see Wallon (1897), pp. 25–28.
88. Évrard de Fayolle (1900), p. 55.
89. Many were struck with an effigy of Louis XV well after his death; the dies stayed in Paris, but at a merchant's house.
90. "...comprit qu'une grande Compagnie devait avoir ses coins et faire frapper des jetons à la mode...," contemporary document quoted per Évrard de Fayolle (1900), p. 54; see also p. 50.
91. "...pour marque de satisfaction de leurs services...," contemporary document quoted in Évrard de Fayolle (1900), p. 54.
92. Charvet (1878), p. 18.

Rouen the chamber of commerce graciously acceded to the request of the visiting royal librarian, Armand-Jérome Bignon, for a set of its jetons and medals for the royal library.[93]

While perhaps epiphenomenal to the historical record, contemporary French chambers of commerce seem neurotically insecure and desirous of pleasing established authority. In Lyon, for example, the chamber of commerce expressed concern that the weight of its jetons not be heavier than for those of the town council (the *consulat*) in order not to offend them.[94] The chamber of commerce in Rouen seems to have been especially inclined to offer jetons and medals to those in power in order to secure its position and the good will of the authorities, at one point speaking of "the need we have for protection in the general and particular affairs of commerce," and the historian of the Rouen chamber remarks that "gold medals and likewise the purse of one hundred jetons were often used by the Chamber to acknowledge or assure the protection of the nobles or magistrates charged with the governance of the province."[95] In what seems almost like a bribe, the Rouen chamber decided to give the local intendant, a M. de Courson, who "having seemed until now favorably disposed to commerce and to merchants and having even given signs of this on several occasions, a purse of one hundred silver jetons to maintain his good disposition and as a small token of the Chamber's gratitude."[96] In 1715 the intendant received a purse of 100 silver jetons simply for the honor he bestowed on the chamber by chairing a meeting.[97] Apparently, some crisis in 1710 shook the Rouen chamber of commerce. In response, it sent a gold medal to Michel-Jean Amelot de Gournay (1655–1724), a king's counselor and president of the national Bureau de Commerce. It gave gold medals to the parents of the leading Norman nobleman, the Maréchal and Pair de France, Henri d'Harcourt (1654–1718), in order to persuade him to accept a gold one himself—he accepted a silver one. The governor of the province, the duc de Luxembourg, received a gold medal in a special box with gold nails and his insignia on it, and a velvet purse of 100 jetons went to a certain M. Chaillou, the duke's secretary. And, likewise in 1710, "as we have urgent need in the present circumstances of the protection and good offices of the Intendant," the chamber gave the intendant a gold medal and sent a purse of 100 jetons to his secretary.[98]

This toadying, particularly toward the local intendant, continued on the part of the Rouen chamber of commerce throughout the rest of the eighteenth century. The new intendant in 1721, for example, received a purse of 100 silver jetons and a gold medal worth 375 *livres* in a lovely presentation case made by the local clockmaker, Mlle Lemoine, for which she was paid an additional 100 *livres*; on the occasion, the subdelegate to the intendant got his purse of 100 jetons, as did the intendant's secretary.[99] Ditto in 1732 when the new intendant was offered a gold medal in a moleskin case garnished with gold nails with his coat of arms on one side and his cipher on the other;

93. Wallon (1897), pp. 40–43.
94. Charvet (1878), p. 18.
95. "...les besoins que l'on a de protection dans les affaires générales ou particulières du commerce"; registers of the Chambre de Commerce of Rouen quoted in Wallon (1897), p. 20 n; "...La médaille d'or fut souvent employée par la Chambre, comme la bourse de cent jetons, à reconnaître ou à assurer la protection des seigneurs ou magistrats chargés de gouverner la province...," Wallon (1897), pp. 18–19.
96. "M. de Courson, Intendant, ayant paru jusques à present dans la disposition d'estre favorable au commerce et aux negocians et en ayant mesme donné des marques en plusieurs occasions, et pour l'entretenir dans cette bonne disposition et luy en donner une petite marque de reconnoissance de la part de la Chambre, il a esté deliberé qu'il lui sera donné une bourse de cent jettons d'argent...," registers of the chamber of commerce of Rouen quoted in Wallon (1897), pp. 16–17.
97. Registers quoted in Wallon (1897), p. 21. The Intendant was nominal head of the chamber and presided when he attended.
98. "...comme on a extrêmement besoin dans la conjoncture presente de la protection et des bons offices de M. l'Intendant....," registers of the Rouen chamber of commerce for June 11, 1710, quoted in Wallon (1897), p. 20.
99. Wallon (1897), p. 29.

a purse of 100 silver jetons went to M. Girault, his secretary, "all in the name of the Chamber as is customary."[100] Again in 1743 and across the rest of the eighteenth century the Rouen chamber of commerce regularly distributed gold medals and jetons as favors to the powers that be.

The Rouen chamber gave out medals and jetons for projects more directly in line with its charge of economic development. A purse of 100 jetons, for example, went to a certain Le Tellier in 1768 for a project to improve the port of Rouen on the Seine. An engineer in the Royal Navy, Jean-Baptiste Degaulle, made a map of the Norman coast for which the chamber of commerce offered him a purse of 100 jetons; Degaulle accepted on the condition that he give the jetons to two poor students at the hydrography school at Rouen, and the chamber naturally accepted this offer. New lighthouses along the Norman coast fell under the auspices of the Rouen chamber, and in 1776 an engineer in charge of a new lighthouse project received a purse of 40 jetons; in 1786–1787, the inspector of the department of bridges and roads of Dieppe was granted a purse of 100 jetons for his services regarding lighthouses, and so on.[101]

The full story of contemporary French chambers of commerce remains to be written, but in the end they show themselves to be deeply involved in the culture of jetons, yet defensive, largely ineffective, and on the periphery of real power and control. That would change.

A handful of private banks, insurance companies, and mining and manufacturing establishments struck jetons through the eighteenth century, but these hardly figure in the larger story either of jetons or the history of capitalist institutions.[102] Granted formal recognition by the central government, for example, the major private bank in France prior to the Revolution, the *Caisse d'Escompte, Besnard et Compagnie*, arose only in 1776; it issued three undated octagonal jetons in silver and bronze in the 1780s that were distributed to shareholders who attended semiannual meetings.[103] *Jetons de compte* were struck for salt mines in the sixteenth century, but none after that. The odd iron or coal mining operation, most from the latter part of the eighteenth century, had jetons that were possibly used as accounting tools, as *jetons de présence*, or even as *méreaux* that could be exchanged for bread, wine, soup, fresh and cooked meat.[104] In the 1720s the state nationalized all mining operations under the *Compagnie Royale des Mines de France*, headed by the duc de Bourbon as grand master of mines and minerals, and this organization issued a handful of jetons in the 1720s and 1730s.[105] Beginning in 1666 and continuing through the first decade of the eighteenth century, authorities struck a number of jetons and medals in connection with the construction of the famous Canal du Midi linking the Atlantic (via the river Garonne) with the Mediterranean; some of these were struck in zinc and distributed as *méreaux* to workers.[106] Occasionally, individual superintendents of mining operations had personal jetons struck, but all of the jetons mentioned here were few in number and issued only sporadically; they are peripheral to the larger story of jetons, and they do not testify to any larger movement towards industrialization in France under the Old Regime.

100. "...une médaille d'or avec un étui en chagrin garni de clous d'or à ses armes d'un côté et à son chiffre de l'autre, et qu'il serait donné une bourse de cent jetons d'argent à M. Girault son secrétaire...le tout au nom de la Chambre suivant l'usage," registers quoted and paraphrased in Wallon (1897), p. 30.
101. These details from Wallon (1897), pp. 50, 51; on Degaulle, see McClellan and Regourd (2011), p. 509.
102. On the handful of jetons struck for pre-Revolutionary maritime, life, and fire insurance companies, most from the later part of the eighteenth century, see Gailhouste (2002), pp. 11–26. For the jetons of insurance companies, see F.4923–4939.
103. Babut de Rosan (1909), especially pp. 5–6. In addition, at the general assembly of April 16, 1778, the bank distributed a special medal to shareholders in attendance; for other jeton of banks, see F.4944–4962, F.6311–6318.
104. On these points see the catalog of Jules Florange (1904), passim and here especially #156.
105. Jules Florange (1904), #169, p. 46.
106. Depeyrot (1982), passim and here p. 83; see also Mukerji (2009).

Notably also, the larger French merchant houses involved in colonial trade and the slave trade, that of the Montaudouin brothers of Nantes or others in Bordeaux, for example, did not produce their own jetons. This absence is telling. Merchant trading involving coffee, sugar, and slaves was a major and growing element of the contemporary French economy and much more on the cutting edge of the historical development of capitalism than anything we have seen regarding jetons.[107] In the end, profit and economic/capitalist development was viewed with suspicion by the reigning powers in pre-capitalist Old Regime society. Nowhere is this sense more evident than in the scandalous token of the united merchants of the *Six Corps* (Fig. 5.6).[108] Feuardent dates this jeton as having appeared under Louis XIV. On the obverse it features seven individuals supporting the globe and on the reverse a beehive and a swarm of bees. The latter was a common trope for jetons whenever collective action was evoked.[109] But the notion embodied in the legends here was that of a united association of merchants combining their forces to support the earthly orb: STAT MUTUIS VIRIBUS (It stands firm through mutual strength) and SOCIETATIS BENE UNITAE (Belonging to a well unified society). Certainly in the era of the Sun King, as Feuardent put it, "It would have been very presumptuous that these merchants would have the idea that they held up the world through their commerce."[110] Being Atlas was the king's job and the job of the monarchy, not theirs.

FIGURE 5.6: Token of the united merchants.

Jetons became well integrated into the fabric of the society and culture of France in the Old Regime. Elsewhere outside of officialdom, we find a number of private organizations issuing jetons, for example, the Renaissance-style academy, the Académie des Lanternistes on the rue Bourglabbé in Paris dating from 1695 and modeled after earlier humanist groups in Italy.[111] Similarly, jetons were struck for the literary and poetic competitions of the Académie des Jeux Floraux of Toulouse. Private music and art schools, such as the *Société des Enfants d'Apollon*, struck jetons every so often that were used as admission tokens. A number of special interest and eleemosynary associations, such as the *Lycée des Arts* or the *Société Philanthropique* arose just prior to the Revolution, and they, too, had their jetons used as tokens for admission (Fig. 5.7). These also share a more modern iconography.

107. On the growing importance of colonial trade to the French economy at the end of the Old Regime, see Tarrade (1972).
108. F.5437–5439; see also Blanchet (1890), vol. 2, p. 426, but he certainly mistakes the date of this jeton.
109. Jambu (2013), pp. 185–86.
110. "Il eût été très présomptueux que ces marchands aient eu l'idée qu'ils soutenaient le monde par leur commerce," Feuardent (1904), p. 429.
111. Chautard (1890), pp. 499–500; on the taxonomy of academies see McClellan (1985), Ch. 1.

FIGURE 5.7: Jeton the *Société Philanthropique*, 1781.

Of the more interesting and revealing jetons of this sort were the ones struck for clubs of archers and marksmen.[112] These "privileged companies" arose in towns all over France from the early sixteenth century, with those for marksmen ultimately replacing those of archers and cross-bowmen by the end of the century. Supported by their local towns, these groups were independent of the army, but were deployed in wartime and called on for the defense of towns and to quash rebellions. They were privileged in being free from certain taxes. They held weekly practices with attendance at least once a month required for continued membership; their jetons served as *jetons de présence* and were given out as prizes to the best shots.[113] Annual competitions took place in Paris, Soissons, and elsewhere, and these events attracted large crowds, including magistrates; the town's mayor took the first honorary shot; members wore swords and badges. Targets were painted cardboard or wooden cutouts of birds (*papegai*) placed atop poles. In some cases competitions lasted four days. The winner was crowned "the Bird King" (*Roi de l'oiseau*) for the year, exempted from the *taille* and lodging soldiers, and in certain areas could sell a certain amount of wine. If the same individual won three years in a row he was declared "emperor" and held these privileges for life. In addition, believe it or not, larger regional and national competitions took place, and these were grand events with formally regulated masses and jetons handed out to participants and the winning company. Nine different companies competed at Bourges in 1619, for example; 14 companies totaling 243 marksmen entered the contest at Tours in 1753; the *Concordat général de l'Arquebuse Royale de France* of 1775 brought delegations from Brie, Picardy, Île de France, and Champagne. The competition of 1778 drew 152 marksmen. The winning company at the competition at Villeneuve-de-Lyon in 1767 was headed by a woman who that year won as *Roi* [sic] *du tir*. The society of Parisian marksmen had the statutory right to go and congratulate the king on happy occasions and to demand a present from him. The past is truly a foreign country, and the case of companies of French marksmen, suppressed during the French Revolution, provides a good example of how jetons can carry us deep into the bygone society and culture of Old Regime France (Fig. 5.8).

Finally, mention needs to be made of Masonic tokens struck in the second half of the eighteenth century coincident with the Masonic movement that swept through contemporary France and that affected a certain segment of enlightened, bourgeois, male society.[114] These coins are a largely separate phenomenon from other jetons discussed to this point. They are iconographically distinct without royal or other official insignia. They may have played a role in initiation ceremonies. They were certainly shared as tokens of sociability, friendship, and recognition among fellow Masons (Fig. 5.9).[115]

112. Blanchet (1930), pp. 295–96; Jules Florange (1899).
113. J. Florange (1889), p. x.
114. For background, see Darnton (1968).
115. Blanchet (1930), pp. 320–22; Brichaut (1884).

FIGURE 5.8: Jeton of the *Arbusquiers* of Lyon, 1741.

FIGURE 5.9: Masonic jeton: *Pont Parfait de l'Or. de Paris, 5760* (1759).

With this, we have surveyed the principal domains where jetons were to be found in the kingdom of France up until 1789: The royal court and the major organs of government and the military; the church; among the nobility; in the academies, learned societies, and certain faculties; in cities, towns and provincial estates; guilds, chambers of commerce, and yet others, such as Masonic groups or the marksmen of Lyon. It is worth underscoring that within these institutions jetons do not stand alone nor are they to be seen simply as material objects of another era, but, rather, they are to be associated with and shed light on certain social practices,...in action, in other words, such as their exchange at the Étrennes or in initiation ceremonies or, still, on counting tables across the kingdom. In all, again, like Blake's grains of sand, the jetons of the Old Regime allow us a view of a whole world we otherwise would not glimpse in quite the same way.

It remains next to examine two key institutions essential to the production and historical evaluation of jetons: the Académie des Inscriptions et Belles-Lettres that played a key role in the design of jetons and the Monnaie des Médailles, the special mint that actually produced them.

FIGURE 6.1:. Jeton of the Académie Royale des Inscriptions et Belles-Lettres.

6.

Aesthetics and the Special Role of the Académie Royale des Inscriptions et Belles-Lettres

Jetons are constructed and exist as material objects. Two different aspects went into their creation. On the one hand, designers concocted the *type* or *devise*: the imagery, legends, insignias, busts, and related iconography and inscriptions that came together in the design of a jeton. On the other hand, jetons had to be fabricated in the minting process at the Monnaie des Médailles. The next chapter looks at how jetons were actually minted there. This chapter attends to what went into the design of jetons and the key role played by the Académie Royale des Inscriptions et Belles-Lettres in this process (Fig. 6.1).

Prior to the establishment of what was known as the Petite Académie, the subgroup chosen from the Académie Française in Paris in 1663, the process of envisioning a *devise* or design for a jeton was entirely haphazard. In the sixteenth century and through the first half of the seventeenth, private, usually high-ranking savants, often with court or other connections, would propose *devises* for jetons. They performed this work in conjunction with a sponsor—a minister, an administrative department, the Cour des Monnaies, and other entities issuing jetons—or sometimes independently as an intellectual game.[1] We saw above that in the first decade of the seventeenth century, Henri IV and his minister, Sully, themselves composed the annual *devises* for the kingdom. Offered along with madrigals or the sycophantic pamphlet, lesser clients would sometimes craft *devises* for patrons, hoping for attention and promotion.[2] By the mid-seventeenth century these private *devisistes* commonly received honoraria for their suggestions first in cash, then as a purse of 100 jetons.[3] Pierre de Montmaur (1576–1648), professor of Greek at the Collège Royal, had a reputation for his *devises*, earning, for example, fifty *livres* in 1646 for the one he created for the city of Paris.[4] Intrigue and avarice were sometimes involved, as well as gaining a reputation as a *bel esprit*. Claude-François Ménestrier, for example, in 1686 complained of having offered a *devise* for the Secrétaires du Roi that was

1. Valin (2016) is especially strong on the creation of *devises* prior to the Petite Académie; see especially vol. 1, pp. 188–95; see also Veillon (1994), passim and pp. 28–29.
2. Loach (1995), p. 24, indicates that prior to the creation of the Académie des Inscriptions, officials in Lyon used local talent for *devises* for local jeton; Roche (1978), vol. 1, p. 123, gives examples of other provincial academies doing the same into the eighteenth century.
3. Veillon (1994), p. 29. Payment for devises began at end of sixteenth century: the successful author of *devises* for Ville de Paris received 50 *livres* in cash in the 1630s and 1640s...300 *livres* in 1660. From the 1670s forward the award was a purse of 100 jetons.
4. Veillon (1994) p. 28; Barnard (1916), p. 67.

rejected by the sponsor only to see the same one appear several years later with someone else getting the bourse.[5]

Louis XIV's great state minister, Jean-Baptiste Colbert, formed the Petite Académie within the Académie Française in 1663 in part to regularize and supervise choices of *devises* for jetons and medals, and as it morphed into the full-fledged Académie des Inscriptions et Belles-Lettres, this institution became nominally responsible for selecting the *devises* used for all French jetons and medals. Closer scrutiny is thus in order for the Académie des Inscriptions and its actions as the design authority for Old Regime jetons.

An August Authority

The official letters patent of the Académie des Inscriptions et Belles-Lettres of 1701 specified the body's role and authority as the commemorative agent for the crown. Article XX stated that "the said Academy, having been principally established to work on inscriptions and other monuments that have been created or that will be created to preserve the memory of great men and their noble deeds, it will continue to work on everything that concerns such works, such as statues, mausoleums, epitaphs, medals, jetons, *devises*, inscriptions on public buildings, and other works of a similar nature."[6] Article XX of a revised set of letters patent in 1787 repeated the point, saying, "The Academy, which has the institutional charge to consecrate for posterity the principal events of the reign of the King, will continue to work with the same zeal that it has up until now on medals, inscriptions and other monuments that His Majesty will consider assigning it."[7] But the letters patent went further in formalizing the academy's authority over *all* jetons, medals, and like commemorative works: "When the provinces, towns, incorporated bodies and even individuals wish to consecrate the memory of some important event worthy of being transmitted to posterity, they will request inscriptions or medals from the Academy, which will apply itself very assiduously to providing prompt and complete satisfaction."[8] The Académie des Inscriptions thus came to have a monopoly over the design of jetons and medals.

Through 1758, one way the academy fulfilled this assignment was, as we have seen, to provide the *devises* for the annual standing orders for the administrative jetons issued and paid for by the crown for various of its governing departments (i.e., Trésor Royal, Parties Casuelles, Ordinaire and Extraordinaire des Guerres, la Marine Royale and les Galères, and the Bâtiments du Roi).[9] But far more often, where funds and the occasion permitted, other court, bureaucratic, civic and cor-

5. Ménestrier (1686b), pp. 12–14, and pp. 15–27 for his unfortunate experiences creating a *devise* for the French clergy.

6. "Ladite Académie estant principalement establie pour travailler aux Inscriptions & autres Monuments qui ont esté faits, ou que l'on pourra faire, pour conserver la mémoire des hommes célébres & de leurs belles actions, elle continuera de travailler à tout ce qui regarde lesdits ouvrages : tels que sont les Statues, les Mausolées, les Epitaphes, les Médailles, les Jettons, les Devises, les Inscriptions d'Edifices publics, & tous autres ouvrages de pareille nature…," AIBL, H&M 1 (1736), Histoire, pp. 13–14, article XX.

7. "L'Académie chargée par son institution, de consacrer à la posterité les principaux évènemens du règne du Roi, continuera de travailler avec le même Zèle qu'elle a fait jusquà présent, aux Médailles, Inscriptions & autres monumens que Sa Majesté jugera apropos de lui ordonner," new letters patent of 1787, Article XX, copied in AI, AIBL archives, A:70, minutes of the meeting of January 9, 1787.

8. "Lorsque des Provinces, des Villes, des Corps, et même des Particuliers, voulant consacrer la mémoire de quelqu'évènement important & digne d'être transmis à la posterité, demanderont des Inscriptions ou des Médailles à l'Academie, elle S'appliquera très particulierement à donner une promte & entiere Satisfaction," AI, AIBL archives, A:70, minutes of the meeting of January 9, 1787, and letters patent of 1787, Article XXVIII. This, *nota bene,* parallels the sanctioned authority of the Académie des Sciences as the governor of science.

9. These standing orders to the Académie des Inscriptions for jetons continued until 1758, except for the Extraordinaire des Guerres, a series which continued through 1779. AIBL, H&M, vol. 1 (1736), pp. 2–4, 324 mentions these. On this activity within the academy, see also Sarmant (2003), pp. 109, 187; Mainguy (1999), pp. 12, 70; Veillon (1994), p. 31; AI, AIBL archives, A:109. "Jetons et médailles, 1715–1730."

porate bodies regularly approached the Académie des Inscriptions et Belles-Lettres for *devises* for their jetons and commemorative items. The list of these jetons is long and the dates extend across the eighteenth century: the Maison de la Reine (the Queen's Household, from 1726) and other royal households, the Menus-Plaisirs du Roi, the Argenterie, the Écuries (stables), Chambre aux Deniers, Artillerie, the Ville de Paris, the *prévôt des marchands*, the assemblies of the clergy and various clerical orders, the bishops of Nîmes and Castres, the Estates of Languedoc, Brittany and other estates, the city fathers of Lyon, Rouen, Toulon and others, the military order of Saint-Louis; chambers of commerce, the *Officiers du Guet à cheval* (mounted watchmen), and so forth.[10] Year in, year out, decade after decade. The academy quickly became aware of its increased workload, remarking in 1736:

> When the Company grew to forty academicians [in 1701], our work on *devises*, inscriptions and medals, far from diminishing, on the contrary became even more extensive than it was before. This was because the Company, being better known, more numerous, and *especially with greater authorization*, requests came in not only for everything having to do with the king or official public matters, but the academy was consulted even more from the depths of the provinces by towns and by individuals having some taste in these matters, and often even by foreigners.[11]

Be it noted that the academy's work on *devises* for jetons and medals formed only part of a larger picture of its collective activity in providing inscriptions and learned recommendations for all sorts of commemorations, and the complete list of solicitations dealt with by the Académie des Inscriptions et Belles-Lettres in responding to requests for its expertise is little short of astonishing in the range of matters involved. In no particular order consider: medals galore; the design of royal tapestries produced by the royal Gobelins dye works; inscriptions for the great gate at Bayonne; an epitaph for Mme the Duchesse d'Arpajon; road markers in the Cévennes; plaques demarcating the meridian from Paris to the Pyrenees as laid out by the astronomers of the academy of sciences and the Observatoire Royal; an inscription for the main door for a military fort the king was having built in the colony of Saint Domingue in the Caribbean; a medal commemorating "the establishment of the port of Louisbourg on Isle Royale at the mouth of the Saint Lawrence River" and a plaque set at the entry into the town of Louisbourg; ditto in 1772 for a medal and a plaque for new fortifications on the island of Guadaloupe; inscriptions for commemorative plates to be set up by French astronomers on their expeditions to Lapland and Peru in the 1730s; the standards for several regiments; for a fountain in Aix; foundations of the new cathedral at Metz; for the mausoleum of Stanislas in Nancy; an *arc de triomphe* at Bordeaux; another for a gate in Saint-Malo; in 1726 the Ville de

10. H&M of the AIBL specifies the *devise* requests the academy received each year and the actions it took, e.g., AIBL. H&M, vol. 1 (1736), p. 324, where we read: "Outre les jettons du Thrésor Royal, des Parties Casuelles, de l'Ordinaire & de l'Extraordinaire des Guerres, des Bastiments, de la Marine & des Galéres, il s'est passé peu d'années où l'on n'ait encore demandé à la Compagnie des devises pour ceux de Madame la Duchesse de Bourgogne Dauphine, pour ceux de la ville de Paris, de l'Artillerie, de la Chambre aux Deniers, des Menus-plaisirs, de l'Argenterie, &c...."

11. "Quand la Compagnie eût esté augmentée jusqu'au nombre de quarante Académiciens...le travail commun des devises, des inscriptions & des médailles, loin de cesser, devint au contraire plus estendu qu'il ne l'estoit auparavant; parce que, la Compagnie estant plus connuë, plus nombreuse, & sur-tout plus autorisée, on continua à luy demander, non seulement toutes les choses de ce genre qui avoient rapport au Roy, ou au public, mais qu'elle fut encore consultée du fonds des Provinces par les villes & par les particuliers qui avoient quelque goust, souvent même par les estrangers...," AIBL, H&M, vol. 1 (1736), p. 325, "Devises, Inscriptions et Médailles faites par l'Académie." Author's emphasis in the translation. Original also quoted in d'Affry de la Monnoye (1878), p. 379.

Paris wanted suggestions for a fireworks display. A reservoir. A new road. A new bridge. A new hospital.[12] The list went on and on.

As the example of the Metz cathedral suggests, not infrequently medals were ceremoniously set into the foundations of buildings, particularly larger public buildings, as time capsules of a sort, and it was not unusual for the academy to be asked to provide some appropriate inscription for one or another for these constructions.[13] The surgeons requested the like for their new surgical amphitheater; the new Hôtel des Monnaies on the quai de Conti had a medal placed in its foundations where it presumably remains today. Commemorative medals were buried in the foundations of the Louvre restorations. Medals were placed in the foundations of the new headquarters of the Compagnie des Indes in Lorient.[14]

The academy was correct to note that requests came from outside of France as well. The Spanish Crown in particular regularly asked to be provided *devises* for jetons and medals for the kings and queens of Spain. The academy received solicitations from and for the Grand Order of Malta; the antiquarian of the Holy Roman Emperor wrote in; Austrian authorities wanted inscriptions for pedestals to be erected in a square in Vienna to mark the marriage of the dauphin to Marie Antoinette. In 1720 Russian authorities did the same for an inscription for a pedestal for Peter the Great in Saint Petersburg.[15]

Americanists will note that in 1785 and again in 1789 the U.S. Congress meeting in New York approached the Académie des Inscriptions et Belles-Lettres in Paris with requests for inscriptions for commemorative gold medals for heroes of the American Revolution.[16] The overture in 1785 came through David Humphreys, a former colonel in the American forces, for congressionally mandated medals for generals Washington, Horatio Gates, and Nathanael Greene. Later that year Congress voted further medals for General Daniel Morgan, Colonel William Washington (1752–1810, a cousin of George Washington) and Colonel John Eager Howard.[17] The request of 1789, this one transmitted through the American ambassador, Thomas Jefferson, was for designs and inscriptions for medals for General Anthony Wayne, Major Walter Steward, and Commodore John Paul Jones.[18]

Generally speaking, the academy did not involve itself with medals, jetons or inscriptions for individuals outside of the academy itself, only for other institutions, but it made an exception in the case of the Marquis de La Galissonnière (1693–1756) whose family wanted an epitaph to

12. "En 1720 on fit une Inscription & une Médaille, demandées par le Conseil de la Marine, au sujet de l'établissement & la construction du port de Louisbourg dans l'Isle Royale...," AIBL, H&M, vol. 5 (1729), p. 362; see further AIBL, H&M, passim, and manuscript minutes of academy meetings. For a similar range of work undertaken in the seventeenth century by the Petite Académie, see Wellington (2015a), pp. 107–23.

13. For a medal to be placed by the king in the foundations of the church at the École militaire, see AI, AIBL archives, A:65, minutes of the meetings of May 6 and 31 and June 14, 1768. For medal to be placed in the foundations of the new Hôtel des Monnaies, see AI, AIBL archives, A:65, minutes of the meeting of 16 July 1770; A:68, minutes of the meeting of July 29, 1783, for a medal to be placed in a dam.

14. See previous notes and Guiffrey (1881–1901), vol. 1, p. 189.

15. AIBL, H&M, vol. 3 (1736), "Devises, Inscriptions et Médailles faites par l'Académie," pp. 288–90.

16. On this episode, see AI, AIBL archives, A:68, minutes of the meetings of April 8, 19, 22, and 26 and May 13, 1785. The academy provided Latin inscriptions; Shattuck (2015) details the medals in question and pictures two of them, clearly the design products of the Académie des Inscriptions et Belles-Lettres; per the website, Medals of America (www.medalsofamerica.com/content--name-History-of-The-Revolutionary-War-Military-Medals; accessed September 23, 2018), these medals were delivered in 1790. See above, Ch. 4, for Benjamin Franklin's earlier contact with the Académie des Inscriptions et Belles-Lettres.

17. AI, AIBL archives, A:68, minutes of the meetings of November 25, 1785 and December 13, 1785.

18. AI, AIBL archives, A:71, *plumatif* and formal minutes of the meetings of January 13 and February 10, 1789.

honor his grave; the academy decided to provide one because La Galissonnière had been a member of its "sister" institution, the Académie Royale des Sciences.[19]

In sum, who cannot feel for the poor academicians of the Académie des Inscriptions et Belles-Lettres to be so widely put upon?

At Work on Emblems and Legends

Forty academicians at any one time made up the Académie Royale des Inscriptions et Belles-Lettres. They were all very learned men, expert in a variety of different fields of humanistic scholarship. They were the inheritors of Renaissance humanism and thereby the literary and scholarly worlds of ancient Greece and Rome. They all knew Latin, of course, and many were Latinists of the highest order.[20] In terms of erudition they were the men most qualified to set the standard for proposing medals, jetons and commemorative inscriptions to carry on into posterity. And it took erudition because of the critical standards involved and the public arena in which work was judged (in certain circles at least).[21] The great French playwright, Jean Racine (1639–1699), for example, was a member of the Petite Académie and composed *devises* for medals and jetons, as did the noted poet Nicolas Boileau (1636–1711) and other of the learned, including several abbés.[22] Some academicians, such as the numismatist Jean-François Simon (1654–1719), were thought to be especially artful when it came to composing *devises*.[23]

Working up *devises*, settling on designs for jetons and medals, concocting inscriptions, and like activities took considerable time at meetings of the academy and constituted a major part of the collective endeavor of academicians of the Académie des Inscriptions et Belles-Lettres.[24] Procedures varied over the course of the eighteenth century. Individual academicians always seemed able to make suggestions for *devises*. At an early date academicians apparently filled out forms for suggesting and voting on *devises* for the standard administrative jetons.[25] For a while in the 1720s and 1730s, with Gros de Boze as permanent secretary, academicians were not as diligent as they should have been, and much of the work on *devises* fell to de Boze who prodded his colleagues for help.[26] Particularly in the second half of the century the academy appointed ad hoc "commissions" of four members to handle *devises* and inscriptions. At other times the academy worked as a committee of the whole, but at least in one instance in 1785, it voted to go back to creating commissions to handle these matters because they took up too much time in regular sessions.[27] Later in 1787, the academy bemoaned the fact that the same four commissioners were usually appointed, and there was fear that there would be no one left in the Academy after a while with the expertise to properly come up with designs and legends. The academy then voted that every *devise* commission be raised to eight members in addition to officers of the academy.[28] When the matter came

19. AI, AIBL archives, A:63, minutes of the meeting of July 21, 1758. The academy accepted involving itself in a bequest for a prize contest, but refused to allow the sponsor's bust to appear on the medal; AI, AIBL archives, A:68, minutes of the meeting of July 8, 1785.
20. Blanchet (1930), p. 225. Blanchet lists notable AIBL academicians and jeton devises and images for which they are responsible.
21. Blanchet (1932).
22. On medals and jetons with devises created by these men, see Jacquiot (1961), (1962), and (1980), p. 860; see also Jacquiot et al. (1970), pp. 33–40; Veillon (1994) p. 28.
23. Veillon (1994), p. 31.
24. Sarmant (2003), p. 187; AI, AIBL archives, A:08, "Notes diverses et dessins de médailles."
25. See AI, AIBL archives, A:08, "Jetons et médailles, 1715–1730."
26. Guichon de Grandpont (1854), p. 7.
27. On this point see AI, AIBL archives, A:68, minutes of the meeting of April 19, 1785.
28. AI, AIBL archives, A:70, minutes of the meeting of July 24, 1787.

up again in 1788, academicians reaffirmed that eight commissioners be appointed for all business having to do with medals, jetons, epitaphs, etc.[29]

Ordinarily, inscription and *devise* committees met in the academy's quarters after the regular meetings, but in 1783 the question arose if they could meet outside the academy and, if so, whether that had to be at the home of an officer or if it could be the home of an ordinary member of the commission. That such an issue would arise bespeaks the suspicious, nit-picking tenor of the institution and other academies like it. The academy voted to allow any commission to meet anywhere it chose.[30] But the issue proved contentious, settled only when the academy again ruled by a formal vote the next year that commissioners could meet whenever and wherever they liked as long as they fulfilled the commissions assigned to them in a timely fashion.[31]

A related incident over a medal to celebrate the queen giving birth in 1779 shows this negative and picayune side of the academy in dealings with jetons and medals. Academician-commissioners were appointed by vote of the academy to come up with proposals. In this case the commissioners chose their own top three and passed these directly to state minister Malesherbes. The rule in the academy, articulated formally in 1754, was that the commissioners should choose from among suggestions made by any academician as well as by the commissioners themselves and that these suggestions were then then supposed to be presented to the Academy itself for ratification, which was not done in this case. Given that in the meantime academicians had proposed two additional possibilities, the question became whether the academy should start the process over again. After some apparently ardent discussion, the group resolved that the three suggestions arrived at by the commissioners, in effect now presented to the Academy, would be approved retroactively.[32]

As various commissions proceeded to do their work year after year, the reasonable fear developed of repeating *devises*.[33] To help avoid this, in 1710 the otherwise scandalous academy treasurer, Jean-François Félibien, whom we encountered earlier absconding with jetons of the academy, compiled a list of *devises* that the academy had used up until then.[34] As late as 1785 the academy was embarrassed to discover that many of *devises* it had concocted for a medal to commemorate the birth of a duke of Normandy had been used previously to commemorate the births of sons and grandsons of Louis XIV. To avoid such pitfalls and the painful research required not to make such mistakes, the academy requested a complete set of the metallic history of the kingdom as necessary for its work.[35]

Proposals for medals and jetons would specify an obverse of a standard bust and inscription. Proposals for reverses stipulated the image or *type* (e.g., Minerva holding a globe) and a suitable legend. These were overwhelmingly in Latin, although some were in French.[36] In 1784 the merchants of Paris (the *Six Corps*), for example, asked the academy for a commemorative medal for the recent peace, and it requested the legend in French.[37] The same year the city of Salon solicited

29. AI, AIBL archives, A:70, minutes of the meeting of March 4, 1788.
30. AI, AIBL archives, A:68, minutes of the meeting of December 19, 1783.
31. AI, AIBL archives, A:68, minutes of the meeting of January 16, 1784: "qu'elle n'a aucun droit d'exiger de Ses Membres qu'ils travailent plutôt dans un endroit que dans un autre, & qu'il doit lui suffire que les Commissaires remplissent leurs missions, et en rendent compte dans un temps convenable."
32. See AI, AIBL archives, A:67, minutes of the meetings of January 8, 12, and 15, 1779.
33. On this point see Veillon (1994), p. 31; Jacquiot (1980), pp. 865–66, n. 11.
34. Josset (1978), p. 337.
35. One this episode, see AI, AIBL archives, A:68, minutes of the meeting of June 17, 1785.
36. See list of over 6,000 legends in Feuardent, vol. 3, pp. 363–431. Note the difficulties/concerns about putting the names of towns into Latin, per Jacquiot (1980), pp. 862–63.
37. AI, AIBL archives, A:68, minutes of the meeting of January 27, 1784; Ménestrier as early as 1686 expressed his indifference as to whether inscriptions are in Latin or in French; see Ménestrier (1686), p. 30,

an inscription in French for the bust of a local notable. The academy formally deliberated and decided by ballot that it would prepare inscriptions in French when so requested.[38] Despite what has been said of the expertise of academicians, some critics complained that the Latin of these legends and inscriptions often strayed so far from classical purity to approach a scene of gibberish Latin in Molière.[39]

The Académie des Inscriptions et Belles-Lettres generally worked on *devises* from mid-November to mid-December, in time so that jetons could then be minted at the Monnaie des Médailles and made available for the first of the year. Virtually all academicians participated in this exercise.[40] As noted, all proposals for medals, jetons, and commemorative inscriptions had to be presented to the academy as a whole. Commissioners would usually present 8–10; 15 members made suggestions for a set of sponsored jetons in 1709.[41] Suggestions by commissions were sometimes modified in full assembly.[42] Everyone always had a hand in the final decisions. The academy voted on *devise* suggestions with decisions made by the plurality of academicians in attendance.[43] The top three proposals were then forwarded to the minister for final approval, and this activity continued across the eighteenth century. The end result in each instance was a written report such as the archival document of 1774, "*Devises* provided by various members of the Company in 1774 for the Extraordinaire des Guerres."[44]

> The 12th, 13th, and 14th suggestions were presented to the minister. The 14th, DONEC TUBA, etc, was accepted.
> #12: An olive tree, symbol of Peace, with different weapons around it in the manner of a hedge and the words: NE QUIS NOCEAT [Lest anyone do harm]. 15 votes.
> #13: A laurel branch and an olive branch with its fruit intertwined with the words: CONSOCIARI AINANT [They love to be partnered]. 11 votes.
> #14. A soldier seated beneath an olive tree with his weapons laid on the ground at his side: DONEC TUBA FUGERE COGAT [Until the trumpet sounds the retreat]. 10 votes.

The minister chose #14, the least preferred in the academy.[45]

38. AI, AIBL archives, A:68, minutes of the meeting of November 16, 1784.
39. Nocq (1910), p. 169, who is clearly exaggerating: "Les termes et la syntaxe s'éloignent souvent de la pureté classique pour se rapprocher avec hardiesse du langage médical de la 'Cérémonie' du *Malade imaginaire*." Murray McClellan, private communication, agrees about the somewhat barbaric quality of Old-Regime inscriptions compared to their predecessors in classical antiquity.
40. Veillon (1994), p. 30.
41. See AI, AIBL archives, A:109: "Jetons et médailles, 1715–1730."
42. See, for example, AI, AIBL archives, A:70, minutes of the meeting of March 11, 1788.
43. AI, AIBL archives, A:66, minutes of the meeting of June 5, 1772.
44. "Devises fournis par divers membres de la Compagnie en 1774 pour l'extraord. des guerres. Les 12e, 13e, et 14e ont été présentées au Ministre. La 14e admise, Donec tuba &. – # 12, Un olivier, symbole de La Paix, entouré de différentes armes, comme d'une haye, avec ce mot: NE QUIS NOCEAT [15 votes] – #13, Une branche de Laurier et un d'Olivier distinquée par son fruit, entrelassées, avec ce mot: CONSOCIARI AINANT [11 votes] – #14, Un soldat assis au pied d'un olivier, avec ses armes posées à terre à côte de Lui: DONEC TUBA FURGERE COGAT [10 votes]," AI, AIBL archives, A:08. See also AIBL, minutes of the meeting of September 5, 1775 for the jeton of the Extraordinaire des Guerres for 1776, where the names and the votes of the commissioners are listed. Other documents of this sort remain scattered in the archives of the academy.
45. This legend was used for four dated and undated 1775 jetons of the Extraordinaire des Guerres in silver and copper with two different busts and two different engravers; Feuardent has the soldier seated beneath a laurel tree, not an olive tree; see F.890–893.

Once officially approved, the exercise became a formal project.[46] The Académie des Inscriptions et Belles-Lettres had its own in-house artist (*dessinateur*) of great caliber who would take the legends and designs suggested by the academy and turn them into a sketch for use by engravers at the mint.[47] Sketches were circulated to the interested parties, and they could be returned for revision.[48] The position of *dessinateur* within the Académie des Inscriptions was a prestigious one obtained through a royal appointment and with its holders coming from the upper echelons of the Académie Royale de Peinture et de Sculpture. Sébastien Leclerc (1637–1714), for example, official royal engraver attached to the Cabinet du Roi and academician at the painting and sculpture academy held the *dessinateur* post within the Académie des Inscriptions and worked with Racine in the 1680s and 1690s in designing jetons and medals, producing sketches in color.[49] The royal history painter and director of the Académie de Peinture et de Sculpture, Antoine Coypel (1661–1722), held this post in the decade after Leclerc's death and was in essence the royal official in charge of royal imagery. According to Nicolas de Launay, director at the Monnaie des Médailles, Coypel was "solely in charge of the sketches of medals and jetons."[50] Regarding a jeton of 1721 for the Argenterie, for example, the academy left instructions that "M. Coypel must be approached for the sketch of this jeton because it's he who ordinarily does them all, and in these small details as well as the larger sweep [of the design] no one is sharper than he is."[51] The royal painter, Louis de Boullogne (1657–1733), succeeded Coypel as director de l'Académie Royale de Peinture et de Sculpture and as *dessinateur* in the Académie des Inscriptions; he, too, was "the only one charged to make the sketches for medals and jetons."[52] Later in the century Louis-Claude Vassé (1717–1772), a royal sculptor of the Académie de Peinture, held the position, and he was succeeded in 1773 by the sculptor Augustin Pajou (1730–1809).[53]

Ministerial Approval

The secretary of the Académie des Inscriptions transmitted the results of the deliberations of the academy to the appropriate minister.[54] No government-sponsored jeton or medal was struck without prior approval by the king himself or at least a royal minister or intendant.[55] Certainly for the administrative jetons, the fact that a minister—generally that of the Bâtiments du Roi, decided on proposals for royal jetons and medals says everything about the academy and its connections to

46. E.g., AI, AIBL archives, A:109, a single sheet: "Projet arresté par l'Academie des Belles Lettres Pour le Jetton de L'Argenterie 1721." Per Valin (2016), vol. 1, p. 16, an important collection of these preliminary designs is Ms. 5331 held at the BnF, Bibliothèque de l'Arsenal.
47. See for example, payments made to various artists, including Sébastien Leclerc; Guiffry (1881–1901), vol. 2, col. 985 [1686]; vol. 3, col. 90. [1688]; col. 991 [1694]; vol. 4, col. 49 [1696].
48. On this back-and-forth process as concerns medals, see Villela-Petit (2015b).
49. Jacquiot (1980), p. 866, n. 18.
50. "[le] seul chargé de faire les dessins des Medailles et des Jetons," Nicolas de Launay, "Journal historique de la Monnoye des Médailles, 1697–1725," CAEF, Ms. F° 80, vol. 2, p. 18.
51. "Il faut s'adresser à M. Coypel pour le dessein de ce jetton, parceque c'est lui, qui, ordinairement les fait tous, et que dans ces petits détails là, comme dans les grands sujets, personne n'a plus d'intelligence ~~que lui~~ [*sic*]," AI, AIBL archives, A:109, single manuscript sheet: "Projet arresté par l'Academie des Belles Lettres Pour le Jetton de L'Argenterie 1721," in folder, "Inscriptions et devises, 1722," itself in folder "Devises de jetons et médailles 1716. 1730."
52. "...il est seul chargé de faire les Dessins des Medailles et des Jetons," Nicolas de Launay, "Journal historique de la Monnoye des Médailles, 1697–1725," CAEF, Ms. F° 80, vol. 2, p. 92.
53. For Vassé, see AI, A:65, membership list heading "Registre des Assemblées Et Délibérations de l'Académie Royale des Inscriptions Et Belles-Lettres pour l'Année 1770." For Pajou's appointment, see AI, AIBL archives, A:66, minutes of the meeting of January 8, 1773.
54. Josset (1978), p. 337.
55. Blanchet (1930), p. 223; La Tour (1897), p. viii; for permissions from Cour de la Monnaie delivered in the sixteenth century for the Chambre aux Deniers and the Maîtres d'hôtel, see Mazerolle (1888), pp. 361, 452.

the upper levels of government and the court. The academy served first and foremost as an arm and agent of royal authority, but the ultimate power was above their pay rank; they were on tap, but not on top. The state secretary of the Bâtiments du Roi, to whom it reported, had both the Académie des Inscriptions and the Monnaie des Médailles in his portfolio, the two units that produced jetons, and the rules required him "to examine *devises*, and should there be a preposterous one among them that does not at all suit the subject or that might have consequences, depending on the circumstances, he will reject them or have them redone."[56] Jetons were scrupulously supervised for fear of something inappropriate coming into play.

Sometimes documents were marked "*agréé par Sa Majesté*," although it may well have been the minister who did this.[57] At other times a scrawled ministerial or royal "*bon*" let the decision be known.[58] High-ranking men of state intervened on the choices for *devises*, and other authorities sometimes overruled *devises* chosen by the Académie des Inscriptions; sometimes decisions did go all the way to the king.[59] In one instance three suggestions were submitted to the king who chose the legend from one suggestion and the *type* or image from another.[60] In yet another case the king chose, but made a change.[61]

But the interaction was not one-way, moving from the Académie des Inscriptions upward to ministerial heights. Things passed in both directions, and in the seventeenth century all three succeeding chief ministers of the king—Richelieu, Mazarin, and Colbert—were known to meddle.[62] Apparently, Colbert was particularly responsible for the design of the jeton of the Bâtiments du Roi for 1683, the year he died. The Abbé Bignon offered suggestions to the academy for appropriate subjects and gave orders for work.[63] Sometimes things could be tricky, as in 1758 when the Artillerie was separated from military engineering (*génie*), but one paying office served both, so one jeton or two? (The answer came back that the king was suppressing standard administrative jetons altogether after 1758, so the point was moot).[64]

Indicative of the back and forth between the Académie des Inscriptions and higher levels of authority were the frantic preparations for royal births and having medals set to go before the event. In September 1781, for example, the Académie des Inscriptions received the following rather direct letter from the head of the Bâtiments du Roi, Antoine-Jean Amelot de Chaillou (1732–1795):

> I thank you, sir, for the information you have provided me concerning the medals that were struck at the time of the late queen's confinements in 1727 and 1729. I have given an account of this to the King. His Majesty wishes that the Academy immediately take up the subject of a medal that we wish to have all ready for the moment of the Queen's delivery in case she gives birth to a Dauphin. His Majesty's intention is that this medal appear the day after the Birth of the Prince. He gave orders to have the dies engraved except for the names and the date to be added, which can be done

56. "Il examine les devises, et s'il y en avoit quelqu'une absurde, qui ne convint point au sujet, où [*sic*] qui put avoir des conséquences, selon les circonstances, il les rejette où les fait reformer...," AN O¹ 2064, 1744 manuscript labelled "Memoire concernant les Médailles." This document makes plain these sensitivities throughout.

57. See, for example, AI, AIBL archives, A:66, minutes of the meeting of September 6, 1774.

58. Ms. labelled "Devises pour le Roy," AI, AIBL archives, A:109.

59. Sarmant (2003), pp. 69, 187–88.

60. AI, AIBL archives, A:67, minutes of the meeting of September 7, 1781.

61. AI, AIBL archives, A:70, minutes of the meeting of August 28, 1787.

62. Barnard (1916), p. 67.

63. AI, AIBL archives, A:109: Letter signed "Rochard de Lextraord.re des guerres. Ruë bavadier, vis avis lhôtel de soubise" and dated "Ce 23 9 1720."

64. AI, AIBL archives, A:63, minutes of the meeting of August 18, 1758.

quickly. I request that as soon as possible you bring His Majesty's intentions before the Academy that it might fulfill them.[65]

Three days later at its next meeting the academy scurried to make three proposals for this medal for the case of a male birth. The king chose the image from one (France seated on an antique pedestal with the dauphin on her knees or in her arms) and the legend from another (FELICITAS PUBLICA, civic happiness).[66] Indicative of the same set of sexist, Salic preferences, back in 1727 Gros de Boze transmitted to the academy on the part of the then head of the Bâtiments du Roi, the duc d'Antin, a request for proposals for *two* medals pending the outcome of the queen's pregnancy. "Its *type* should vary as the fruit of this delivery. If it's a princess, we only need a simple medal celebrating fecundity; if it's a Prince, as we wish, one can mark that his birth not only brings happiness to France but to the whole of Europe."[67] The queen gave birth to twin girls, and the fecundity medal issued, but the academy held in reserve the design for the medal for the prospective dauphin.

In the end, the way to see the interaction of the Bâtiments du Roi, the Académie des Inscriptions, and to include the Monnaie des Médailles is as three units coordinating in the design and production of jetons and medals under the close supervision by powerful authorities at the very center of government and monarchical rule.[68]

Two Unhappy Clients

The work of the Académie des Inscriptions et Belles-Lettres was hardly limited to these dealings with central government, but involved the whole range of clients mentioned above and many others besides. The academy constantly had dealings with a variety of outside parties and their agents who sought *devises* sanctioned by the academy. The record suggests that these interactions generally worked well, but not always.

One unhappy example of such dealings concerns the first jetons of the Bordeaux chamber of commerce mentioned earlier, an institution that seemed unaware of the authority of the Acadé-

65. "Je vous remercie, Monsieur, de renseignements que vous m'avès donnés sur les Medailles qui ont eté frappées lors des couches de la feu Reine en 1727 et 1729. J'en ai rendu Compte au Roi. Sa M. desire que L'Academie s'occupe très incessament du sujet d'une Medaille qu'on Voudroit avoir toute prête pour le moment des couches de la Reine, en cas qu'Elle accouchât d'un Dauphin. [¶] L'intention de Sa M. etant qu'elle paroisse le lendemain de la Naissance du Prince. Elle a donné ses ordres pour faire graver les coins de maniere qu'il ny'a aura plus que les noms et la date du jour à ajouter, ce qui seroit promptement fait. Je vous prie de mettre le plutôt possible les intentions de Sa M. sous les yeux de L'Academie afin qu'elle puisse les remplir," AI, AIBL archives, A:67, minutes of the meeting of Tuesday, September 4, 1781, that transcribe Amelot's letter of August 28, 1781; see also minutes of the meeting of August 31, where this matter comes up for the first time. The event in question, in fact, concerned the birth of Louis-Joseph, the Dauphin of France, who was born on October 22, 1781 and died of tuberculosis on June 4, 1789.

66. For the academy's three proposals and the king's choice, see variously AI, AIBL archives, A:08, Ms. "1781. Medailles arrêtées dans la séance du 7 Sept. 1781 dont l'une doit etre présentée à la Reine le lendemain de son acouchement si elle mit au jour un Dauphin," A:08, packet labelled, "1781. Types d'une medaille pour l'accouchement de la Reine, si elle donne un Dauphin"; and, A:67, minutes of the meetings of September 4 and 7, 1781, where the three choices are described further "en cas qu'Elle accouchât d'un Dauphin."

67. "en 1727, Mardi 17 juin Deux médailles demandées, que l'on voudroit avoir prêtes pour le lendemain des Couches de la Reine. Le type en doit être varié autant que le peut être le fruit qu'elles produiront. Si c'est une princesse, il ne faut qu'une simple médaille sur la fécondité; si c'est un Prince, comme on le souhaite, on pourra marquer que sa naissance fait le bonheur non seulement de la France, mais encore de l'Europe entiere (demande faite par M. de Bose de la part de M. le Duc d'Antin)," AI, AIBL archives, A:08, undated single sheet, "Notes diverses et dessins de médailles." See similar activity surrounding a royal birth in 1753, A:08, folio headed "1754. Sujets d'une Médaille pour la Naissance de Mgr. Le Duc d'Aquitaine. Né le 8 sept. 1753 cette année."

68. Blanchet (1930), p. 231, makes this point. The academy was supposed to receive a finished medal for each project it worked on, but this was not always the case; see complaints AI, AIBL archives, A:67, minutes of the meetings of December 7, 1781 and March 15, 1782.

mie des Inscriptions.[69] It's true that it sometimes took the academy months or even years to come up with the appropriate Latin.[70] But Paris twice rejected all of the suggestions from Bordeaux for *devises* for prospective jetons and medals, including a view of the port of Bordeaux that the chamber wanted. Possibly hinting at the true cause of the delay, the chamber expressed the worry that "it would be a shame for the Chamber if the *devises* were not to the tastes of the connoisseurs we have in the province."[71] The Paris academy then dragged its feet, and after two years the Bordeaux chamber was still without jetons or medals. The Bordeaux chamber got annoyed, writing to its agent in Paris on 23 February 1706:

> It seems very strange the *Messieurs* of the Academy have not settled on our medals and jetons.... We didn't realize that we had to have recourse to them for this. If the ideas behind our *devises* are faulty, it's easy for them to correct them, and we are completely ready to give in to what they decide. But something has to be decided. If the matter has become serious as you tell us, we cannot see through to the bottom of it.[72]

A follow-up letter three weeks later made the same complaint:

> We are beyond astonishment to see the matter drag on so long. We don't know by what exigencies the thing depends so absolutely on Monsieur Anisson that we have to have recourse to him for this.[73] If *Messieurs* of the Academy have reservations, they have only to come up with some other idea. We're happy to defer to them, but it seems that more than three months having passed since our agent arrived in Paris, that should be plenty of time for these gentlemen to make up their mind. Since we are at the end of our year we emphatically beg you to leave nothing undone to bring this matter to an end.[74]

When the jetons finally did arrive in Bordeaux, they were adjudged too small.[75] For some unknown reason, neither their suggestion (HIC VENIUNT FELICIUS UVAE—Here the grapes grow more abundantly) nor the one chosen in Paris Académie des Inscriptions (CLARIOR ADJUNCTIS SURGET RECTORIBUS AEDES—The edifice will rise more distinguished than its builder) appeared on the 1706 jeton of the Bordeaux chamber of commerce, but rather QUO NON HAC DUCE (Whither

69. On this episode, see Évrard de Fayolle (1900), pp. 22–27.

70. Jacquiot (1980), p. 858.

71. "...Il sera bien fâcheux pour la Chambre si les devises ne sont pas du gout des connoisseurs que nous avons en province," contemporary document quoted in Évrard de Fayolle (1900), p. 26.

72. "Il paroit bien étrange que Messieurs de l'Académie ne se soient pas encore déterminés sur nos médilles et jettons... nous n'avions pas cru être dans la nécessité d'avoir recours à eux pour cella. Si les idées de nos devises sont fausses, il leur est aisé de les rectifier, nous sommes touts prêts à nous soumètre à ce qu'ils en décideront; mais enfin, il faut se déterminer à quelque chose. Si l'affaire est devenue sérieuse comme vous nous le dites, nous n'en pouvons pas pénétrer la cause...," letter of Bordeaux Chambre de Commerce, dated February 23, 1706, quoted in Évrard de Fayolle (1900), p. 24.

73. This would be Jean Anisson (1642–1721), director of the royal press.

74. "...nous sommes du dernier étonnement de voir traîner si longtems cette affaire; nous ne sçavons pas par quel endroit la chose dépend si absolument de Monsieur Anisson qu'il faille avoir recours à luy pour cela. Si Messieurs de l'Académie trouvent à redire sur quelque chose, ils n'ont qu'à fournir quelque autre idée. Nous voulons bien y déférer, mais il semble que plus de trois mois, qui se sont écoulés depuis que notre argent est à Paris, est un tems plus que suffisant pour déterminér ces messieurs; comme nous voila à la fin de notre année, nous vous prions très instamment de ne rien oublier pour terminer incessamment la chose," letter of Bordeaux Chambre de Commerce dated March 13, 1706, quoted in Évrard de Fayolle (1900), p. 25.

75. On rare occasions jetons and medals were returned because poorly struck or the wrong size; for examples, see A. Barthélemy (1875), p. 231; AI, AIBL archives, A:65, minutes of the meeting of December 11, 1781.

without this guide?). The gentleman from the Collège de Guyenne who suggested the original devise that wasn't accepted was supposed to get a gold medal for his trouble, but because his suggestions weren't used, he received six *louis d'or* instead.

FIGURE 6.2: Jeton of the Estates of the Artois.

A similar dustup occurred over the 1713 jeton for the estates of Lille. (Fig. 6.2). Lille and French Flanders had been part of France from 1668, but in the War of Spanish Succession (1701–1714), Lille was occupied by the Dutch from 1708 to 1713, and the city had just returned to French rule. The situation was tense, and the Lille estates/town council (the *Loy*) itself may have been divided. Initial proposals for jetons and medals from Lille were rejected by academicians at the Académie des Inscriptions in Paris. Ultimately, Paris accepted one of the proposals from a Lille alderman, that of a sunflower turning toward the sun, but changed the legend to UNI SEVIVIT AMOREM (It kept its love for one).[76] The Lille authorities had only to accept what had been decided in Paris and furnish the necessary funds to cover the costs. Still, it took four months of back and forth between Lille and Paris to settle on the *devise*, and it took another month for the medals and jetons to be struck and sent to Lille.[77] The Monnaie des Médailles engraved the dies and struck the jetons and 45 additional gold medals that were handed out to the councilmen of Lille and certain widows.[78]

The Science and Art of the Devise

A jeton's *devise* consisted of two things: an image and a legend. The image or *type*, as it was known, was considered the "body" of the jeton, the legend its "soul."[79] The creativity involved in concocting a *devise* lay in artfully combining the two—the body and soul of the *devise*—in ways that were apt, elegant, and sophisticated in their references and allusions. *Devises* were not made up out of thin air, and academicians of the Académie des Inscriptions put much thought into creating the designs for the jetons they sanctioned. Several modern historians and numismatists have studied the *devises* of seventeenth- and eighteenth-century jetons and medals, notably Marie Veillon, who in her 1994 study of *devises* for jetons struck in the reign of Louis XIV, speaks explicitly of "the art of the *devise* for jetons."[80] Sabrina Valin's 2016 dissertation likewise probes the interpretative subtleties surrounding the *devises* of royal jetons, particularly those of Louis XIII.[81]

76. Bordeaux (1905), tells this story, p. 20. This jeton is F.7234.
77. Jacquiot (1980), p. 858, comments on the long time it took the Académie to respond to clients.
78. The gold medals alone cost 6,665 *livres*, a not insignificant sum; Bordeaux (1905), pp. 15, 17. See also Van Hende (1858), pp. 2–3, who portrays the Académie des Inscriptions as more aloof in this incident.
79. On this point, see above Ch. 2, n. 79; La Tour (1897), p. xviii; Guichon de Grandpont (1854), p. 10; see also Mainguy (1999), pp. 190–96.
80. "...l'arte des devises de jetons," Veillon (1994); Veillon (1994), appendix, lists and translates the *devises* of 200 jetons taken mostly from royal administration. On the character and importance of *devises*, see also Fontenay (1854), pp. 54–63; La Tour (1897), pp. xviii–xx; Blanchet (1930), p. 206; Lerouge and Lerouge (1977), pp. 62–64.
81. Valin (2016), 3 vols.; see in particular vol. 1, pp. 165–88, 213–29.

In the 1960s, '70s, and '80s Josèphe Jacquiot of the Département des Monnaies of the BnF pioneered the scholarly examination of the *devises* of jetons and medals and their Latin inscriptions.[82] She and other scholars make the point that it is hard to decode jetons because we need to know the reasons authors chose in picking a *type* or legend, and these reasons are not easy to come by.[83] Jetons were often intrinsically and deliberately enigmatic even to experts in the seventeenth and eighteenth centuries, and in so many cases they are certainly so today. Furthermore, what the Latin inscriptions actually say usually only means to suggest another sense which is not expressed directly.[84] All is allegorical, figurative, and metaphorical…almost hermetic and certainly requiring hermeneutics! A "double language" destined to outlive time is always involved. The choice of a legend follows the choice of the *type*. Words must be noble, forceful and even on occasion sweet. Jetons and their *devises* always reflect the glory of the sovereign, of course, but otherwise jetons can be opaque. Mme Jacquiot gives examples of the complex thinking that went into the choice of Latin legends. In considering their choice of imagery and inscriptions, she relates the creators of *devises* for jetons to painters, but one could also add not just the creators of *devises*, but the *dessinateurs* and the engravers who sketched them and carved them in metal. The result was an object and form of expression conveyed in a language the great majority of the population could not understand at all.[85]

The academicians of the Académie des Inscriptions and others connected to jetons and their *devises* had at their disposal a surprisingly large and intellectually rich literature to guide them in their choices.[86] These are works largely of the seventeenth century, but their influence was felt well into the eighteenth century. They derived from and served the strong contemporary Latinist culture we have seen. These works offer a body of theory as to how *devises* should be constructed.[87] An early work of this sort was by the printer Henri Estienne who published *L'Art de faire des devises* in 1645, dedicating the work to the Cardinal Mazarin.[88] The *De l'Art des devises* by the Jesuit Pierre Le Moyne (1602–1671) appeared in 1666. At some point in the 1670s or 1680s, the literary giant, Charles Perrault (1628–1703), secretary of the Petite Académie as well as head of the Bâtiments du Roi, contributed a *Discours sur l'art des devises*.[89] The Jesuit scholar and heraldist, Claude-François Ménestrier (1631–1705), was undoubtedly the most influential of such authors. We've had occasion to mention his 1689 work, *Histoire du roy Louis le Grand par les medailles, emblêmes, devises, jettons, inscriptions, armoiries, et autres monumens publics*, but Ménestrier wrote and published other titles that can be translated as: *The Philosophy of Images, Composed of an ample collection of* devises *and an evaluation of all the works that have been done on this topic* (1682–1683) and his *Science and art of the* devise *based on new rules, with six hundred* devises *on the principal events*

82. See Jacquiot œuvre cited in Bibliography.
83. Longpérier, p. 425; Blanchet (1950), p. 200, remarks that what contemporaries were going for in their mottoes and iconographic materials was clear to them, but often not to us. Blanchet (1932), p. 234; Blanchet (1930), pp. 220–21; Veillon (1994), p. 28 make this same point.
84. Jacquiot (1980), p. 857.
85. Josset (1978) makes this important point, p. 337.
86. On this point, see also Sarmant (2003), pp. 115–16, 187–89; Mainguy (1999), pp. 39–56.
87. These points per Josset (1978), p. 337.
88. See Estienne (1645); an English translation of this work as *The art of making Devises* appeared the following year in London in 1646.
89. A reference now lost to the author points to BnF, Bibliothèque de l'Arsenal, Ms. 3328, Recueil, C. Perrault, "Discours sur l'art des devises"; see mentions of these and related works in Veillon (1994), pp. 22–23, 27; Mainguy (1999), pp. 316–19.

in the life of the King and four hundred sacred devises *with the words all drawn from Holy Scripture* (1686).[90]

As Ménestrier's reference makes plain, a complicated set of rules governed what constituted a successful *devise*. Ménestrier elaborates many of these.[91] For example, in an image/*type* one should not use many bodies, but only one to represent a common action. Legends should not mention the image/*type* directly. Of two items in an image/*type*, one should not appear literally or concretely and the other abstractly or allegorically. The look of the Latin legends was a consideration furthermore, not just the play of the words. Ménestrier gives a hundred references from Horace that can be used for legends. To guide preparers, he gives specific examples for specific cases, such as showing a ship with a map and a compass and the legend DOCTUS ITER MELIUS (Taught a better way) for a heretic who converted through reading the Gospels and the Church Fathers.[92] Or, when the king rewards men of letters, to depict an orange tree laden with fruit and the words from Horace's *Ars Poetica*, DIVITIIS ANIMAT COLENTES (With riches he animates his cultivators).[93]

Earlier Italian humanists began this sort of thinking in evaluating ancient coinage and medals, and much of that background carried over to French consideration of the medals they produced, but previous Italian work did not apply directly to jetons because Italians didn't have a jeton tradition. So in this sense the French were on their own in creating these rules and developing a sensibility regarding jetons and their *devises*.[94] The Académie des Inscriptions itself reflected on the state of the literature in 1739 and its own implied superiority in understanding, applying, and adjudicating rules for jetons and medals:

> And finally, we undertook an ongoing metallic history of the principal events of the king's reign. The subject matter was ample and magnificent, but it was difficult to get it underway. The ancients, who left us so many remaining medals, bequeathed us no other rules than the medals themselves which until our day have only been examined for the beauty of the workmanship and only studied from the point of view of history. Moderns have had many medals struck over the last two centuries, but they have been little bothered by rules. They haven't followed any, and they have pronounced none. In compilations of the sort one can hardly find three or four instances where genius was happily complemented by method.[95]

Academicians were attentive to these rules and sensibilities. The various proposals they considered in 1773 for the jeton of the Extraordinaire des Guerres for 1774 provides a nice example.[96]

90. See Ménestrier (1682–1683), *La philosophie des images, composée d'un ample recueil de devises, et du jugement de tous les ouvrages qui ont été faits sur cette matière*, and (1689b), *La Science et l'art des devises, dressez sur de nouvelles règles, avec six cens devises sur les principaux événemens de la vie du Roy et quatre cens devises sacrées dont tous les mots sont tirés de l'écriture sainte*. In this connection see the five-volume collection of *devises* complied by Chassant and Tausin (1878–1895).

91. Ménestrier (1689b), separately paginated, "Discours ou il est traité de la nature des devises et de leurs principales Règles," pp. 1–74.

92. Ménestrier (1689b), p. 49.

93. Ménestrier (1689b), "La Vie du Roy en Divises," pp. 75–379, here pp. 231–32.

94. On these points see Veillon (1994), pp. 22–23; Lerouge and Lerouge (1977), pp. 98–105.

95. "Enfin, on entreprit de faire par Médailles une histoire suivie des principaux événements du regne du Roy. La matiere étoit ample & magnifique, mais il estoit difficle de la bien mettre en œuvre. Les anciens, dont il nous reste tant de médailles, n'ont laissé sur cela d'autres regles que leurs médailles mesmes, qui jusques-là n'avoient guères esté recherchées que pour la beauté du travail, & estudiées que par rapport aux connoissances de l'Histoire. Les modernes qui en avoient frappé un grand nombre depuis deux siécles, s'estoient peu embarrassez des regles ; il n'en avoient suivi, il n'en avoient prescrit aucune ; & dans les recueils de ce genre, à peine trouvoit-on trois ou quatres piéces où le génie eûst heureusement suppléé à la méthode...," AIBL, H&M, vol. 1 (1736), p. 5.

96. "Devises fournis en 1773 pour l'extraord. des guerres," AI, AIBL archives, A:08.

Among the four considerations (or four *essais de jetton*) was one with a sword entwined with an olive branch that surmounted it and the legend PAX NITIBUR ARMIS (Peace will sustain itself through arms). Regarding this proposal one of the commissioners noted the following: "This *type* has two problems. One, the arms are represented concretely, whereas Peace is there only symbolically. Second, it is difficult on the small surface of a jeton to distinguish olive branches from laurel, etc. Perhaps contrary to the rule of jetons, a thousand times violated, it would be better to represent the Goddess of Peace leaning on a pile of arms to be covered by a shield."[97] Another commissioner made the same point about this jeton: "One will remark that the arms are represented concretely, peace being only as a symbol, as should be the case."[98]

In composing *devises*, part of idea was to incorporate a sly reference to some event of the previous year...to point to some notable something in the last 12 months pertaining to the issuing department or the nation for that matter that was to be indelibly stamped into the jeton of current year.[99] (In this, again, jetons mark the more ephemeral in contrast to the supposedly timeless aim and character of events commemorated by medals.) Thus, for example, a proposal for the jeton of the Ordinaire des Guerres for 1714 suggested roses intertwined with stalks of lilies, "in connection to the recent peace signed with England, as roses signify England as lilies do France," a reference to the Treaty of Utrecht of the year before.[100] For the administrative jetons annually commanded by the king, the suggestion has been made that the various departments themselves put forward to the Académie des Inscriptions the event of the last year they wanted to see incorporated into the *devise* of their jetons.[101] A treaty like Utrecht, a victory, a coronation, a marriage, a birth,...for 150 years the designers of jetons imaginatively wove in a clever reference to such happenings for those in the know to appreciate.[102]

In a like vein, if the *devise* didn't make direct reference to something in the immediate past, it often alluded to the function or character of the entity sponsoring the jeton. This sense comes across in the case of the jeton of the Chambre des Comptes du Roi of 1608 with the legend NEC FALLIT.NEC.FALLITUR.UNGUAM (Never deceived, never deceiving). The more standardized jeton of the Académie des Sciences, again, with the *devise* of Minerva surrounded by scientific instruments and the legend INVENIT ET PERFICIT (It discovers and perfects) summarized the essence and aim of the institution. Images and legends signifying abundance flowed from the Trésor Royal; widespread use of naval and sea imagery is seen not just in the series for the royal navy and the galleys.[103] Jetons were not necessarily subtle in getting their point across, as, for example, with the jetons of the Parties Casuelles for 1682. The Parties Casuelles, again, was the formidable bureaucracy that kept track of offices and office holders and collected the annual tax on same, and,

97. "1°. une épée debout entortillée d'une branche d'olivier qui la surmonte. PAX NITIBUR ARMIS. (Deux défauts dans ce type. 1°. les armes sont représentées réellement, la paix ne l'est que dans son symbole. 2°. il est assés difficile dans le champ étroit d'un jeton de distinguer l'olivier d'avec le laurier &. Peut-être contre la règle des jettons, mille fois violée, vaudroit-il mieux représenter la Déesse de la paix appuyée sur un monceau d'armes qui seroient couvertes d'un bouclier.)," AI, AIBL archives, A:08, "Devises fournis en 1773 pour l'extraord. des guerres."

98. "...on remarqua que les armes sont représentées réelement, la paix ne étant que dans son symbole, comme cela doit être," AI, AIBL archives, A:08, small unbound booklet marked "jettons/1774."

99. On this point see, for example, Valin (2016), vol. 1, p. 75 and above, Ch. 2, nn. 79 and 82.

100. "Les roses designent langleterre comme les lis, la France," "Projet de devises pour l'année 1714," AI, AIBL archives, A:109, with the legend QUAM DULCE SPIRANT (How sweetly they breathe).

101. Guichon de Grandpont (1854), p. 6.

102. Mainguy, pp. 90ff; Blanchet (1930), pp. 228–29; Veillon (1994), p. 27 gives an example from the Trésor Royal of 1679 that alludes to the treaty of Nijmegen; Berry (1977) gives several examples of jetons of Louis XIII and their allusions/connections to current events in that reign.

103. Guichon de Grandpont (1862), pp. 3–15, gives a list of 144 jetons with naval images exclusive of those for La Marine or Les Galères from 1556–1786.

as reported in the *Mercure galant*, its *devise* for 1682 featured "a field of poppies on the reverse with the words around PLACIDOS DAT DUCERE SOMMOS (It permits continued peaceful slumbers), which signifies that as poppies give a restful sleep to those who take them, one can similarly sleep in peace after one has paid the annual tax that is due for keeping an office."[104] Another of the jetons of the Parties Casuelles in 1723 depicted an elephant, "which, according to naturalists, breaks its tusks to avoid being pursued."[105] The appeal to the science of the naturalists is a nice touch here.

The Connection to Antiquity

The importance of antiquity and ancient Rome in particular hardly needs to be pointed out, but it well deserves emphasis. The historian Peter Gay long ago drew attention to the fact that Enlightenment thinkers were deeply influenced by the world of classical antiquity and steeped in its literature.[106] Even if by the eighteenth century the so-called Moderns had won the literary battle of the Ancients versus the Moderns, everyone was shaped by deep immersion in ancient authors. In their jetons and medals contemporaries were not imagining a future categorically different from their present. They were not looking forward to a different time ahead—certainly not the one that would come with industrialization and the French Revolution. They were still backward looking, the keepers of a status quo that imitated antiquity in seeking to project their present onto an imagined future.[107] Scholars who have studied jetons have identified the ancient sources contemporaries relied on and tapped in creating their jetons and medals.[108] Virgil was a principal inspiration for many jetons and medals.[109] With Virgil came Horace, Ovid, Juvenal, Cicero, and other Latin authors.[110] Many of the players involved undoubtedly knew Greek, and divinities and architectural forms from classical antiquity are featured images on jetons, but Greek and the Greeks are essentially absent from among the *devises* of jetons.[111]

The academicians of the Académie des Inscriptions were self-conscious and explicit about their links to the ancient world.[112] Along with men in the Cabinet du Roi, they were numismatic and historical experts who knew and studied the medals of antiquity, such as, for example, one mentioned for Vespasian.[113] For a medal for the embassy of the Turks in 1720 academicians proposed the legend REGI CRESCENTI (For the king who prospers) and added the comment: "The REGI CRESCENTI is a happy imitation of several Roman emperors, among others those of the young

104. "Pour les Parties Casuelles. Plusieurs Pavots sont au Revers, et les paroles suivantes sont autour, PLACIDOS DAT DUCERE SOMNOS; ce qui marque que comme les Pavots donnent un sommeil tranquille à ceux qui en prennent, on peut de mesme dormir en repos, apres qu'on s'est acquité du Droit annuel pour la conservation des Charges qui sont obligées de la payer," *Mercure galant*, January, 1682(1), pp. 58–59; also quoted in Guichon de Grandpont, 1862, pp. 17–18. Guichon de Grandpont gives similar extracts of the *Mercure galant* and descriptions for fourteen other jetons of 1682. Bonnet (1899) in Depeyrot (2007), pp. 91–127 provides further examples and explications.

105. "Les parties casuelles. Un elephant qui selon les naturalistes rompt ses dents pour eviter d'estre poursuivi," AI, ABIL archives, A:109, "Jettons 1723."

106. Gay (1966–1969), especially vol. 1: "The Rise of Modern Paganism."

107. See Wellington (2015a), whose subtitle, *Artefacts for a Future Past*, captures this point perfectly, and in particular pp. 127, 145–54.

108. Longpérier (1863), for example, pp. 427 ff., lists references and ancient sources of allusions made in *devises* elaborated by Sully from 1589 to 1601.

109. Jacquiot (1989b) has devoted a study to the use of Virgil, notably the *Aenied*, as legends on medals and jetons.

110. Blanchet (1930), pp. 227–28. The Académie des Inscriptions would itself often list the ancient authors used for legends; on these, see, for example, AI, AIBL archives, A:08.

111. As noted, Feuardent (1904–1915), vol. 3, pp. 363–431, lists approximately 6,000 different legends on jetons in Latin and French, but only twenty-one of the legends listed are in Greek.

112. On this point, see also Loach (1995), p. 27.

113. AI, AIBL archives, A:08, Packet: "Medaille sur la mort du feu Roi [Louis XV]." See other references AIBL, H&M, passim.

Gallienus on which one reads JOVI CRESCENTI. Connoisseurs will greatly appreciate the imitation and even more the wisdom that changed JOVI [Jove] into REGI [King], which says nothing less for us."[114] An image of 1776 featured "France seated on an ancient pedestal."[115] Another of 1779 pictured public felicity "as in ancient monuments."[116] These resonances were all well and good, but who could possibly know that behind the armed Hercules and the legend ASSERIT OTIA MUTIS (He defends peace from brutes) suggested for the Extraordinaire des Guerres for 1776 were allusions to Hercules Musagetes of the Greeks (leader of the muses) or the Temple of Heracles erected in Rome in 187 BCE by Marcus Fulvius Nobilior and restored by Marcius Philippus in 29 BCE?[117]

Hercules was an especially prominent image, and the iconography of jetons features the gamut of Roman and Greek Olympians.[118] For the medal struck in anticipation of the treaty of Seville of 1729, for example, Wisdom and Prosperity unite with War and Peace, as Minerva extends her hand to Mars who holds an olive branch.[119] Apollo appears in various guises. Poseidon is there, as is Atlas. Goddesses and idealized personifications representing Victory, Justice, Peace and the like litter the landscape of Old Regime jetons and medals. These usually robed figures appear with their attributes, such as Minerva's owl, attributes which only underscore how much they speak a symbolic language known only to an elite. The palm tree was a symbol of fecundity, the olive tree of peace, and so on.[120] "A trophy of arms held high by the goddess of peace to show that what could be an object of horror can serve as an ornament to the state."[121] "A lion enchained to show that the king's love of peace holds back, so to say, the valor of his troops."[122] This was a language requiring its own semiotics. Symbolism, personification, allegory, antiquity...all folded together in an artful jeton.[123] Consider as a single example the 1775 proposal for *devises* for the standards of a regiment of dragoons suggested on request by the Académie des Inscriptions.[124] On one side: the arms of the king's brother whose regiment it was. On the other: Bellerophon mounted on a winged horse (Pegasus) combating a chimera vomiting flames. The Académie proposed two legends: NEC MONSTRA PAVET PAVESCIT (Of monsters it neither fears nor grows afraid) and VEL MONSTRA LACESSIT (It assails monsters). The proposal went on to explain the *type* and how to portray the imagery. "The winged horse indicates the speed and celerity proper for dragoons. The chimera being fought signifies the character of a dragoon to never be fearful of anything. For painting and costuming

114. "Le Regi crescenti est heureusement imité de quelques Médailles d'Empereurs Romaines, entr'autres de celles du jeune Gallien sur lesquelles on lit JOVI CRESCENTI. Les connoisseurs sentiront bien l'imitation, et plus encore la Sagesse qui a fait changer le JOVI en REGI qui ne dit pas moins pour nous," AI, AIBL archives, A:109, "Medaille sur l'Ambassade des Turcs." The date is probably 1720 marking a formal envoy from the Ottoman sultan that year.

115. "la France assis sur un socle à l'antique," AI, AIBL archives, A:08, "Pour 1776. Jetton pour l'Extraordinaire des guerres."

116. "L'allegresse publique representée, comme dans les Monumens Anciens," AI, AIBL archives, A:67, minutes of the meeting of January 15, 1779.

117. "Allusion à l'hercule Musagete des Grecs, et à l'hercule musarum de Fulvius Nobilior et Marius Philippus," AI, AIBL archives, A:08, "Pour 1776. Jetton pour l'Extraordinaire des guerres. Types fournis avec leurs dévises, en 1775 pour l'Année 1776."

118. See Veillon (2005); Gombert (1974); Jacquiot (1971).

119. Bonnet (1899) in Depeyrot (2007).

120. See reference, AI, AIBL archives, A:68, minutes of the meeting of June 17, 1785.

121. "Un trophée d'armes elevé et soutenue par la Pais, pour montrer que ce qui pourroit faire un objet d'horreur sert d'ornement a l'etat...," AI, AIBL archives, A:109, manuscript. labelled "Devises pour le Roy."

122. "Un Lion Enchaisné...L'amour du Roy pour la paix enchaîne, pour ainsi dire, la valeur de ses troupes," AI, AIBL archives, A:109, manuscript labelled "Devises pour le Roy."

123. On the use of allegory in jetons and medals see La Tour (1897), p. xviii; Jacquiot (1976); on the allegorical character of Lyon consulate jetons, see Loach (1995), especially pp. 28–45. Mainguy (1999), Part III, pp. 183–301 is especially strong here, as is Valin (2016), passim.

124. AI, AIBL archives, A:08, Ms. "1775: Le Régiment de Dragons de Monsieur demande une dévise pour ses Etendards ou Guidons."

Bellerophon, Pegasus and the Chimera one can consult Pindar, Ausonius and other poets."[125] So deep was the attachment to classical antiquity that the creators of jetons looked to Pindar and his peers to see how to clothe their figures.

In crafting their *devises* the academicians of the Académie des Inscriptions strove to avoid repetition and clichés that might displease. Thus for the 1775 medal and jeton for the coronation of Louis XVI, the academy proposed Religion anointing with holy oil the head of the king who is on his knees at the altar and the legend, DEO CONSECRATORI (Unto God, He who halloweth). The proposal went on to comment: "This medal would be in the antique style. Nothing is more ordinary than the legends JOUI, APOLLONI, HERCULI, CONSERVATORI, PROPAGUATORI, SORPITATORI, TUTATORI, etc. CONSERVATORE is not from Cicero. It is in St. Jerome, in Tertulian, and in Julius Firmicus, but it is not certain that this word can be used for a Christian ceremony."[126]

Political considerations made themselves felt in the choices for *devises* offered for jetons and medals. Thus, regarding a 1779 proposal for a medal to commemorate the birth of the eldest daughter of Louis XVI and Marie Antoinette, depicting Austria offering a child to France with the legend NOVUM CONDORDIAE PIGNUS (A new pledge of harmony), a marginal note presumably from the minister emphasized that "Austria must not appear here, much less the word *concordia*."[127] For plaques mounted in Ecuador by the astronomical expedition of the Académie des Sciences in the 1730s, a sheet labeled "Doubts and Remarks on the Subject of the Proposed Inscription," notes: "One would have liked to be able to use REGE CATHOLICO [Catholic king] for Spain in conjunction with REGIS CHRISTIANISSIME [Christian king] for France, but it was thought better to use HISPANIARUM ET INDIARUM REGE [King of Spain and the Indes], and thus CATHOLICO seemed superfluous and to avoid saying the [Most] Catholic king of Spain."[128]

Beautiful and exquisitely crafted, if occasionally enigmatic, silver and gold jetons were the result of all these considerations often enough in the eighteenth century. Many have been depicted here already. The silver one minted for the Artillerie for 1731 expresses all these virtues (Fig. 6.3). Two female figures in antique garb promenade along a path. The jeton catches them as they are surprised to encounter cannons and discarded armaments beneath a tree. The legend is ET MUTA MINANTUR (And silent they threaten). The jeton seems not to connect to any current event of 1730, but only to the fearsomeness of the Artillerie itself.

125. "Le cheval ailé désigne la vitesse et la célerité propre aux Dragons. La chimere a combattre signifie quil est du caracter du Dragon de n'avoir peur de rien. Pour la peinture et le costume de Bellerophon, pégaze et la chimere on peut consulter pindare; ausone; et d'autres poétes." AI, AIBL archives, A:08, Ms. "1775: Le Régiment de Dragons de Monsieur demande une devise pour ses Etendards ou Guidons." Ausone would be the late Latin poet and rhetorician from Gaul, Decimus Magnus Ausonius (310–95). For other examples of this sort, see AI, AIBL archives, A:109.

126. "Cette médaille seroit dans le gout antique. Rien n'est plus ordinaire que des légendes, JOUI, APOLLONI, HERCULI, CONSERVATORI, PROPAGUATORI, SORPITATORI, TUTATORI, &c. CONSERVATORE n'est pas de Ciceron – mais il est dans S.t Jerôme, dans Tertullien, dans Julius Firmicus, et il n'est sure que ce mot peut estre employé pour une cérémonie chrétienne," AI, AIBL archives, A:08, undated [1774] folio: "La répétition du meme type ou a peu près de la meme légende pourra déplaire."

127. "(on observa que l'Autriche ne doit pas paroître ici, non plus que le mot Concordia.)," AI, AIBL archives, A:08, Folded loose sheet,"Arrêtés dans la séance du 15 janvier 1779 et envoyés le même jour," "Types d'une Médaille pour la naissance de Madame de France fille du Roi."

128. "On auroit voulu pouvoir mettre *Rege Catholico* pour opposer a *Regis Christianissime* mais on a crû quil etoit a propos de mettre hispaniarum et Indiarum Rege et alors Catholico a parû superflu. n'ayant point connoissance quon dise le Roy Catholique d'Espagne," AI, AIBL archives, A:109, undated single sheet labelled: "Doutes et Remarques au sujet de l'inscription proposée." On this expedition and the plaques ultimately set up in South America, see Safier (2008).

FIGURE 6.3: Jeton of the Artillerie, 1731.

By contrast, the iconographic and stylistic emphasis on antiquity and the "tools to think with" that antiquity provided the men of the Académie des Inscriptions and their compeers meant that jetons with other sorts of images were uncommon, or so it seems. One hardly finds images reflecting contemporary life or even images from earlier periods of French history. Unusual and more exceptional visually, for example, are these two jetons of the Extraordinaire des Guerres of 1755 and 1772 (Figs. 6.4 and 6.5).[129] The first shows two medieval mounted knights jousting; the second a man in contemporary dress with a rifle beside a cannon and a plow.[130]

FIGURE 6.4: Jeton of the Extraordinaire des Guerres, 1755.

FIGURE 6.5: Jeton of the Extraordinaire des Guerres, 1772.

129. See also the related choices for the 1776 jeton of the Extraordinaire des Guerres, AI, AIBL archives, A:65, minutes of meeting of September 2, 1766 (#3).

130. See further descriptions, List of Figures and Tables below.

Royal busts often graced the obverse or front side of jetons.[131] The bust of the king appeared on most jetons, although the jetons of the Royal Navy and the galleys often featured the bust of the current *Amiral de France*. Ministers apparently paid great attention to the portraits of their masters on jetons.[132] The busts reproduced in the Feuardent catalog give an almost complete view of these. Paradoxically, they allow us to see the king aging—for both Louis XIV and Louis XV we see them from early childhood to old age; Louis XVI had another fate. Numismatists make distinctions in categorizing busts of the king as "laurelled," "with headband," or "in military garb," and such like. In flattery, the Académie de Marine, as we saw, once complained that the likeness of the king on its jeton made him look old, and apparently in 1770 Louis XV sent one back for that reason.[133]

The stock legends that graced the bust likewise require little attention. Only Louis XIV was MAGNUS REX (Great King), but he and the others were regularly D.G. FR. ET NAR. REX (By the grace of God, King of France and Navarre [the tiny kingdom in the Pyrenees]) and more commonly, REX CHRISTIANISSIMUS (Most Christian King). The interesting exceptions, showing his various royal guises, present the king as father of his country, protector of the Académie Française, protector of commerce, protector of the fine arts, and, in the case of mirror makers, protector of the crafts ("arts").[134] If it were present, the exergue, or the space for inscriptions at the bottom on the reverse of jeton, usually gave the name of the institution for which the jeton was struck. These appeared with and without dates, but the date on a jeton is no reliable indicator of when it was struck. In some odd instances jetons and medals incorporated chronograms where the letters of a legend hid a date in Roman numerals.[135]

Jetons as Fine Art

"The artistic importance of coinage requires no emphasis, since it has always been recognized" is how Philip Grierson, the historian and numismatist encountered earlier, put the point in 1951, and what he says about coins in general turns out to be especially true of Old Regime jetons as *objets d'art* and the products of talented artists and critical artistic traditions.[136] In other words, jetons need to be evaluated as works of art and in terms of visual culture. This point has never been in dispute among numismatists and others who have taken to jetons and who have always called them *petits monuments* (little monuments). Adrien Blanchet said this about the matter in 1930: "In the esteem given to the output of numerous medalists and engravers, French art has truly been graced and honored by these *petits monuments* whose evolution can be followed over several centuries."[137] For Serge Boutin in 1966: "From the end of the sixteenth century, jetons became more and more beautiful. Their busts engraved with

131. Lerouge and Lerouge (1977) discuss seventeenth- and eighteenth-century busts, pp. 67–72; see also Wellington (2015a), pp. 155–82; Feuardent (1995), vol. 4, plates I–XXII.

132. Blanchet (1930), p. 229.

133. Belaubre and Colllin (1992), pp. 162–63.

134. Mainguy (1999), pp. 183–301 mentions these attributes and their symbolic functions on jetons.

135. Bordeaux (1905), p. 7, gives an example of a chronogram, where the legend, LVDoVICo Magno LIberatorI, contains the letters LDCLLVVIII representating in Roman numerals the date of 1713; see also Barnard (1916), p. 55, and Blanchet (1892), p. 217, who notes that the 1722 coronation jeton/medal has a chronogram.

136. Grierson (1951), p. 7. The art historical approach to numismatics has been especially strong, and the point is no less true when it comes to seeing jetons as art objects; see, for example, the dissertations by Mainguy (1999) and even more Valin (2016).

137. "L'Art français a vraiment été illustré et honoré par ces petits monuments dont on peut suivre l'évolution pendant plusieurs siècles, en admirant les œuvres de médailleurs et de graveurs fort nombreux," Blanchet (1930), p. v.

infinite care by the most highly celebrated artists. Most of the reverses are as finely crafted as the obverses. With a delicacy of execution much greater than the coin of the realm, jetons are really little works of art."[138]

The fine art and visual art aspects of jetons stemmed from two quarters. The first we have just encountered in the design suggestions for the *devise* and the initial sketches made by the royal *dessinateur* of the Académie des Inscriptions. Sketches are already sophisticated works of art produced by the top royal painters. We will turn to the artist-engravers and the process involved in turning initial sketches into dies and finished jetons in the next chapter, but we owe the skill and talent of engravers a word here. These were not mere mechanics, but sophisticated artists who brought nuance to their work. Within the tiny circumference of a jeton they exercised their will and expressed an artistic vision. On the unforgiving surface of steel they chiseled enduring works. Busts are often regal portraits. On the reverse artist-engravers brought faces to life and captured motion, fleeting gestures and expressions, as in the Arcadian women encountering the cannon of the Extraordinaire des Guerres above (Fig. 6.3). The scene of the Valenciennes town council depicted in its jeton of 1726 (Fig. 3.10) is fabulously dynamic with activity everywhere; the two figures at the back beside the empty royal chair, one looking, one pointing tell whole tales. These gifted engravers reproduced in metal the tactile quality of cloth and fur, as in the buttoned up coats of the gentlemen corduroy-makers (Fig. 5.1) or the mantle that bedecked Peter I of Russia in the medal Jean Duvivier carved with his image (Fig. 7.5). French jetons of the seventeenth and eighteenth centuries were unmistakably the products of a high art tradition. The fact that jetons were usually signed by the engraver is another indicator that jetons need to be considered in this light.

The art styles found in jetons span French classicism, the Baroque, and the Neoclassical. With somewhat too broad a brush back in 1927 Charles Florange and Stéphane Strowski summed up this sweep of art styles and the place of jetons in the history of art more narrowly considered. Their emphasizing jetons more than medals as vehicles for thinking about the history of art in the *Âge classique* is to be noted:

> Although the art behind medals reflects the evolution of styles less than does painting or sculpture, one can follow on our jetons the transformations that have characterized the great styles of French art over the course of these two centuries. The first truly belongs to this transitional art that links the Renaissance on the decline with the art of the grand century being born. The sweep of the somewhat solemn and well-regulated style of Louis XIV characterizes the fine examples of the seventeenth century. Here and there one even makes out hints of the future pretentious style. On the other hand, the style of Louis XV hardly shows itself except in drawings of coats of arms or on certain representations of female bodies, transparent beneath their veils. Finally, Benjamin Duvivier (1728–1819), royal engraver and member of the Académie Royale de Peinture et de Sculpture, who chiseled in metal the profile of Louis XVI in the manner of his forebear, knew how to give to his work the incomparable style of the monarchy at its twilight.[139]

---

138. "À partir de la fin du XVIe siècle, les jetons se font de plus en plus beaux ; ce sont des portraits gravés avec un soin infini, par les artistes les plus célèbres ; la plupart des revers sont aussi soignés que l'avers : d'une finesse d'exécution beaucoup plus grande que les monnaies, ce sont vraiment de petits objets d'art," Boutin (1966), p. 241. The lesser artistic value of coined money is a point indeed to be noted in this context.

139. "Bien que l'art des médailles reflète moins que la peinture ou la sculpture l'évolution des styles, l'on peut suivre, sur nos jetons, les transformations qui ont caractérisé les grands styles français, au cours de ces deux siècles. Les premiers appartiennent bien à cet art de transition, qui rattache la Renaissance déclinante à l'art du grand siècle qui va naître. L'ampleur, un peu solennelle et bien ordonnée, du style Louis XIV caractérise les belles pièces du XVIIe. De ci,

One is not surprised to learn, then, that from the end of the seventeenth century through 1789 jetons (along with medals and sketches of same) were exhibited at the art salons held annually in the great Salon d'Apollon at the Louvre. These renowned exhibits began with the Académie de Peinture et de Sculpture and the associated art school in the later seventeenth century; they presented the sanctioned and very best in every aspect of French art for the year, including jetons and medals, and the Paris salons were the loci of much critical discussion.[140] Jetons and medals were shown in the Paris salon 19 times from 1699 to 1789.[141] Indicative of the level of art of French jetons, in 1771 authorities in Flanders (Belgium) complained about the low aesthetic quality of their *jetons d'étrennes* compared to the French.[142]

While not a point for contemporaries, for us today a final aesthetic characteristic of jetons deserves notice, that is, their patinas. The few surviving gold and gold-plated ones have remained unspoiled, of course. Other jetons in silver and copper and various alloys come to us bearing the mark of time, and some of their patinas are breathtaking in the depth and richness of their tones. They have suffered the ravages of oxidation and the oil of human hands. The jetons in question here are hundreds of years old, and they carry with them their own history since the day of their striking. In this connection, the author made a point of collecting every 1746 jeton he could find for the Estates of Burgundy. Coincidentally, that jeton is illustrated above (Fig. In.5) in the introduction to the present work. What is not illustrated is what you see when several of these jetons are collected back together, each with a common origin on a particular day at the Monnaie des Médailles, yet coming down to us today each with its own long trajectory etched uniquely into its patina. History and numismatics find a curious union here.[143]

Pressures for standardizing jetons increasingly made themselves felt from the early part of the eighteenth century, and the withdrawal of government funding for the administrative jetons (excepting the Extraordinaire des Guerres) in 1758 negatively impacted the vitality of jeton culture in the second half of the century.[144] Compare this to the plain and boring jetons of the Société Royale de Médecine of the 1780s (Fig. 4.3) as indicating a different aesthetic sensibility regarding jetons than is true for the arcane and baroque ones of the era of Louis XIV, not to mention jetons from earlier in the sixteenth and seventeenth centuries. Still, given what has been said in this chapter about the detailed work of the Académie des Inscriptions et Belles-Lettres and all that went into creating

de là, on y entrevoit même quelques soupçons du futur style pompier. Par contre le style Louis XV ne se manifeste guère que dans le dessin des armoiries ou dans certains modèles de corps féminins, transparents sous leur voiles. Enfin, Benjamin Duvivier, graveur du roi et membre de l'Académie royale de peinture et sculpture (1728–1819), qui a gravé dans le métal le profil de Louis XVI, après celui de son aïeul, a su donner à son œuvre, le style incomparable de la monarchie finissante," Florange and Strowski (1927), p. vii.

140. Notable especially are Denis Diderot's commentaries on the salons of the Académie de Peinture et de Sculpture from 1759–1781.

141. L. Gillet (1906), pp. 2–8; jetons were exhibited in 1699, 1737, 1739, 1740, 1742, 1746, 1750, 1755, 1765, 1769, 1771, 1773, 1775, 1777, 1779, 1781, 1783, 1785, 1789; the next salon was in 1795. See also Desavis (1933), pp. 31, 32.

142. Witte (1899), p. 11.

143. Grierson (1951), p. 12 is sensitive to fate of coins after they left the mint. On these Burgundian jetons, see Rossignol (1851), chapter 23.

144. Mainguy (1999), pp. 88, 181; Lerouge and Lerouge (1977), p. 64; Lapassade (1970), p. 368; Mazerolle (1888), p. 371. The Extraordinaire des Guerres was the only office to maintain annual production; see AI, AIBL archives, A:65, minutes of the meeting of January10, 1769, where the purse received from the Extraordinaire des Guerres is said to be "la seule que subsiste." See also similar receipts noted in the minutes of AIBL meetings of January 9, 1770, and January 8, 1771.

devises for jetons down to the end of the eighteenth century, it *cannot* be said that 1750 marks the *terminus ad quem* for jetons or their history. Rather, although stylistic and other changes are to be acknowledged, the jetons of the French Bourbons again present themselves as a revealing socio-cultural phenomenon spanning two centuries and running past 1750 and all the way down to the French Revolution. For the decades that it operated across the eighteenth century, the Académie des Inscriptions et Belles-Lettres was a big part of that story.

Great care was devoted to jetons. Once a *devise* was worked up and decided upon within the Académie des Inscriptions et Belles-Lettres, once it was approved by the minister, and once the royal *dessinateur* had drawn a sketch, the minister then turned the project over to the special mint for jetons and medals, the Monnaie de Médailles.[145] The Monnaie des Médailles housed the artist-engravers who engraved the dies that stuck the blanks that were minted into jetons. The next chapter explores this institution and its activities.

145. See example of this hand-off, AI, AIBL archives, A:68, minutes of the meeting of July 5, 1785.

FIGURE 7.1: Jeton struck by the Monnaie des Médailles for the Rouen Mint, 1787.

7.

The Monnaie des Médailles

The Monnaie des Médailles was the special mint that, with the merest handful of exceptions, minted *all* French jetons and medals from 1552 through 1789, a span of time pushing two and a half centuries.[1] This institution has been mentioned many times in earlier chapters, and the centrality of the Monnaie des Médailles to the story of jetons and to any appreciation of jetons and their place and role in Old Regime society and culture should already be apparent. The Monnaie des Médailles was the means of production and the place where the artistic vision of the *devise* through the further art of engraving was turned into an actual, material object, a jeton.

The very materiality of jetons calls out for investigation. To this point, Philip Grierson overstated the case, perhaps, again in 1951 when he wrote that, "coins throw more light on their manufacture than on the uses to which they were put."[2] But Grierson is right again that coins and jetons in particular do shed a critical light on the processes and, indeed, the world of their manufacture. This technical dimension of the story of jetons and the Old Regime has not received its just due from numismatists, and it is certainly not to be overlooked.[3] As loci of production, all minting operations in Old Regime France connected to the contemporary world of advanced technology, and in the eighteenth century minting, like printing, for example, was a complicated and characteristic technology of the times. But, even here, the Monnaie des Médailles stands out and deserves closer scrutiny because it was in its day simply the best mint in the world...the most technologically advanced, artistically inclined, and sophisticated minting operation anywhere.[4]

From 1609 the Monnaie des Médailles was located within the precincts of the Louvre. It was not originally to be found there, however, having been first installed in 1552 on the other side of the Seine at the tip of the nearby Île de la Cité.

1. On the Monnaie des Médailles, see Guiffrey (1884–1888); Mazerolle (1903); Mazerolle (1907a); Rondot (1904); Mainguy (1999), pp. 133–41; Lerouge and Lerouge (1977), pp. 33–34; Lapassade (1970). See also the key document, CAEF, Ms. F° 80: Nicolas de Launay, "Journal historique de la Monnoye des Médailles, 1697–1725," 2 manuscript vols. Similarly see Dubois de Saint-Gelais (1717), vol. 2, pp. 152–63, and materials in AN, O¹ 2064, especially 1744 document labeled "Memoire concernant les Médailles."
2. Grierson (1951), p. 5.
3. Guiffrey (1884–1888) and Lapassade (1970) are the obvious exceptions.
4. Mazerolle (1907a), p. 29.

A Separate Mint for Jetons and Medals: The Étuves

Prior to the establishment of the Monnaie des Étuves in 1552 the manufacture of jetons was haphazard.[5] Engravers would cut dies and someone, somewhere would literally hammer out jetons.[6] The minting of money, as we have seen in the Paris Mint, was another matter, strictly controlled as a matter of state, as was the nascent art of French medals, but indicative of the circumstances, the position of royal engraver-general of the mint dates only from 1547 and the ascension of the new French king, Henri II (r. 1547–1559).[7] Henri II took a further interest in upgrading the country's coinage and minting operations by ordering the establishment of two new ones in Paris in addition to the main royal mint located on the aptly named rue de la Monnaie on the Right Bank. The one begun in 1550 near the Nesle tower had an ephemeral existence. Henri established the other, more long-lasting mint in 1552 at the western tip of the Île de la Cité, near the royal baths, whence its name, the Monnaie des Étuves (sauna); the establishment was also known as the Monnaie du Moulin (mill) because it was powered by a waterwheel turning in the Seine.[8]

Henri II had a particular reason for creating the mint at the Étuves—he wanted the best and most up-to-date technology for the kingdom's mints.[9] From his ambassador in Augsburg in Germany, Charles de Marillac, he had heard of the invention of the new coin press; this would be the screw press or *balancier* as it came to be known in French.[10] Possibly with the printing press as inspiration, a German silversmith named Max Schwab had invented the machine for the mechanical stamping of coins and jetons, and German coin presses were in operation at this time in Augsburg and Nuremberg.[11] Henri sent his confidant, Charles' brother, Guillaume de Marillac (1521–1573), along with a mint master from Lyon, and a skillful smith, Aubin Olivier, to investigate the new German technology for coining money.[12] Olivier, on his return, constructed an improved machine that not only stamped the faces of the flans but their edges at the same time. Henri set up the Monnaie des Étuves, imported other equipment and materiel from Germany, appointed Guillaume de Marillac to run the Étuves administratively, and had Olivier to work the machines. As a result of this technology transfer, and thus possessing the day's most sophisticated, most highly mechanized minting machines, the Monnaie du Moulin at a stroke became a leading center for the art and industry of striking coinage, jetons, and medals.[13] This development had far-reaching consequences for the history and production of jetons.[14]

The Monnaie des Étuves began striking coin of the realm and jetons, and as head of the new mint, Guillaume de Marillac was empowered "to make known to whomever it may concern by his trumpet and public cry that they may bring to the said mint any sort of bullion, worn out coins, or other materials to be minted."[15] Guillaume de Marillac resigned in 1556 to move on

5. On the mint at the Étuves, see Valin (2016), vol. 1, pp. 51–53, 56; Parsons (2014), pp. 27, 48, 109–113; Mazerolle (1907a), pp. 26–31; (1902–1904), vol. 1, pp. 10–36; Dubois de Saint-Gelais (1717), vol. 2, pp. 157–58.
6. See Mazerolle (1902–1904), passim and vol. 1, p. 554, for example, regarding jetons pre-1550.
7. Guiffrey (1884),) p. 486.
8. Near Place Dauphine in Paris today. See Darnis (1996/1999), pp. 15, 160–61; Sauval (1724), vol. 2, p. 504.
9. Valin (2016), vol. 1, part 1, "Les enjeux de la frappe mécanique en France entre 1610 et 1661," especially pp. 99–120, now presents the most up-to-date account of this aspect of the story.
10. See eighteenth-century *balancier* depicted and described further below, Fig. 7.3.
11. Blanchet (1890), vol. 1, pp. 2–3; Droulers (1989), p. 186; Belaubre and Collin (1992), pp. 158–60.
12. Barnard (1916), pp. 55–56. On these developments, see also Babelon and Jacquiot (1951), pp. 14–15.
13. Mazerolle (1903), p. 113; see Mazerolle (1902–1904), vol. 1, pp. 276–81 for a detailed inventory of the Monnaie des Étuves; also Bouclier (1924), pp. 105–06.
14. Mazerolle (1903), p. 125, points out that the new equipment allowed for medals in particular to be struck and not cast. Historical medals were no longer cast after this period.
15. "...a faire signifier à son trompe et cry publicq à toutes personnes qu'ilz ayent à apporter à ladicte Monnoye toute sorte de billon, monnoye rongnée et aultre matière pour y monnier," contemporary document quoted by Mazerolle (1903), p. 116.

to bigger and better things in royal administration, and Aubin Olivier took over the Monnaie du Moulin.[16]

Henri II's initiative to modernize minting in France ran into fierce opposition from the Cour des Monnaies and the regular Monnaie de Paris.[17] At the royal mint, coinage and jetons were still hammered by muscular *monnoyeurs* whose longstanding privileges inured them from the incursion of the machine and mechanized production; they were the "John Henrys" of their day.[18] Until its own eclipse a century later, the Cour des Monnaies worked with the Monnaie du Moulin and approved dies used there.[19] The Chambre des Comptes had given its blessing as well, but the Monnaie de Paris was largely successful in resisting its more up-to-date competitor. The Monnaie du Moulin at the Étuves always minted medals, jetons, and gaming pieces, but only copper currency after 1563. In 1585 it was forbidden to coin money altogether, and in the same legislation the Monnaie de Paris succeeded in outlawing the use of the mechanized coin press in its own operations, effectively maintaining the technology of hammering new coins throughout the realm. An effort in 1606 to have the Paris mint shift to the *balancier* proved similarly unsuccessful, and it wasn't until 1645, under Cardinal Mazarin and the royal engraver Jean Warin, that French mints finally adopted the screw press to mint royal coinage.[20] The struggle lasted almost a century before the old *monnoyeur* who hammered the blanks became the *monnoyeur* who sat at the machine feeding it blanks and removing the stamped coinage.

If in 1585 the Monnaie du Moulin was prevented from minting money, hence becoming an entity entirely separate from the general French mint, it gained a lot in the same legislation when King Henri III granted it the exclusive right to produce medals, jetons, and gaming pieces.[21] This small privilege greatly solidified this little mint and gave it a place in society and a remarkable history that extended over the next two centuries and, in fact, down to today. In 1586 the Cour des Monnaies registered the letters patent from Henri III to Alexandre Olivier, the son of Aubin Olivier, conferring on him the charge of "superintendant, operator and custodian of the machines of the Monnaie du Moulin in the main building of the Étuves with permission to engrave, make, and mint *all, but only* ancient and modern medals, gaming pieces, and jetons, but not that he should make any coined money of copper, silver or gold."[22] The privilege and the monopoly was maintained over the next several generations.

The Monnaie du Moulin was weak in its early years and suffered on again, off again operations. Business was slow in the 1550s, and the mint was essentially closed from 1563 to 1569. The Oliviers sought to develop the market for their products, royal orders for jetons and historical medals came in, orders mounted, and a special room was set aside at the Moulin for engravers to work on their dies.[23] The *balancier*, as we will see further, was powered by humans, but other machines

---

16. Darnis (2005), p. 9. On this background history, see also Blanchet (1890), vol. 1, p. 3; Rachline (1992), p. 33.
17. Valin (2016), vol. 1, pp. 39–51; Darnis (1996/1999), p. 160; Mazerolle (1903), pp. 117, 120; Bouclier (1924), p. 103–05.
18. Rachline (1992), pp. 31–32; Blanchet (1888), p. 88.
19. See examples given in Mazerolle (1902–1904), vol. 1, pp. 54–58.
20. Rouyer (1893), p. 176; Bouclier (1924), pp. 105–06.
21. Wellington (2016), p. 296; Lapassade (1970), p. 354; Blanchet (1890), vol. 1, p. 3; CAEF, Ms. F° 80, vol. 1, pp. 5–6; Dubois de Saint-Gelais (1717), vol. 2, p. 158.
22. "…maître garde, conducteur et gouverneur des engins de la Monnaie du Moulin, au logis des Etuves [sur l'Île de la Cité] avec permission de graver, fabriquer et monnayer seulement toutes les médailles antiques et modernes, pièces de plaisir et jetons sans qu'il puisse faire aucune monnaie de billin, argent ou or," quoted in Lapassade (1970), p. 352; author's emphasis for clarity; see also Mazerolle (1907a), p. 30. On Alexandre Olivier, see especially Mazerolle (1903), pp. 130–35.
23. Darnis (2005): pp. 884–85; Mazerolle (1903), p. 126.

in the Monnaie du Moulin were powered by a large waterwheel turning with the flow of the River Seine. In 1557 the Cour des Monnaies issued an edict prohibiting anyone from interfering with the flow of the river for the mill at the Monnaie, so apparently, there was some conflict over this.[24] But one way or another, by the turn of the seventeenth century the Monnaie du Moulin had secured itself. In 1609 the Monnaie des Médailles as it had come to be known, moved to new and better quarters in the Louvre. We will examine these new accommodations momentarily, but first, we should review the men at the helm of the institution in its early days as it solidified its place in society and in Parisian urban space.

Early Directors

The Monnaie des Médailles remained in the hands of a succession of Oliviers: Aubin, Alexandre, Baptiste, Gilbert, another Aubin, and René Olivier through the 1620s, and by that date the Monnaie des Médailles had moved from the Étuves to the Louvre, as we will see further, so the story of the Monnaie des Médailles in the seventeenth century concerns the mint at the Louvre.[25] Typical of successions in the trades, a Pierre Régnier married into the Olivier family, as did Jean Warin, a silversmith who married René Olivier's widow, Jeanne Desjours, in 1629. Jean Warin (or Varin, 1596–1672) turned out to be an eminent artist and a more than able and politically well-connected administrator who considerably upgraded the caliber of the art and output of the Monnaie des Médailles.[26] Presumably along with Desjours, Warin ran the Monnaie des Médailles from 1629 through 1636; for reasons unknown Jean Régnier, son of Pierre Régnier, had charge of it from 1636 to 1639, but in 1639 Warin was again at its head with a royal appointment as its superintendent (*maître-garde*), a post he held for 33 years until his death in 1672.[27] In the 1640s he became chief royal engraver (*tailleur général des monnaies*) and controller-general of the royal image, and his position was further upgraded to royal privy councilor, intendant of the Bâtiments du Roi, and in 1665 artist-academician at the Académie de Peinture et de Sculpture. Warin was himself an engraver of the highest order, and he is known for founding a veritable school of engravers who brought the art of engraving medals and jetons at the Monnaie des Médailles to a high degree of perfection.[28] The Monnaie des Médailles now set the absolute standard for the production of commemorative historical medals and jetons.

The Monnaie des Médailles and its director fell under the authority of the royal Cour des Monnaies who had its own commissioner for the Louvre mint. All along, the Monnaie des Médailles dealt with the Cour des Monnaies and followed its administrative procedures. Nevertheless, as the authority of that monetary court became restricted in the second half of the seventeenth century and especially when the metallic history project under Louis XIV began, the Monnaie des Médailles was placed under the control of the minister in charge of the Bâtiments du Roi, to whom directors more often reported.[29]

The directors of the Monnaie des Médailles who followed Warin in the last quarter of the seventeenth century were solid leaders. There was one exception, Warin's son François, who nomi-

24. Mazerolle (1903), 116.
25. Darnis (2005) gives a complete list of directors from 1551 to 1803; see also Valin (2016), vol. 1, p. 106; Lapassade (1970), pp. 358–59.
26. On Warin see Blanchet (1888); Mazerolle (1932); Rachline (1992), pp 17–19; Charvet (1907–1909), pp. 62–63; Lapassade (1970), p. 358; Droulers (1989), p. 186.
27. Blanchet (1888), pp. 87–88; Lapassade (1970), p. 358.
28. On Warin and his work, see also Lerouge and Lerouge (1977), p. 65; Jacquiot et al., pp. 78–93.
29. CAEF, Ms. F° 80, vol. 1, p. 11, article xx of the 1696 regulations; Lapassade (1970), pp. 356, 358.

nally ran things from 1672 to 1676, but had the job taken away from him. Claude Ballin, a highly regarded royal silversmith, followed but unfortunately died after only a short time at the helm; the Abbé Pierre Bizot, a competent and connected administrator and himself a numismatist, had charge of the institution from 1678 to 1684.[30] Nicolas Petit, a royal secretary, followed as head for 12 years from 1684 to 1696. Let us recall that at this time Louis XIV was in his full glory, historical medals marking every event, and jetons becoming more and more part of the Étrennes and the esoterica of the parlor games of the *Mercure galant*. With the jetons of the administrative departments ordered year after year, the Monnaie des Médailles was a place of considerable activity. The heyday of jetons and the centrality of the Monnaie des Médailles would all continue well into the eighteenth century, but in 1696 the Sun King reorganized the administration of the Monnaie des Médailles, installing Nicolas de Launay as director, and with that, a new chapter opened for the Monnaie des Médailles. We will examine these developments momentarily, too, but it would do well to pause first to look more closely at the monopoly enjoyed by the Monnaie des Médailles.

The Monopoly of Monnaie des Médailles

The legislation of 1585 granted the Monnaie des Médailles an absolute monopoly on the production of jetons and medals, a right confirmed again and again by legal authority, as, for example, in the royal edict of 1672, "prohibiting the fabrication of jetons elsewhere other than at the *balancier* of the Louvre."[31] The *Mercure* put it this way in 1722: "The Monnaie des Médailles should not be confused with the other mints in France. The Monnaie des Médailles is very old and the only one permitted to manufacture medals, jetons, and other curios."[32] No one else could strike jetons or medals, and only those made at the Louvre could be sold or used in the kingdom. None could be imported. Silversmiths were forbidden from selling silver jetons and could only sell or keep copper ones in their shops and only those that came from the Monnaie des Médailles. All other merchants were banned from trading in them at all.[33] Shops and homes of suspected violators could be searched, and they were to be treated as counterfeiters and their tools confiscated.

Especially through the first half of the seventeenth century jetons were certainly struck elsewhere on occasion. These seem mostly to have been *jetons banaux* or common tokens for the counting table. Tournai was a known site for their manufacture, although the Monnaie des Médailles in Paris also produced *jetons banaux*.[34] At one point the mint at Nancy received permission to strike some jetons on a freelance basis, and at other times authorities in Rouen, Lyon, Dijon, and Sedan struck the odd ones, and even the Paris mint itself did so at least once.[35] But these were

30. On Ballin and Bizot, see also Dubois de Saint-Gelais (1885), vol. 2, pp. 159–60; Chautard (1881), p. 334, calls Bizot, author of *Histoire métallique de la République de Hollande* (1687), "très versé dans la science des médailles et auteur de l'histoire métallique de Hollande."

31. "...défendient de fabriquer des jetons ailleurs qu'au balancier du Louvre," per Blanchet (1912a), p. xxxiii, and (1912b), p. 281.

32. "...il ne faut pas confondre la Monnoye des Medailles, avec les autres Hôtels des Monnoyes de France. La Monnoye des Medailles est très-ancienne, & la seule où il soit permis de fabriquer des Medailles, des Jettons, & autres Pieces de curiosité," *Mercure de France*, 1722(2), p. 116.

33. See "Arrest de la Cour des Monoyes: qui défend a tous orfèvres de vendre des Jettons d'argent, & à tous autres Marchands d'en vendre de cuivre, autres que ceux qui auront esté fabriquez en la Monoye des Galeries du Louvre. Du 14 Juillet 1685," CAEF, Ms 8° 3: Ordonances des Monnoyes depuis 1680 Iusqu'en 1699, pp. 103–08; see also CAEF, Ms. F° 80, vol. 1, pp. 16–17, and royal edict quoted in Lapassade (1970), p. 356.

34. Schodt (1873), p. 14; Blanchet (1930), p. 216, (1912b), pp. 281-82. The inventory published by Mazerolle (1903-1907) makes clear that the Monnaie des Médailles struck *jetons banaux*.

35. On these exceptions, see Blanchet (1930), p. 215; Blanchet (1924); Loach (1995), pp. 46, 50; Rossignol (1851), p. 51; Barnard (1916), pp. 59–60; Blanchet (1909a), pp. 399–400.

decidedly exceptions proving the rule. The overwhelming majority of French jetons emerged from the workshops of the Monnaie des Médailles.[36]

Needless to say, those who held it sought to enforce this monopoly. Jean Warin was notoriously litigious in guarding his rights.[37] In 1672, for example, he instituted a successful legal attack on a competing minting activity that had arisen on the grounds of the Palais Royal, not far from where he was in the Galeries du Louvre.[38] In 1685 the then head of the Monnaie des Médailles, Nicolas Petit, secured an edict from the Cour des Monnaies reaffirming his monopoly. Silversmith shops indicated by Petit were to be searched; anything inappropriate confiscated, and merchants and silversmiths found in violation fined 1,000 *livres*, half of the fines going to the Hôtel-Dieu and half to poor prisoners. On the other hand, the edict allowed merchants and silversmiths one week to bring all silver jetons in their possession to Petit to exchange for the current value of silver. "As for copper jetons, the said Petit has full authority, and merchants who trade in these can only sell those furnished to them by said Petit and with his permission with the same penalties for violation to apply, unless the royal prosecutor should think better advised otherwise."[39] The edict from the Cour de Monnaies was posted in the guildhall of the silversmiths in the presence of its officials and all other interested parties. That would have been an interesting scene to witness.

By the same token, monetary legislation constantly railed against the importation of German jetons. The Cour des Monnaies or royal authority repeatedly reaffirmed the exclusive privilege of the Monnaie des Médailles to mint jetons, as happened in 1663, 1672, 1685, 1696, 1699, and again in 1709, all of which suggests that such legislation was to a degree at least ineffective.[40] These challenges to the monopoly of the Monnaie des Médailles seem somewhat less operative in the course of the eighteenth century, but they never went away.

The Monnaie des Médailles Moves to the Louvre

The construction of a new bridge—the Pont Neuf—connected the left and right banks of the Seine across the Île de la Cité.[41] The "new bridge," begun in 1578 and completed in 1607 and still spanning the Seine as the oldest bridge in Paris, necessitated the relocation of the Monnaie du Moulin, and in 1609 Henri IV ordered its removal and emplacement on the right bank of the Seine along the Galeries du Louvre—a stretch of arcades near the old Louvre palace itself and the Tuilleries gardens. This move was an early step toward what the Louvre was to become later in the seventeenth century as an enclave of royal artists and the academies. The stunning 1739 map of the parish of Saint Germain L'Auxerois that included the Louvre (Fig. 7.2) shows the new headquarters of the Monnaie des Médailles as well as its old emplacement where the Pont Neuf now stood.

Fronting another rue des Monnaies, the Monnaie des Médailles occupied four full arcades at #16 along the Quai des Galeries du Louvre. Its emplacement stood next to a building known as *Lesdiguières* (the Dikers) with the royal presses (the Imprimerie Royale) on other side. The placement of the Monnaie des Médailles next to the Imprimerie Royale is to be noted, in that both were similar, technical and mechanical operations. The royal presses, too, were involved with metal

36. Blanchet (1930), p. 215.
37. Mazerolle (1932), vol. 1, pp. 44–48.
38. Lapassade (1970), p. 354.
39. "...l'option demeureroit referée audit Petit purement et simplement quant aux jettons de cuivre, les marchands qui en font commerce ne pourroient debiter que celuy que ledit Petit leur fourniroit et avec sa permission sous les memes peines sauf au Procureur general a prendre telles conclusions qu'il avisera bon estre," CAEF, Ms. 4° 316 (1 et 2) "Introduction Au Droit des Monoyes," fol. 284r–287v.
40. Lapassade (1970), p. 354. On this point, see also Barnard (1916), p. 60.
41. See above and Mazerolle (1907a), pp. 31–45.

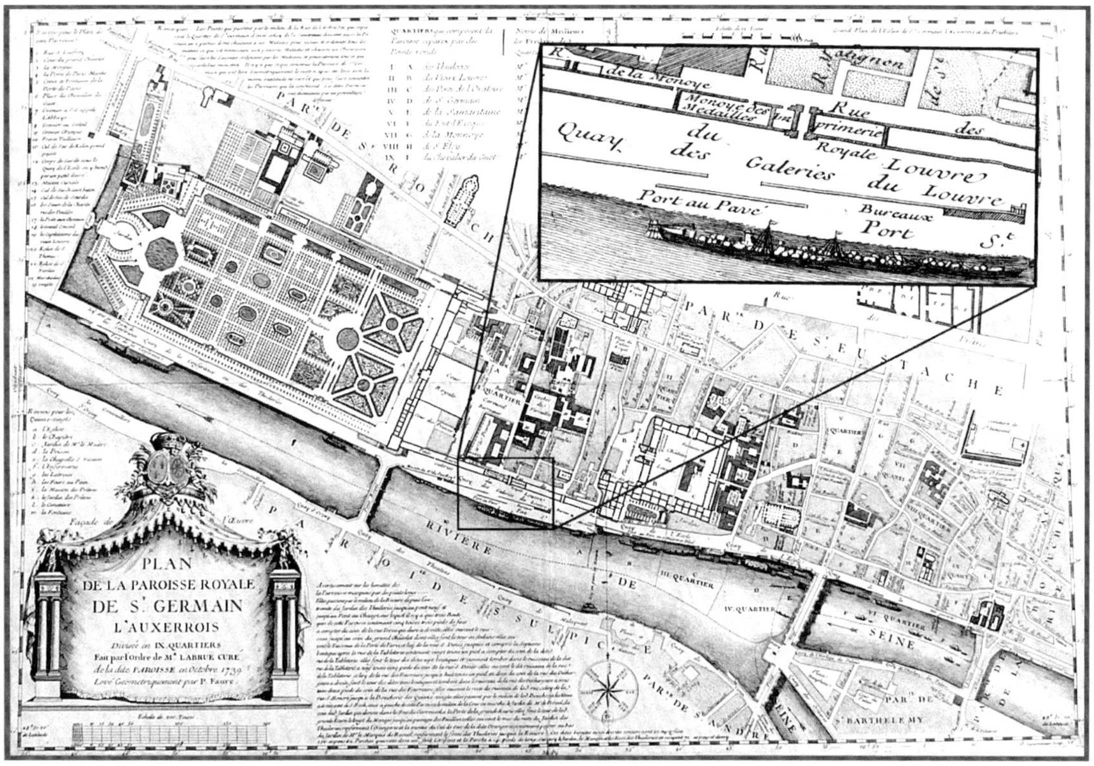

FIGURE 7.2: Engraving: The Parish of Saint Germain L'Auxerois, 1739, showing the Monnaie des Médailles in the Galeries du Louvre.

work, die making, and casting type, and at various points the Monnaie des Médailles and the Imprimerie Royale may have shared tasks and resources.

An inventory made in 1696 allows a precise description of the quarters of the Monnaie des Médailles.[42] The establishment took up 50 m of frontage and 2,500 m^2 of total space.[43] The establishment existed on four levels. Stables, a machine room, a forge and a kitchen were to be found at the bottom level by the docks along the Seine.[44] Reception rooms and offices—a big office and a smaller one for deliveries—fronted the rue des Monnaies on the other side, and there were two stories with eight rooms above that served for lodging and workshops and later a display gallery. Until de Launay moved out, the director lived there.[45] The engravers had their own room with a view of the street. Elsewhere could be found workrooms for the *balanciers*, machines to cut blanks, and other equipment including additional forges, rolling mills, reverberatory furnaces, other ovens, copper boiling-pots, hammers, tongs, tripods, pokers, pincers, sieves, and bellows. At its origin in 1609, by the main door to the Monnaie des Médailles a plaque of black marble with gold lettering indicated to the world that this was the "royal mint for the manufacture of medals, jetons, and gaming pieces in gold, silver, bronze, and copper," underscoring that this mint was a separate entity from the Monnaie de Paris and other mints responsible for royal

42. CAEF, Ms. F° 80, vol. 1, "Procès verbal d'Installation de Monsieur De Launay à la Monoye des Medailles fait par M. Jean Favieres Conseiller dans la Cour des Monnoyes," pp. 33–34.
43. In addition to the view depicted in Fig. 7.3, architectural renderings of the Monnaie des Médailles can be found at BnF, Département des Estampes, Va 440(a) Ft 6; Va. 218e, A16854; and Va 218e, 1356, A16852bis.
44. Lapassade (1970), p. 366.
45. Lapassade (1970), p. 363.

coinage.[46] At some point this sign broke and was replaced with a more modest one that said only, *Monnoie des Médailles*.[47]

1696 Legislation and the Reign of Nicolas de Launay

In 1696 Louis XIV created the hereditary office of director of the Monnaie des Médailles. He may have done so because he coveted the income from the sale of the office or because the notion bubbled up from other quarters, perhaps from de Launay himself or other patrons. One way or another, earlier that year—in June, 1696—Louis XIV had at first decreed the establishment of two new positions to run and oversee the Monnaie des Médailles.[48] One was that of director of the mint (*Directeur du balancier du Louvre*), and the other was controller of the Monnaie des Médailles. The former was nominally in charge of the administration and operation of the mint, the latter nominally superintending its finances and accounting operations.[49] In November, 1696 another royal edict fused the two positions into one overlordship for the Monnaie des Médailles.[50] The position was hereditary and subject to taxation by the Parties Casuelles. The holder of this office had to be at least 25 years old, a Catholic, of good moral character, and to swear an oath.[51]

The Monnaie des Médailles was never an independent institution and was nominally tied to government and more and more to the court through the Maison du Roi or the Bâtiments du Roi, whence flowed *devises* from the Académie des Inscriptions and orders for the production of jetons.[52] This reporting structure remained in place with the reforms of 1696.[53] But with the creation of this formal office in 1696 the Monnaie des Médailles became evermore a cog in the machine of royal administration and its head evermore a powerful personage. In essence, with this new legislation Louis XIV put the Monnaie des Médailles on same administrative footing as other royal manufacturing establishments.

In 1696 Nicolas de Launay (1647–1727) purchased the office of "royal counselor, director, controller and keeper" of the Monnaie des Médailles for the hefty sum of 75,000 *livres*.[54] De Launay was a talented silversmith, and by the time of his appointment as head of the Monnaie des Médailles he had been a royal silversmith for two decades providing fancy goods to the court, including, for example, a seven-foot silver cross destined as a present from Louis XIV to the cathedral in Strasbourg. De Launay had married the niece of a previous director, Ballin, and had worked with Ballin at the Monnaie des Médailles prior to Abbé Bizot taking over in 1678. De Launay was brother-in-law to the sculptor Corneille van Clève and apparently possessed or had access to the

46. "Monnoye du Roy pour la fabrique des Medailles, Jettons, et Pieces de plaisir, d'or, d'argent, de bronze et de cuivre," CAEF, Ms. F° 80, vol. 1, p. 7.
47. Dubois de Saint-Gelais (1717), vol. 2, p. 158; Mazerolle (1907a), p. 32.
48. CAEF, Ms 8° 3: "Ordonances des Monnoyes depuis 1680 Iusqu'en 1699," pp. 755–70.
49. CAEF, Ms. F° 80, vol. 1, pp. 10–18, "Extrait de l'Edit du Roy." The director was to receive 1,500 *livres* a quarter for three quarters and an extra yearly honorarium of 2,000 *livres*, in addition to housing, specified fees for each run of jetons and medals produced, and the same privileges and exemptions as accorded to officers and mint workers at the regular mint. The controller was to be paid 1,000 *livres* for three quarters with an honorarium of 1,333 *livres*, 6 *sols* and 8 *deniers* a year with similar privileges and exemptions. See articles xxi, xxiv, xxv of the 1696 edict; CAEF, Ms. F° 80, vol. 1, pp. 13–16.
50. CAEF, Ms. F° 80, vol. 1, pp. 18–21.
51. CAEF, Ms. F° 80, vol. 1, pp. 24–30, gives the letters patent for de Launay as Director of the Monnoie des Médailles; also Mazerolle (1907a), p. 35.
52. See previous notes and Guiffrey (1885), Lapasssade (1970), p. 356; Blanchet (1890), vol. 1, p. 4. See also CAEF, Ms. F° 80, vol. 2, passim, and CAEF, Ms. 4° 316 (1 et 2) "Introduction Au Droit des Monoyes," vol. 2, fols. 270v–279r.
53. J.-J. Guiffrey, 1885: director reports to "Secrétaire d'État ayant le département de la Maison de Sa Majesté."
54. "Conseiller, directeur, contrôleur et garde de la Monnaie des Médailles," CAEF, Ms. F° 80, vol. 1, p. 15. On de Launay, see also Jacquiot (1989a).

75,000 *livres* required to purchase the office. The Cour des Monnaies registered his letters patent in 1696.[55]

De Launay held the post and ruled over the Monnaie des Médailles for 31 years until his death in 1727. Contemporaries saw him as "one of the most famous personalities of his time," with an important art collection that included a Raphael, a Poussin, and a Carracci in addition to bronze and marble vases.[56] As his wealth and fame increased, lodgings in the Monnaie des Médailles proved cramped. De Launay married his daughter Suzanne to Robert de Cotte, and in 1717 de Launay and his son-in-law bought from the *prévôt des marchands* for 45,600 *livres* a parcel of land across the river near the quai de la Grenouillière. In 1722 they built two houses designed by the first royal architect there at the corner of rue du Bac and rue de Bourbon (presently the rue de Lille), a site later torn down to make way for the d'Orsay railway station, today's Musée d'Orsay.[57] It's easy to imagine de Launay and his son-in-law making their way back and forth from their homes across the new Pont Royal (1689) to the Galeries du Louvre and the Monnaie des Médailles. The engraver Jean Duvivier fashioned medals after his benefactor and his wife, and François Chéreau engraved a portrait of de Launay in 1719.

De Launay and then his son-in-law, Robert de Cotte, followed by Robert de Cotte's son and grandson, Jules-Robert de Cotte and Jules-François de Cotte, exercised total authority over the Monnaie des Médailles for all of the eighteenth century. The position formally paid very little—seemingly in the neighborhood of 2,500 *livres*, but it was extremely lucrative.[58] An anonymous complaint of 1727 had it that de Launay raked in 150,000 *livres* a year, truly an enormous sum.[59] The government paid for all of the administrative jetons and those of the academies, and doubtless de Launay and his successors profited from these payments. There was a minting fee for each run of jetons that applied to the government and private clients alike; this fee was doubtless sufficient to cover the standard pay of the mint workers. De Launay was nominally required to bear the expense of the upkeep of the machines, but he passed that expense off to the government.[60] He priced and sold medals for purchase by connoisseurs, and he decided on how much to pay engravers for their work, and at each step of the way there were opportunities for an extra *sol* or *livre*.[61] Even more, de Launay was an entrepreneurial free agent who took on independent commissions and made money from the manifold ranks of customers throughout the kingdom who desired and ordered jetons. We have seen that they were legion, and it was wholly up to de Launay to decide if ministerial involvement or approval was necessary.[62]

55. CAEF, Ms. F° 80, vol. 1, pp. 24–30.
56. Dubois de Saint-Gelais (1717), vol. 2, pp. 160 and 152 n, where he calls de Launay "...un des plus fameux curieux de son tems & un des plus habiles orfèvres de Paris."
57. See BnF catalog entry, "Cabinet Robert de Cotte."
58. CAEF, Ms. A-1, 9 (8): "Suite de la Liquidation des Finances" [circa 1770]; CAEF, Ms. A-3, 9(2) (9) 1788 (payments to various mints and officers around the country): "Monnoie du Louvre: Jules François de Cotte, Directeur et Contrôleur du Balancier des Médailles...2250#." Lapassade (1970), p. 359, puts the salary at 5,000 *livres*.
59. See anonymous memorandum circa 1727, "De la Monnoyes des Médailles," AN K 903 (#2), 92; the complaint called for more oversight by the Cour des Monnaies, suggesting that the rumor of 150,000 *livres* ("J'ai toujours oui dire que cette monnoye rapportoit jusqu'a 150000# de benefice...") was probably a politically inspired exaggeration; also, Mainguy (1999), p. 305.
60. Compare CAEF, Ms. F° 80, vol. 1, pp. 13–14 (Article XXI of the edict of 1696) and de Launay's view to the contrary, CAEF, Ms. F° 80, vol. 2, p. 4. Lapassade (1970), p. 360, makes this point, too.
61. Lapassade (1970), p. 359.
62. CAEF, Ms. F° 80, vol. 2, p 4: "Il [le Directeur] leur [aux ministres] communiquoit les sujets de celles que le Public demandoit, lorsqu'il était necéssaire qu'ils fussent approuvez."

Right away, de Launay renovated the Monnaie des Médailles, spending twelve to fifteen thousand *livres* to do so.[63] Thirty-five hundred *livres* paid for new display cases, but the largest expense, almost 8,000 *livres*, went for two new *balanciers*.[64] Both were large; the larger of the two was used to strike medals; the smaller, jetons.[65] De Launay's *balanciers* were famous.[66] They were cast by the royal arsenal from cannons furnished by the royal arsenal presumably with direct royal authorization. Both were emblazoned with royal arms and the same Latin inscription provided, naturally, by the Académie des Inscriptions et Belles-Lettres: RERUM GESTARUM FIDEI ET AETERNITATI (For eternal faith in lasting deeds).[67] The smaller press carried the date, MDCXCVIII, the larger MDCXCIX. Bas-relief of allegorical figures decorated both: Justice, Abundance, Force and Religion were pictured on the larger; Music, Sculpture, Engraving and Minting the other. De Launay designed the bas-relief, and the artist who engraved them received 500 *livres* for his work. (These insignia were effaced in the Revolution by vandals or patriots, depending on how you look at it.) Possibly because of a fear of something being struck during off-hours, the *balanciers* were kept locked, and only de Launay had the keys.[68] The bigger of the two coin presses at least remained in use into the twentieth century.

FIGURE 7.3: Engraving: The *Balancier.*

63. Lapassade (1970), p. 361.
64. On these *balanciers* see also Guiffrey, 1884–1888 (1885), pp. 97–101; Mazerolle (1897), pp. 6–7; Jacquiot et al., pp. 399–417.
65. In the eighteenth century a third *balancier* was also used to stamp jetons.
66. Lapassade (1970), pp. 363–67, 388; Lapassade reproduces images of two of these *balanciers*, one especially constructed for striking jetons.
67. Mazerolle (1897), pp. 6–7; (1907a), pp. 39, 165–66: The inscriptions came from the AIBL on April22, 1698. On a related medal with a similar inscription of the AIBL, see Wellington (2015a), pp. 8–9. The Monnaie des Médailles had another *balancier* dated MDCXXIX (1629) and marked AERE ARGENTO AURO FLANDO FERIUNDO (For striking and casting bronze, silver and gold), part of the official title of moneyers in the Roman Republic.
68. CAEF, Ms. F° 80, vol. 1, p. 15, 1696 Extrait de l'Edit du Roy, Article xxiv.

Nicolas de Launay died in 1727 at the age of 89. He was succeeded by Robert de Cotte who ran the Monnaie des Médailles until 1735. Robert de Cotte put together an elaborate manuscript catalog of the jetons he held at the Monnaie des Médailles, but it was never published.[69] The cessation of annual standing orders of administrative jetons from the government in 1758 doubtless had an impact on the vitality of the Monnaie des Médailles, and it seems fair to say that the mania for jetons that ran rampant under Louis XIV and the first part of the reign of Louis XV lessened, and this diminished the vigor of the Monnaie des Médailles somewhat in the later part of the eighteenth century. But we should not lose sight of the fact that the jetons of the academies continued to be churned out until the Revolution and orders from everywhere still flowed in and were faithfully fulfilled year in, year out. Robert de Cotte's son, Jules-Robert de Cotte, took over and ran the institution until 1767, and his son, François de Cotte, then led it all the way through the turmoil of the Revolution and until 1803. At that point, the Monnaie des Médailles, still at the Louvre, passed to the French Ministry of the Interior. In 1807 it was transferred to the regular mint across the Seine at the Hôtel des Monnaies and fused administratively with it in 1832.[70] Even now, the Paris Mint remains a center of artistic expertise and craftsmanship in producing art and commemorative coins and medals. The irony of the history of the Monnaie des Médailles is that when the industrial production of French coinage exceeded the capacity of the Hôtel des Monnaies in Paris, requiring a new factory to be established in Pessac, France in 1973, the artistic production of the Monnaie de Paris remained on the Quai de Conti and became once again an institution distinct from the national mint.[71]

Staff and Engravers

In addition to the director who lorded over the establishment, something like two dozen other people worked at the Monnaie des Médailles.[72] A baker's dozen were the mint workers themselves, *monnayeurs, ajusteurs, ricochons, recuiteurs,* and *tailleresses*, the men and women who tended the forges, prepared the blanks, and actually operated the machinery. Their names remain unknown. They were all members of the guild of mint workers, of course, and they had all sworn their oaths.[73] Their status as guild members probably created a degree of worker-management tension within the mint, despite any shared pride, and everyone's adherence to the religious calendar, in addition to the demands of their own *confrérie*, impacted the nature and pace of work carried on at the Monnaie des Médailles.[74]

The *serrurier*, effectively the blacksmith, was an important person in the Monnaie des Médailles.[75] There was plenty of metalwork that involved him in mint activities. The *serrurier* seems not to have been salaried, but was paid for each bit of piece work. He supplied the engravers with the steel blanks they used to cut their dies, and for each he received 30 *livres* either from the

69. CAEF, Ms. F° 82. "Collection Robert de Cotte (1656–1735)." The title page of de Cotte's manuscript appears as the frontispiece to the present work.
70. For these details, see Mazerolle (1907a), p. 174; Darnis (2005), pp. 884–85; on this organizational history, see also Blanchet (1890), vol. 1, p. 3. Per Lapassade (1970), p. 368, the holdings of the Monnaie des Médailles became the basis for the *Musée monétaire* in Paris.
71. https://www.monnaiedeparis.fr/en/1-150-years-of-history; accessed September 23, 2018.
72. No source gives this figure; it is the author's estimate of mint workers, engravers and other known staff. Valin (2016), vol. 1, pp. 85–116, surveys this personnel.
73. See above, Ch. 5 and CAEF, Ms. A-1, 3a for these oaths.
74. On these points see Lapassade (1970), here pp. 351, 362; Darnton (1984) famously explores this tension in the case of contemporary print workers.
75. On the serrurier, see especially Valin (2016), vol. 1, pp. 87–89.

director or the engravers themselves.[76] He was responsible for making repairs to the machines used in the mint, i.e., rolling mills, a mechanical punch to cut blanks for jetons, the *balancier*, and lesser equipment.[77] In 1727, for example, one Pierre Bagüiet received 2,346 *livres* for these sorts of repairs.[78] Jean de Rondelle took over as *serrurier* in 1732 and was paid 3,003 *livres* for that year and similar sums in subsequent years for like work. The *serrurier* tended his forge on the lowest level of the Monnaie des Médailles. A fire broke out there in 1707 that was not devastating, but required his temporary relocation.[79]

The position of *commissaire*, perhaps a royal commissioner or a connection to the Cour des Monnaies, existed within the Monnaie des Médailles.[80] Subaltern to the director, a *commis* helped run things. The Monnaie des Médailles employed its own coachman and paid someone 200 *livres* to dust the medals and dies in the display cases. At one point the widow of an engraver received a small sum. The Monnaie des Médailles remained at the Louvre for almost two hundred years, and every so often major repairs became needed and outside contractors called in, and in this way the Monnaie des Médailles had dealings with the wider world of merchants and craftsmen, as it would in other connections as well. We find in 1733, for example, repairs to the foundations for the *balanciers* required a mason named Ozé who was paid 2,988 *livres* for his work; in 1736, a certain entrepreneur named Martin earned 3,092 *livres* for masonry and carpentry work to restore balances, forges and rolling mills.[81] Other outsiders occasionally did specialty work for the Monnaie des Médailles in supplying copper hardware or making repairs to the display cases. In 1720 seven individuals, aside from the artist/engravers, received small sums on the account of the Monnaie des Médailles.[82]

Engravers (*tailleurs*) existed as a separate but key group within the Monnaie des Médailles.[83] They made the dies that were struck in the *balancier* to create jetons and medals. They were artists, and they had a certain status. Many were renowned and sought after for their work, so much so that artistic rivalries arose among them.[84] They are decidedly to be distinguished from ordinary engravers, although they may have risen through those ranks.[85] Their positions came with royal appointments and royal pensions, and many were academicians at the Académie Royale de Peinture et de Sculpture.[86] As Jules Guiffrey remarked in 1884, around the Monnaie des Médailles

76. Lapassade (1970), p. 362.
77. CAEF, Ms. F° 80, vol. 2, pp. 6–7.
78. AN, O¹ 2064, "Registre des ordres de Paiment, 1711–1736."
79. See AN O¹ 2064, folder labeled "Monnaie, Fabrication des Médailles, 1692–1744" and letter signed "Chamillard" and dated "A fontainebleau ce 27 septembre 1707."
80. Lapassade (1970), p. 362. The exact nature of this position or the commissioning authority is unclear.
81. See AN, O¹ 2064, "Registre des ordres de Paiment, 1711–1736" for these years.
82. Nocq (1908), p. 425.
83. On the engravers of the Monnaie des Médailles, see Feuardent's list (1915), pp. 355–62; Nocq (1908) and (1911); Blanchet (1920a); Jacquiot, et al. (1970), pp. 69–261; Mazerolle (1902–1904). Desavis (1933), pp. 28–33 gives a list of engravers who crafted jetons for the Faculté de Médecine. Lerouge and Lerouge (1977), pp. 127–30 list the principal engravers of notary jetons through the nineteenth century.
84. Lerouge and Lerouge (1977), p. 65; Mainguy (1999), pp. 109–15, 119. Blanchet (1930), p. 229, reports that in 1715 ministers set up a competition between two engravers, Le Blanc and Jean Duvivier.
85. Mazerolle (1902–1904), vol. 1, pp. 501–84 lists some early Parisian engravers; see vol. 2, pp. 182–205 for some of the jetons by these engravers.
86. See, for example, the early commission afforded Guy Brucher in "l'office de tailleur et graveur de la Monnaie des Étuves" approved by the Cour des Monnaies; Mazerolle (1902–1904), vol. 1, pp. 560–61, 564. On royal pensions for engravers, see Guiffry (1881–1901), vol. 1, col. 1203, with the example of Joseph Roettier's pension in 1679 of 1,500

arose "a sizable school of distinguished engravers, little remarked upon until now, a school that nonetheless merits holding an honorable place in the history of French art."[87]

The engravers had their own workroom, as mentioned, that was fully stocked for their needs. De Launay and his successors, as directors of the Monnaie des Médailles, would choose the engraver for any particular order, paying him according to the type, complexity, and quality of the dies he produced at a price determined by de Launay.[88] Within the Monnaie des Médailles the workshop (the *cabinet des graveurs*) grouped six or seven in-house engravers; de Launay lists seven of them attached to the Monnaie des Médailles in 1715: Dollin, Du Vivier, Le Blanc, Roettiers, Rög the Dane, and De Rosnel.[89] Other royal engravers sometimes did contract work outside of the Monnaie des Médailles when demand was high, but this core group of engravers within the Monnaie des Médailles is to be singled out. Whereas the *serrurier* was called by his first and last names, it is revealing that engravers were known by their last names and the honorific "Monsieur." Thus Pierre Bagüiet was the *serrurier*, while Monsieur Duvivier was the royal engraver.[90]

From Jean Warin onward across the seventeenth and eighteenth centuries, a long line of accomplished royal engravers at the Monnaie des Médailles pursued their art and craft. Once again, the pattern of the arts and trades passing down in families appears here. Thus, Joseph Roettiers (1635–1703), for example, drawn to Paris from Anvers by royal liberality, joined Warin at the Monnaie des Médailles and became an eminent engraver under Louis XIV.[91] An inventory at his death shows him to have become a rich man.[92] His son, Joseph-Charles Roettiers (1691–1779) became royal engraver at the Monnaie des Médailles in 1720, and his son, Charles-Norbert Roettiers (1720–1771) succeeded at some point. A nephew of the patriarch Joseph, also named Norbert Roettiers (1666–1727) was active in the craft, too. Most were academicians at the Académie Royale de Peinture et de Sculpture.

Jean Mauger (1648–1722) was another notable engraver at the Monnaie des Médailles in the reign of Louis XIV.[93] But the most creative and important family of engravers at the Monnaie des Médailles was decidedly the Duviviers. The founding father, Jean Duvivier (1687–1761), became *graveur du roi* in 1719 succeeding Maguer. This Duvivier had a prolific output and was an artist-academician at the Académie de Peinture et de Sculpture. Lodged in the Louvre, he quarreled with another painter and refused to engrave a bust of the king, the result of which was that he was deprived of royal commissions for five years.[94] Jean Duvivier had many children; his son Thomas became an engraver and his daughter Jeanne married one and apparently practiced the art herself. But another son, Benjamin Duvivier (1730–1819), surpassed them all.[95] He succeeded his father in the Monnaie des Médailles in 1762, and in 1774 he received appointment as royal engraver-general of the Monnaie des Médailles, succeeding Joseph-Charles Roettiers. He held the post until 1791, the last person to do so. He, too, was a member of the Académie de Peinture et de Sculpture from 1764; he exhibited regularly at the annual Paris Salon, and after the Revolution he

---

livres paid for through the Bâtiments du Roi.

87. "...une école nombreuse de graveurs distingués, peu remarquée jusqu'ici, et qui mérite cependant d'occuper une place honorable dans l'histoire de l'art français," Guiffrey (1884), p. 477. Lapassade (1970), p. 358, echoes this point, speaking about Warin as responsible for "une véritable école française de gravure en médailles."

88. Lapassade (1970), p. 361.

89. CAEF, Ms. F° 80, vol. 2, p. 6: "Noms des Graveurs attachez à la Monnoye des Médailles."

90. See various entries, AN, O[1] 2064, "Registre des ordres de Paiment, 1711–1736."

91. On the Roettiers, see Villela-Petit (2015a); Lerouge and Lerouge (1977), p. 65; Desavis (1933), pp. 29–30.

92. Blanchet (1954).

93. Guiffrey (1884), pp. 480–81; Jacquiot et al. (1970), pp. 175–20; Chautard (s.d.), p. 2.

94. Desavis (1933), p. 31 n.

95. On the Duviviers, see especially Nocq (1911); Darnis (1988), p. 68; Lerouge and Lerouge (1977), p. 65.

was elected to the Institut de France and taught at the École des Beaux-Arts. Duvivier recruited Jean-Pierre Droz (1746–1823), a peerless artist from Switzerland for the next generation.[96] Other notable names included Nicolas-Marie Gatteaux (1751–1830), who received an appointment as royal engraver in 1781 and was founder of another family of engravers and artists that continued into the nineteenth century.[97] All of these men worked at the Monnaie des Médailles.

Making Dies and Planchets

Three things had to come together to make a jeton or medal: a prepared coin blank or planchet, a die, and the striking force provided by the *balancier*.[98] The engravers prepared the dies; the mint workers produced the blanks and planchets and struck them in the *balancier* to yield up jetons and medals. The following section details die production.

To repeat, in general and certainly for the administrative jetons, the director of the Monnaie des Médailles coordinated with the state minister having the portfolio of the Bâtiments du Roi and with the Académie Royale des Inscriptions et Belles-Lettres before the Monnaie des Médailles undertook any project to manufacture jetons or medals.[99] Once the *devise* had been chosen and a sketch done by the *dessinateur* of the academy, the director would receive the commission, and he assigned the task of crafting a die or dies to one of the engravers.[100]

The engraver would sometimes start his work by carving a large mock-up in wax.[101] Using a mechanical reducing device this initial version could then be scaled down to the size die itself.[102] Engravers would begin with a small block of steel, obtained from the *serrurier*, amalgamated onto an iron punch, and they would chisel and carve their way from there.

Engravers at the Monnaie des Médailles cut two kinds of dies, *poinçons* and *carrés*, and the complexity, costs and functions of the two varied considerably. The one kind, called a *poinçon* (or "punch"), was a die in relief (*en relief*), that is, a "positive" die that looked way the jeton or medal in question would look when minted.[103] The purpose of a *poinçon* was to then strike an intaglio or "negative" die (*en creux*) on another block of steel. These dies were the *carrés* (also known as *coins*) that were used to strike the actual jeton or medal itself creating an impression in relief. (Medals were usually struck in high relief; jetons in low relief.) A *poinçon* permitted any number of *carrés* to be taken from it.[104] Sometimes engravers cut only *carrés*, particularly for jetons in low relief, but an engraved *carré* itself was good for only as long as it lasted.[105] Put another way, *poinçons* made *carrés*, which in turn were used to strike medals and jetons. Alternatively, a *carré* could be created from scratch and then used in the coin press.

96. Darnis (1988), p. 36.
97. Feuardent (1915), pp. 355–62 gives a list of 160 different engravers of jetons, including German engravers.
98. On manufacture, see Lerouge and Lerouge (1977), pp. 35–38; La Tour (1897), pp. x–xiii; A. Barthélemy (1875).
99. Lapasssade (1970), p. 356; Blanchet (1890), vol. 1, p. 4.
100. CAEF, Ms. F° 80, vol. 2, p. 4; on this point, see also Sarmant (2003), p 187.
101. La Tour (1897), p. x; see also Valin (2016), vol. 1, pp. 89–90.
102. Joly (1970) provides an overview of the process for medals, but the same steps apply to the manufacture of jetons; see also Mainguy (1999), pp. 128–30.
103. All this per Joly (1970), pp. 378–79.
104. One suspects that *poinçons* were carved for the obverse of jetons because the royal bust would be used on strikings of many different jetons, requiring many *carrés*.
105. On these processes see Lerouge and Lerouge (1977), p. 35; Joly (1970), pp. 378–88; Blanchet (1890), vol. 1, pp. 5–6; Mainguy (1999), pp. 124–32.

The *cabinet des graveurs* was a large room in the Monnaie des Médailles with workbenches that could accommodate a number of engravers and apparently had an ingenious arrangement for lighting the quarters.[106] To make a *poinçon* or a *carré*, engravers chiseled and worked the blank provided by the *serrurier*. At their disposal were literally hundreds of different chisels and assorted burins, hammers, rasps, and scissors. Engravers used separate punches to stamp the lettering or even standard body parts, for example, onto their die, and the *cabinet* was equipped with over 400 punches for letters of the alphabet.[107] Engravers tested a die's progress by stamping it on blanks of lead or a mixture of lead and tin, light metals on which a die would leave an impression without damaging the die itself.[108] Once the engraver was satisfied, the dies—*poinçons* or *carrés*—were hardened by first heating them to red-hot and then quenching them. After that, they were thoroughly cleaned. To create a *carré* from a previously engraved *poinçon*, a blank *carré* was heated to red-hot and then struck by the *poinçon* in the *balancier*. The engraver would work on the *carré* further with burins and punches, making tests in lead, and the result then tempered and likewise painstakingly cleaned.

FIGURE 7.4: A die.[109]

The completed die (Fig. 7.4) was turned over to the director, thereby becoming the property of the Monnaie des Médailles.[110] In principle, an engraved die could be refused if the commissioning authority was not satisfied.[111] Particularly for medals, but for jetons as well, the rule was that six proof strikings be pulled from the completed *carré*. The chief engraver was paid six *livres* a day to supervise these proof strikings.[112] Once this was done, for the administrative jetons the director prepared a memorandum and sent the proofs to the minister for a formal go-ahead.[113] Once that

106. Lapassade (1970), p. 363.
107. Lapassade (1970), p. 361. Jacquiot et al. (1970), pp. 390, 402–03 show some of the specialized punches involved; the illustrations for the article "Monnoyage" in the *Encyclopédie* similarly show some of the tools used by engravers.
108. Lacronique (1902a), p. 415. The result was a uniface coin. Ones in lead have been found in the Seine where they were presumably tossed.
109. The die pictured here is a nineteenth-century one. Blanchet (1909a) plate VII shows some more contemporary, eighteenth-century dies.
110. Engravers reportedly kept a proof in tin for submission to the Académie des Inscriptions et Belles-Lettres; Joly (1970), p. 386.
111. La Tour (1897), p. xxx.
112. Lapassade (1970), p. 362.
113. See, for example, 1744 manuscript labelled "Memoire concernant les Médailles," AN O¹ 2064.

was received, jetons and medals were struck using the dies crafted in the *cabinet des graveurs*. The director set a price for public sale of medals pegged to the quality of the engravers' work; jetons were always destined for the clients who had ordered them.[114] Again for the administrative jetons, the director would prepare for the minister's attention the bill for the work of the engraver, which the minister would then order paid by the Trésor Royal to the director, who would then, finally, pay the engraver.[115] Otherwise, private parties arranged payment to the director and the engraver.

Costs for cutting dies varied greatly, mostly depending on whether a medal or a jeton was involved, but as well whether the engraver was cutting a *carré* or a *poinçon*.[116] In the early eighteenth century the standard rate seems to have been around 200–250 *livres* for a *carré* for a jeton, 300–350 *livres* for a *carré* for a medal, with the cost rising over the century.[117] *Poinçons* cost more—often, but not always, considerably more. The costs of a *poinçon* and the first *carré* could easily rise to 1,000 *livres*.[118] Thus, in 1715, for example, one of the Roettiers was paid 250 *livres* for his *carré* for the Bâtiments du Roi, whereas his colleague Dollin earned 1,400 *livres* for a *poinçon* for a medal depicting the choir of the church of Notre-Dame in Paris.[119] Similarly, in 1729 a certain Ruck took home 1,400 *livres* for a medal marking the birth of the dauphin, while in 1721 Jean Duvivier earned 300 *livres* for a *poinçon* and a *carré* together for a jeton.[120] Benjamin Duvivier seems an extortionist in asking for 1,500 *livres* for a medal for the Rouen chamber of commerce in 1752; he said it would take him a year to prepare, although he could speed up the work for 2,000.[121]

The Trésor Royal annually paid for 15–16 *carrés* and *poinçons*, spending an average of about 6,500 *livres*.[122] Private clients paid for their dies directly to the director. Engravers were sometimes tipped for their work, e.g., the 78 *livres* the engraver Roussel received as a gratification from the Bâtiments du Roi "for the silver jetons he struck for the Bâtiments for 1699 for which he made the *poinçon* and the *carré*."[123] The total costs of producing jetons and medals did increase over time, about doubling over the course of the eighteenth century, not counting inflation.[124]

The dies were one essential part of the minting process, the blanks or planchets struck in the *balancier* were another. These were prepared by the mint workers in their various capacities. Weight was always strictly controlled to avoid pilferage, so the process began by first weighing the metal

114. "suivant la qualité de leur travail," CAEF, Ms. F° 80, vol. 1, p. 3. De Launay stimulated the private market through the series of medals for the kings of France that he struck, medals only available through the Monnaie des Médailles.
115. CAEF, Ms. F° 80, vol. 1, p. 15; vol. 2, p. 4; d'Affry de la Monnnoye (1878), p. xvii.
116. For these details see Lapassade (1970), p. 362, and AN O[1] 2064, manuscript "Registre des ordres de Paiment, 1711–1736."
117. These rates were much lower in the seventeenth century; in 1608, for example, a *poinçon* and the first *carré* earned the engraver only 75 *livres*; see Mazerolle (1902–1904), vol. 1, p. 579; see related entries for these early costs, pp. 572, 577.
118. Lapassade (1970), p. 362. Pierre Crinon, *ALDE Numismatique*, sale catalog (Paris), 16–17 June 2011 at #181, notes 1,000 *livres* spent on a die for a large and elaborate metal.
119. Guiffry (1881–1901), vol. 5, col. 896 [1715].
120. Per dates, "Registre des ordres de Paiment, 1711–1736," AN O[1] 2064.
121. Wallon (1897), pp. 36–37.
122. "Registre des ordres de Paiment, 1711–1736," AN O[1] 2064. Other sources confirm these numbers; see CAEF, F° 80, vol. 2, pp. 7–10, where de Launay reports 9,600 *livres* spent on 38 *carrés* and *poinçons*; also, Guiffry (1881–1901), vol. 4, col. 977 [1703], col. 1077 [1704]; vol. 5, col. 144 [1707], col. 241 [1708], col. 433 [1710], col. 512 [1711], col. 612 [1712], col. 697 [1713], and col. 789 [1714].
123. "pour des jettons d'argent qu'il a frappez pour les Bâtimens pour 1699, dont il a fait le poinçon et le carré," Guiffry (1881–1901), vol. 4, col. 485 [1699]; see related entries col. 626 [1700]; col. 738 [1701; col. 856 [1702].
124. Lerouge and Lerouge (1977), p. 43.

or metals to be used. The metal (or metals in the case of alloys) was then melted and cast into thin ingots in flat molds.[125] The *ajusteurs* then heated the ingots and squeezed them through the bronze rollers of the rolling mill.[126] At the Étuves the power to turn this mill came from the waterwheel turning in the Seine, but the Monnaie des Médailles at the Louvre used horses and a turnstile-like device down below to drive the rolling mill on the floor above. One way or another, the motive force was limited. From the resulting thin plate another screw-style machine (the *découpoir*) punched out blanks. These blanks were then annealed in a reverberatory furnace. The blanks were then set in another machine that rimmed them and that raised their edges slightly to better take the imprint of the dies. Thus blanks were turned into planchets. Along the way, *ajusteurs* and women *tailleresses* filed them and put on any finishing handwork required.[127] The finished planchets were put in a mildly acid bath for 10 minutes to clean them, and once washed and dried, they were ready to be used in the *balancier*.[128]

Planchets were made from various metals and metal alloys.[129] Gold was essentially reserved for the king and the highest of ministers, and planchets in gold were undoubtedly produced in small numbers, although still in the hundreds. The laborers at the Monnaie des Médailles turned out planchets by the thousands in silver, and ever more common beginning in the era of Louis XIV, copper (*cuivre rouge*) and brass (*cuivre jaune*, an alloy of copper and zinc).[130] Planchets in bronze (the bell metal alloy of copper and tin) were not unknown. From the fourteenth century, a mixture of silver and copper known as billon was a common alloy used for coinage and jetons. Planchets for lesser jetons, especially *jetons banaux*, were frequently prepared using an alloy known as *laiton*, which was 60% copper, 30% tin, and 10% lead.[131] Tin and lead and perhaps arsenic figured in other alloys that went into making jetons and medals.[132]

The supply chain for metals used to make blanks was monitored and regulated. In the sixteenth and first half of the seventeenth century the Cour des Monnaies played a major role in verifying the quality of every melt and in keeping records of metals supplied to the Monnaie des Médailles and other mints. These functions probably continued into the eighteenth century, although, again, a fire in 1776 destroyed most of the records of the Cour des Monnaies, so we really only know about the earlier period.[133] The Monnaie des Médailles itself could easily melt down gold and silver and make its own alloys, and alternatively it would contract with suppliers for raw material needed to strike jetons. Parties who ordered jetons would sometimes supply the metal needed. Registers were kept of metal inventories going in and out of the Monnaie des Médailles.[134] Much of the silver and probably the gold was recirculated from metal handed in to the mint by individuals for a

125. The article "Monnoyage" in the *Encyclopédie* of Diderot and d'Alembert details the processes involved in creating the planchets; see Gillispie, ed. (1959) (vol. 1, plates 198–208); see also Valin (2016), vol. 1, pp. 93–99; Delcamp (2015a), pp. 493–94; Lapassade (1970), p. 354; Joly (1970), p. 372.
126. Lapassade (1970), p. 366.
127. On *tailleresses*, see again Delcamp's pathbreaking study (2015a).
128. Blanchet (1890), vol. 1, p. 5, and Dubois de Saint-Gelais (1717), vol. 2, p. 161 provide these details.
129. On the metals involved, see Blanchet (1890), vol. 1, pp. 14–15; vol. 2, p. 401; Blanchet (1930), p. 212; Boutin (1966), p. 241.
130. Barnard (1916), p. 31.
131. Barnard (1916), p. 91 n.; 91n; Elliott-Smith (1952), p. 1175.
132. See, for example, Mitchiner (1988–1998), vol. 2, pp. 727–29. Sources report that jetons were sometimes made out of mother-of-pearl, leather, and other materials, but, per Blanchet, these are not considered to be part of numismatics; see Corre (1986), pp. 475; d'Affry de la Monnoye (1878), p. xviii; Barnard (1916), p. 56.
133. de Truckheim-Pey (1997), p. 41; see also above, Ch. 1, n. 122.
134. Lapassade (1970), p. 360.

regulated price or in-house unused metal. In 1698, the city of Paris used old jetons as the source of silver for new ones; in 1493 silver money was melted down to make jetons.[135]

Planchets gained weight as time went on. In the early seventeenth century jetons weighed 5–6 g, rising to 8–9 g before 1726 and stabilizing at 10–11 g thereafter.[136]

Jetons were mostly round in shape.[137] A limited number were octagonal, fewer hexagonal, and one reference is for an oval jeton.[138] The size of planchets for jetons became more or less standardized in the eighteenth century, too, with a diameter of 27–29 mm. A fad broke out for octagonal jetons from mid-eighteenth century.[139] These were made by clipping the corners of the blanks. The city of Bordeaux ordered octagonal jetons in 1780 precisely because they were more fashionable and elegant.[140] The agent of the poor Bordeaux chamber of commerce offered his advice from Paris as follows in 1779:

> I should nevertheless advise you that round jetons are hardly made today. The whole world wants them cut on eight or ten sides. But I am not suggesting that the Chamber become fashionable in this regard. I know its means, and this new form would augment the expense by a third, not counting the costs of a new die, which would be something on the order of 600 *livres*.[141]

In 1779, the Bordeaux chamber of commerce stuck with round jetons, but in 1783 it adopted octagonal ones.[142]

Production

With dies and planchets in hand, jetons and medals were minted using the *balancier*.[143] The dies—obverse and reverse—set in square iron blocks (hence "squares" or *carrés*) were fitted into the upper and lower striking portions of the *balancier*. As depicted in Fig. 7.3, the *monnoyeur* sat in a lower pit, put in planchets one by one and retrieved the struck jeton. The mechanical force was delivered by the *balancier* itself as the inertial force of its swinging arms would turn the central screw and deliver the striking force that would imprint the dies onto the planchets. *Ricochons*, the men who toggled the weighted arms of the *balancier* back and forth, provided the muscle-power that drove the machine. This took skill and deft experience, both to set up a rhythm amongst themselves working both arms of the *balancier* that would deliver just the right striking force and to develop a timing with the *monnoyeur* who fed the *balancier* and recovered the finished

135. Barnard (1916), p. 31.
136. Depeyrot (2007), pp. 7–9.
137. Barnard (1916), p. 33. The size of jetons today is measured in millimeters; an obsolete scale, the Mionnet scale, has been used to size jetons; see the scale reproduced and preferred by Feuardent, Feuardent (1904–1915), vol. 1, p. xiii and following p. 502.
138. CAEF, Ms. F° 80, vol. 2, entry for 1720: "de grands Jetons d'une forme ovale pour Mad. La Duchesse d'Orleans" Unusually, as mentioned, the jeton for the assembly of the clergy for 1755 was struck on a small rectangular bar; this jeton (F.3971v) appeared on eBay in May 2015.
139. See anonymous document with this complaint circa 1727, "De la Monnoyes des Médailles," AN K 903 (#2), 92.
140. According to Évrard de Fayolle (1902), p. 155.
141. "Je dois pourtant vous observer qu'aujourd'huy il se fait peu de jettons ronds, tout le monde en veut de coupés sur huit ou dix pans; mais je n'invite pas la Chambre à se mettre à la mode à cet égard; je connois ses moyens et cette nouvelle forme augmenteroit la dépense au moins d'un tiers, sans compter celle des frais d'un nouveau coin qui seroit une affaire de 600 *livres*," letter of M. Trouvé, dated December 18, 1779, reprinted in Évrard de Fayolle (1900), pp. 49–50.
142. Évrard de Fayolle (1900), p. 51.
143. On the operation of the *balancier* and the mechanics of production, see Valin (2016), vol. 1, pp. 103–04; Belaubre and Collin (1992), pp. 156–67 ff.; Darnis (1982), p. 1004: Joly (1970), pp. 376–77; d'Affry de la Monnoye (1878), p. xvi.

pieces. This manufacturing process meant that once everything was set in motion, unless something interrupted the flow, minting occurred until there were no more planchets to strike. (The engraver was not present when jetons were struck.) The stamping of jetons and medals generated considerable heat, and operators sprayed water on the *carrés* to keep them cool. Note, too, that this could be dangerous work; *monnoyeurs* had fingers chopped off, and in one instance at least a worker died from being struck in the head by the swinging arms of the *balancier*.[144] In operation the *balancier* could strike several hundred pieces an hour.[145] Mint runs, as we have seen, ran into the thousands. A major difference between jetons and medals was that each jeton was struck only once, whereas a medal might be struck delicately many times over to perfect the resulting artifact. Although the rule was not always respected, another difference between jetons and medals is that for jetons the front and back sides were coordinated to turn along the vertical 12–6 axis of the clock; this was known as a *frappe monnaie*, whereas medals were struck to swivel front to back about the horizontal 9 to 3 axis (*frappe médaille*).[146]

The legislation of 1696 set the manufacturing costs of minting jetons and medals, and those costs remained fixed throughout the eighteenth century.[147] A later document from the 1720s specified those costs in somewhat greater detail, as presented in Table 7.1.[148] By setting the labor cost (*tarif de la main-d'œuvre*) as a function of the weight of the original metals that went into the melt, the number of jetons or medals produced in each run was not relevant. These costs were charged to the government and to private clients for each run on the *balancier*.

The mint workers received tips, and these were an additional cost. An order from the city of Paris, for example, in 1709 included "eighteen *livres* for wine for the *monnoyeurs*."[149] Another in 1780 was for thirty *livres*.[150] Earlier in the seventeenth century an assembly of the clergy set aside an *écu* "for master Olivier's boys."[151] The use of the word boys/*garçons* here speaks volumes about the social status, one might say class status, of the mint workers.

For almost 30 years from 1697–1726, as director of the Monnaie des Médailles, Nicolas de Launay kept a running log of the work of the mint he ran.[152] One after another, month after month, year after year, he recorded information about every striking, and a summary of his reports shows the workings of the Monnaie des Médailles over a substantial period of time. A total of 747 production runs took place over the 30-year period, which works out to an average of 25 strikings a year.[153] The workload was uneven over the course of a year. De Launay's statistics show that two-thirds of the strikings were made in the fall, with December alone accounting for 40% of the work done on the *balancier*; only a third of the work occurred from January through May, with no strikings at all

144. Jambu (2013), p. 188.
145. Belaubre and Collin (1992), p. 159. Dubois de Saint-Gelais (1717), vol. 2, p. 161 says that a single man can produce 20,000 jetons a day, but this seems to concern rimming them, not minting them.
146. Amandry (2001); Barnard (1916), pp. 56–57.
147. CAEF, Ms. F° 80, vol. 1, pp. 13–14, Edict of 1696, article xxi; see also Blanchet (1890), vol. 2, pp. 401, 405–06.
148. AN K 903 (#2), 92, memorandum headed: "De la Monnoyes des Médailles."
149. "18# pour le vin des ouvriers monnoyeurs," order reprinted in d'Affry de la Monnoye (1878), p. 282.
150. Per Bonnet (1899) in Depeyrot (2007), pp. 85–86.
151. "Plus donné aux garçons dudict maistre Olivier un escu," quoted in Florange and Strowski (1927), p. 12.
152. CAEF, Ms. F° 80, vol. 1, pp. 51–354, is the annual list by month of jetons and medals struck, 1697–1715; vol. 2, pp. 19–139, is for 1716–1726.
153. More precisely 24.9 strikings a year; 1720 saw the most strikings (48); 1726 the fewest (16). Blanchet (1930), p. iv, remarks that in years of political or financial stress production dropped off: "Cet exemple nous montre comment l'histoire financière peut tirer parti de la connaissance de modestes disques de métal."

Monnaie des Médailles, Paris
Fixed Labor Costs
1696–1789

	Metal	Type	Cost per *marc** *Livres–sols*	
Medals:	gold		40	
	silver		16	
Jetons:	gold		20	
	silver	ordinary round jetons	4	10
		round gaming pieces	12	
		octagonal up to 18 lines**	8	
		octagonal above 18 lines	16	
	copper	round (including materiel)	2	10
		octagonal of 12 lines	3	
		octagonal of 12 to 14 lines	4	

* measure of weight; ½ of a *livre*

** measure of length or width; 1/12th of a Paris inch

Table 7.1. Monnaie des Médailles, Paris. Fixed Labor Costs (1696–1789).

in June and July, at least in 1716–1717.[154] Orders generally came in late in the year in preparation for New Year's and the Étrennes, which explains the variable pace of work.[155] Engravers may have operated under a slightly different schedule than *monnoyeurs* because the dies had to come first, but engravers were doubtless under no less time pressure to get out the administrative jetons if not others by the first of the year. The editor of the *Mercure galant* commented on this circumstance in 1709:

> I give you the jetons for this year. They were struck at the Monnoye des Médailles, whence nothing issues that is not perfect, notwithstanding the little time often had to complete it. People always wait until the end of the year to provide *devises* in order to decide the events of the year on which *devises* are to be based. Thus the work is always hurried.[156]

In his study of the workings of the Monnaie des Médailles Robert Lapassade made the point that any examination of the clients of the Monnaie des Médailles will quickly turn into a catalog of the exergues of jetons, and how right he was.[157] De Launay's journal every year indicates the standing orders for nine administrative jetons (Trésor Royal, Bâtiments du Roi, etc.) from the king.[158] One also discovers work being done and quarterly payments going to de Launay and the Monnaie des Médailles for the jetons of the three big royal academies (Française, Inscriptions, Sciences) for a total of 16,000 *livres* a year. Royal funding for the administrative jetons ceased in 1758, except

154. Per CAEF, Ms. F° 80, and de Launay data for 1716–1717, vol. 2, pp. 19–131.
155. La Tour (1897) makes this point, p. xi.
156. "Je vous envoye les Jettons de cette année ; ils ont esté frapez à l'ordinaire, à la Monnoye des Medailles, dont il ne sort rien que d'achevé , nonobstant le peu de temps que l'on a souvent pour le perfectionner. On attend toujours à la fin de l'année pour fournir les Devises, afin d'en voir les évenemens sur lesquels les Devises roulent ordinairement ; ainsi ce travail est toujours precipité," *Mercure galant,* 1709 (1), p. 326; also quoted in d'Affry de la Monnoye (1878), p. 321.
157. Lapassade (1970), p. 361.
158. See again CAEF, Ms. F° 80 and payment orders, AN, O[1] 2064, "Ordonnances" for the "Monnoye des Medailles."

for the Extraordinaire des Guerres, but funding for the academies and work done on their jetons at the Monnaie des Médailles continued all the way through to the Revolution. Beyond that, as Lapassade suggested, the list of clients using the services of the Monnaie des Médailles itself rolls on and on: many units in the royal court, various other government bureaus and offices, individuals of high nobility or ministerial rank, the city of Paris (in what became a standing order) and an assortment of bureaus connected with Paris, several provincial estates, a number of *confréries* and religious organizations, a variety of towns, guilds, chambers of commerce, individual intendants and provincial governors, provincial academies, merchants, foreign clients, and families and others ordering marriage medals and gaming pieces. The author's favorite in this connection might be the jeton struck by de Launay and the Monnaie des Médailles for the *Capitainerie des chasses du roy en la Varenne du Louvre* (Captaincy of the royal hunt in the Paris region, established in 1594). Not everyone ordered jetons every year, but all did on a regular basis, and all this work passed through the Monnaie des Médailles year after year both before and after de Launay's reign as director.

Striking jetons and medals was a violent process, and *carrés* sometimes—even often—broke in the *balancier*.[159] Engravers nominally guaranteed their dies for the first 200 jetons struck, but after that clients could incur further nontrivial costs when dies broke.[160] The *carré* for the 1743 jeton of the surgery academy cracked when being tempered; Jean Duvivier offered to reengrave it for an extra 150 *livres*, which rose to 200 *livres* because of other changes the surgeons requested; another of Duvivier's dies for the surgeons broke in the *balancier* in 1751.[161] As a result of these potential pitfalls, engravers came to offer clients insurance on their dies. For example, for 400 *livres* a year the engraver Gatteaux provided a guarantee even if *two* of his *carrés* broke, and in 1783 the city of Marseille took him up on his offer.[162] In this connection its agent in Paris wrote again to the chamber of commerce in Bordeaux in 1786:

> In order to protect them from such an expense [of engraving new dies in case of breakage] most towns make arrangements with the engraver who in exchange for a fee will assume the upkeep of *carrés*, and in this arrangement he assumes any eventuality. A while back I proposed this to the Chamber, but it felt that the 300 *livres* that was being asked annually was too onerous a sum. Bordeaux city officials [*les jurats*] have nevertheless subscribed and you can find out how much they paid for this. They just experienced the same misadventure with one of their *carrés*.[163]

159. See, for example, AN, O[1] 2064, "Registre des ordres de Paiment, 1711–1736" for 1728 and 1736. Évrard de Fayolle (1901), pp. 20–21, reports an incident of a broken carré of the École de chirurgie at Bordeaux; Mazerolle (1902–1904), vol. 1, pp. 572 and 574 for instances of broken dies from 1569 and 1578.

160. Lerouge and Lerouge (1977), p. 44.

161. Lacronique (1902a), p. 414.

162. Évrard de Fayolle (1902) pp. 155, 156, 158.

163. "...c'est pour se mettre à l'abri d'une pareille dépense que la pluspart des villes s'arrangent avec le graveur qui, au moyen d'une rétribution, se charge de l'entretien de ces coins, et par cet arrangement il prend sur lui les évènemens. J'eus l'honneur de proposer cela dans le tems à la Chambre, elle trouva que 300 *livres* qu'on demandoit pour cela par an étoit une somme trop forte. MM. les jurats se sont néanmoins abonné, et vous pourés vous faire informer de ce qu'ils donnent pour cela. Ils viennent d'éprouver la même avanture pour un de leurs coins," letter of 12 December 1786, quoted in Évrard de Fayolle (1900), p. 57.

The Bordeaux chamber of commerce wrote back, saying, "We are surprised by the frequency that you say these sorts of events occur, and we believe it reasonable to suspect that this could be a scam [*insinuation artificieuse*] on the part of the engraver to lead us to subscribe at a rate we find excessively expensive at 300 *livres* a year, "given the costs of a new *carré* to be only 400 *livres*."[164] In fact, the *carré* for the jeton of the Bordeaux chamber of commerce broke in 1787, and the chamber took the insurance at 200 *livres* a year. But not for long.

Once they were manufactured at the Monnaie des Médailles, jetons had to be packaged and either picked up or shipped to customers directly by the mint or through agents in Paris. Shipping and customs duties figure in interesting ways here. In 1780 it cost the Estates of Languedoc, for example, two *livres* for a box for their jetons and 12 *livres* in shipping fees.[165] But then, because Languedoc was appended to the kingdom as a "foreign province" (*province reputée étrangère),* the jetons of the estates were subject to customs duties; these charges started out in the range of 45–50 *livres*, but ended up at 200–300 *livres* in customs duties, and in 1773 it amounted to an incredible 538 *livres* to import jetons into Languedoc.[166] The Estates of Brittany, by feudal contrast, were exempt from customs duties on its jetons, but the charges for the Languedoc estates remained in place until 1789. The city of Bordeaux ran into similar problems with the regional tax farmers (*Fermiers généraux*) who exacted customs duties on jetons as they came into Bordeaux. In 1735 the city objected, saying that it was overcharged; it wanted all customs duties removed in part because the city had built a beautiful building in Bordeaux for the tax farmers. In 1739 jetons were allowed to pass through customs without charges, but then became the subject of further conflict in the 1740s.[167] These circumstances seem so sadly characteristic of Old Regime France.

Several aspects of using dies in minting and of what went on in making jetons and medals at the Monnaie des Médailles may not be immediately apparent, but they shaped the operation and the resulting numismatic legacy of the Monnaie des Médailles. Recall that it took two dies to stamp a jeton, and these were not always the same pair.[168] Jetons made for the royal navy, for example, or the galleys often had the same reverse, but different busts on the front. Similarly, the same die, for royal busts, for example, could and would be used in striking completely different jetons, hence the notion that probably more *poinçons* were made for busts because of reuse.[169] Dies could be paired and reused in a variety of ways to produce jetons with disparate and iconoclastic fronts and backs, as for example a 1756 silver jeton of the Bâtiments du Roi (Fig. 3.2) that featured a bust of a mature Louis XVI, reigning in the 1780s, but who was only four in 1756. The same set of dies

164. "... Nous sommes surpris de la fréquence que vous nous annoncer de ces sortes d'évènements, et nous nous croyons fondés à soupçonner que ce peut être une insinuation artificieuse de la part du graveur pour nous mener à la composition d'un abonnement que nous trouvons excessivement cher à 300 livres par an," response of the Bordeaux Chambre de Commerce of December 23, 1786, quoted in Évrard de Fayolle (1900), p. 58.

165. Bonnet (1899) in Depeyrot (2007), pp. 82, 85–86. These costs had dropped significantly from 1677 when the Languedoc estates paid over 32 *livres* to ship its jetons from Paris to Montpellier.

166. Bonnet (1899) in Depeyrot (2007), p. 87.

167. Évrard de Fayolle (1902), pp. 152–55, 179–80.

168. Barnard (1916), pp. 58–59.

169. On reusing dies, see also Fontenay (1854), pp. 42–43.

could be used over and over again for an institution with the same date.[170] Thus it is plain and needs emphasis that the dates found imprinted on jetons are not reliable indicators of when they were struck. Hybrid jetons, known as mules, had a front and back of different institutions, as for example, the jeton with one face featuring the ordinary reverse of the Académie des Inscriptions et Belles-Lettres and the other face the ordinary reverse of the royal surgery school and the date of 1775.[171] Such jetons are not unknown in the eighteenth century and seem simply oddities or perhaps shared between institutions? Uniface jetons existed, struck only on one side of the planchet with one die, and more than one of these was ordered to reduce costs.[172] Rare bimetallic jetons exist, such as the 1600 jeton of the Chambre des Comptes.[173]

Then, there is the matter of real restrikes, i.e., minting replicas or reproductions of jetons—accurately or not in terms of their obverses and reverses—at a later date. The Monnaie des Médailles certainly restruck seventeenth-century historical *medals* in the eighteenth century.[174] No evidence suggests that the Monnaie des Médailles ever restruck contemporary jetons to produce duplicates of earlier runs, although, again, dies were constantly reused, and the same client would order another batch of jetons which could then be made using old or newly-cut dies. Later in the nineteenth century, however, with no restrictions until 1900 the successor to the Monnaie des Médailles at the national mint turned out new strikings of old jetons left and right using old, seventeenth- and eighteenth-century dies.[175] To satisfy collectors jetons continued to be restruck in the twentieth century and today, as well. Many of these later restrikes, such as the series of the French Canadian colonies, are well-crafted and authentic replicas and are stamped as such. A notable oddity in this regard are copies of eighteenth-century jetons struck in aluminum sometime after 1888 when industrial production of aluminum came into existence. In a word, numismatically speaking, today one confronts a mishmash of jetons old and new that needs to be sorted out in order to be clear in thinking about the Monnaie des Médailles and French jetons of the Old Regime.

Visitors

The Monnaie des Médailles was not only a manufacturing site, but under de Launay it became a tourist attraction. The marvelous *balanciers* were part of the attraction. Another part was the gallery de Launay set up to display royal *poinçons* and *carrés*.[176] The sizable room featured portraits of the Bourbon kings from Henri IV. It was fitted out with eighteen wooden armoires, especially commissioned by de Launay, with glass panes to better show off *poinçons* and *carrés*.[177] According to a visitor in 1717, the glass "permitted objects to be seen without fearing that their being touched or breathed on should spoil their polish, which is what happens typically."[178] Vases further decorated the space which de Launay would open to selected visitors to the Monnaie des Médailles.

170. Mazerolle (1907b), p. 17, reports that the Avocats du Conseil used the same jetons dated 1762 from 1762 all the way through 1789.
171. F.4401; see also Barnard (1916), p. 57–58.
172. See Blanchet (1890), vol. 2, p. 404, and example, AI, AIBL archives, A:108, Ms: Letter from Marville to "M. Dupuy Secretaire perpetuel de l'academie des belles lettres à Vincennes," dated "a Paris ce 3. Novembre 1778."
173. BnF, DMMA.
174. Seventeenth-century medals have rounded edges, whereas eighteenth-century medals and restrikes have straight ones; personal communication Daniel Renaud.
175. On later restrikes see Blanchet (1930), p. 219; Depeyrot (2007), p. 5.
176. Lapassade (1970), p. 361; Mazerolle (1897), pp. 2–3; Dubois de Saint-Gelais (1717), vol. 2, p. 155.
177. Lapassade (1970), p. 367; AN, O^1 2064, "Registre des ordres de Paiment, 1711–1736" for 1728.
178. "Les poinçons & les carrés du Roi...sont...enfermés dans des armoires à panneau de glace qui laissent la liberté de les regarder sans crainte que le toucher ou l'haleine en gâtent le poli, ce qui arrive ordinairement," Dubois de Saint-Gelais (1717), vol. 2, p. 160.

Coincident with his becoming director of the Monnaie des Médailles, a detailed inventory of 1696 cataloged all the *poinçons* and *carrés* being handed over to de Launay. Under the authority of the Cour des Monnaies, the inventory is 420 folio pages, and it took seven months—from September 1696 to March 1697—to complete. It lists 947 *poinçons* and 1611 *carrés* for jetons and medals making a total of 2,758 items.[179] By 1706 the number of *poinçons* and *carrés* at the Monnaie des Médailles and available for display had risen to over 4,000.[180] The entire collection was said to be worth over two million *livres*.[181]

In the eighteenth century the Monnaie des Médailles became a tourist attraction for high society and aristocratic visitors passing through Paris. Already by 1703 the editor of the *Mercure galant* wrote, "I will say nothing of this place that draws the curiosity of all foreigners of good taste and the best of Paris society."[182] Two decades later in 1722 the *Mercure* observed that de Launay "decorated this lovely place the way it is today, and he formed the rare gallery that displays the numberless *poinçons* and *carrés* of the king in such a way that it's one of the singular curiosities in Paris."[183] Dubois de Saint-Gelais in 1717 likewise called the Monnaie des Médailles one of the newer attractions of Paris.[184] He went on to say that the *poinçons* and *carrés* on exhibit themselves formed "a rarity in Europe that garners the admiration of tourists as much as the site itself."[185] The diarist Saint-Simon similarly pointed to the Monnaie des Médailles. The upshot was that notable tourists visited regularly, and we know about a few of them.

The visit of the king, the nine-year-old Louis XV on Wednesday, August 2, 1719, is one well-documented visit.[186] In the morning the head of the king's bodyguards arrived to inspect the site, as was customary. At three in the afternoon, royal French Guards and Swiss Guards were posted in front of the building and formed two rows beside the Imprimerie Royale, and bodyguards and Swiss Guards took over the entrance and the vestibule. An hour-and-a-half later, the king, who earlier had attended a meeting of the Académie Royale d'Architecture, crossed the small passageway by the presses and came on foot to the Monnaie des Médailles accompanied by Monsieur le Duc, the Maréchal de Villeroy, the former Bishop of Fréjus (chief minister, the Cardinal de Fleury), the Marquis de Bellegarde (*surintendant des Bâtiments en survivance*) and many other lords and persons of distinction. His majesty was first shown the room with the *balanciers*, where a gold medal was struck in his presence with the legend DUM SUAM NUMISMATTUM FABRICAM INVISERET (While he visits his own mint), which M. de Launay offered to the young king.[187] De Launay then showed him the instruments used to create medals and how they worked. Louis was

179. AN, KK 960; Mazerolle (1903–1907) reprints this inventory; see also Mazerolle (1907a), p. 36; J.-J. Guiffrey (1884), p. 485; Jacquiot et al. (1970), p. 404. The *poinçons* and *carrés* were held in three large oak armoires, at least one of these dating from 1507, contained many trays (*plateaux*).

180. Lapassade (1970), p. 368.

181. Contemporary source quoted in Lapassade (1970), p. 366.

182. "Je ne vous dis rien de ce lieu qui attire la curiosité de tous les étrangers de bon goust, et de tout ce que Paris a de plus distingué," *Mercure Galant* 1703(1), p. 241; also quoted in d'Affry de la Monnoye, p. 317; see also Mazerolle (1897), p. 7 n.

183. "...c'est [de Launay] qui a décoré ce beau lieu de la maniere qu'il est aujourd'hui, & qui a formé ce rare Cabinet, qui expose à la vuë les Poinçons & les Quarrés du Roy, qui sont sans nombre; en sorte que c'est une des plus singulieres curiosités de Paris," the *Mercure de France*, 1722(2), p. 117; also quoted in d'Affry de la Monnoye, p. 325.

184. "Les poinçons & les carrés du Roi ne se voyoient point auparavant, au lieu qu'ils sont présentement à découvert dans un grand cabinet,...un lieu nouveau & une des curiosités de Paris," Dubois de Saint-Gerlais (1717), vol. 2, p. 160.

185. "...est-ce une rareté en Europe, & qui ne fait pas moins l'admiration des étrangers que le lieu même"; see previous note and also Mazerolle (1907a), pp. 38–40.

186. CAEF, Ms. F° 80, vol. 2, pp. 58–60; Darnis (1996/1999), p. 151.

187. CAEF, Ms. MC-1—1717, "Visites à la Monnaie."

then escorted up to the gallery where the *poinçons* and *carrés* were on view, and he saw other rarities he reportedly considered with great attention. The royal silversmith, Nicolas Besnier, showed off some handsome pieces of that art.[188] In an adjoining room two trays were set out with silver and gold plated medals in two sizes like the gold one just presented to the king. De Launay remarked to his majesty that he had them struck to commemorate the honor he was bestowing on the Monnaie des Médailles, and he implored his majesty to dispose of them as he pleased. The king accepted them and distributed them over the following days to lords and persons of consideration. As soon as the king left and arrived back at the palace of the Tuilleries, where he was staying, de Launay ran over to thank his majesty for his visit. The young Louis supposedly expressed his satisfaction and assured de Launay that he wished to return again to the Monnaie des Médailles.[189]

As rich and revealing as was this royal visit, that of Peter the Great of Russia earlier in 1717 was an even more impressive event.[190] Peter I (1672–1725) became tzar of Russia in 1696 with the goal of modernizing the country. Peter was always interested in advanced technology and had previously traveled, notably to Holland and England, to study, gain practical knowledge, and bring back the best of the West to Russia. His trip in 1717 had the same purpose.

Peter arrived in Paris on May 7. The very refined French court was somewhat put off by the rough-and-ready party that they had on their hands, but none of that deterred Peter as he made his rounds. One stop included the Académie des Inscriptions et Belles-Lettres. This academy would not ordinarily have held a meeting on Saturday, May 29, but informed of the tzar's visit that morning, it arranged one for that afternoon.[191] The tzar spoke through an interpreter. On learning of the academy's role in inscriptions and medals, he asked to see a medal and was shown the historical series of Louis XIV, some 300 medals.[192] The tzar studied them attentively. He pulled out a purse of his own medals, but only to show in principle that such could be had in Russia, not to measure the quality of the work. Everyone lauded everyone else. The record holds no explicit

188. On this, see as well the report of the king's visit of 1719 in the *Gazette de France*, August 5, 1719.

189. This is a paraphrase/translation of de Launay's report, CAEF, Ms. F° 80, vol. 2, pp. 58-60: "Le Mercredi second d'Août fut un jour bien honorable pour la Monnoye des Médailles et pour M.r de launay qui eut l'honneur d'y recevoir le Roi. Le matin le Major des Gardes du Corps vint visiter ce lieu par tout selon l'usage, et trois heures après midi les Gardes François et Suisses furent postées devant, formant deux haies depuis l'Imprimerie Royale et les gardes du Corps avec les Suisses s'emparèrent de la porte et du vestibule. Une heure et demie après le Roi qui avoit traversé la petite galerie et passé par l'Imprimerie se rendit à pied à la Monnoye des Medailles accompagné de Monsieur le Duc, du Maréchal de Villeroy, de l'ancien Evèque de Frejus, du Marquis de Bellegarde [en marge :] – surintendant des Batimens en survivance – et de plusieurs autres Seigneurs et personnes de distinction. Sa Majesté fut d'abord conduite dans l'endroit où sont les Balanciers, et l'on frappa devant elle une Médaille d'or que M. de Launay lui présenta ; Ensuite il fit voir au Roi tous les Instrumens qui servent à la fabrique des Médailles et il lui en explique l'usage par la démonstration. Le Roi monta après au cabinet où l'on garde les Poinçons et les Carrez, et il vit les autres curiositez de cette maison qu'il considéra avec beaucoup d'attention. Dans le cabinet de tableau il a y avoit sur une table deux petits plateaux d'argent et d'or moulu remplis de Medailles semblables à celle qu'avoit le Roi, et M. de Launay dit à Sa Majesté qu'il les avoit fait fraper pour être le monument de l'honneur qu'elle avoit fait à la Monnoye des Médailles, et qu'il la supplioit d'en disposer. Le Roi les accepta et les distribua les jours suivans aux Seigneurs et aux personnes de consideration. Dès que le Roi fut arrivé aux Tuilleries, M. de launay l'alla remercier et Sa Majesté lui témoigna sa satisfaction en l'assurant qu'elle voulait retourner une autre fois à la Monnoye des Médailles."

190. This visit, too, is well documented. See CAEF, Ms. MC-1 – 1717, "Visite du tsar Pierre Le Grand à la 'Monnaie des Médailles'," CAEF, Ms. F° 80, vol. 2, pp. 42–43; Dubois de Saint-Gelais (1885, 1893); the *Mercure* for June, 1717; Piossens (1729), vol. 1, pp. 318–24; Duclos (1791), vol. 1, pp. 311–28; Mazerolle (1897), pp. 6–7; Darnis (1996/1999), p. 151; and, report in AIBL, H&M, vol. 5 (1729), p. 3. A recent work, Firmin et al. (2017), is to be consulted.

191. See report of this visit and related subjects, AIBL, H&M, vol. 5 (1729), pp. 1–7.

192. See manuscript "Inscriptions des Médailles de l'histoire du Roy rangées par ordre alphabétique," AN O[1] 2064.

mention of whether jetons or medals were offered to the tzar on this occasion, but they inevitably were. Additionally, Peter visited the Académie des Sciences where a similar scene undoubtedly unfolded, but more tellingly perhaps, he visited the Gobelins dye works, not once, but twice. It was in this context that the tzar's visit to the Monnaie des Médailles was arranged for June 12, 1717.

The tzar and his party were received at the Monnaie des Médailles by the duc d'Antin (head of the Bâtiments du Roi), several French nobles, and de Launay. Unbeknownst to the tzar, d'Antin, probably at de Launay's urging, had commissioned a medal to commemorate the visit of the tzar. It was designed in secret. The royal engraver, Jean Duvivier, followed the tzar around for a few days to see what he really looked like and made sketches for a portrait. In a short period Duvivier rushed to engrave the one die for the obverse; the Dane Rög did the reverse because time was short. The two *carrés* were in the large *balancier* when the tzar arrived. At the very start of the visit de Launay had the workers work the *balancier* and strike a planchet of gold set in the machine. Out came the medal with the bust of the tzar magnificently engraved on the front side along with the magnificent title PETRUS ALEXIEWITZ TZAR MAG. RUSS. IMP. (Peter Alexiewitz, Great Tzar of Imperial Russia). The reverse features Fame with two trumpets flying from the north and the words Virgil used to describe fame, VIRES ACQUIRIT EUNDO (It gains strength in going), an allusion both to the tzar himself and to the knowledge the prince was acquiring in his travels. The exergue marked the

FIGURE 7.5: Medal: Visit of Peter the Great to Paris, 1717.

date and place for history, Lutetia, Paris, 1717: *LUTET PARISI M.D.CC.XVII* (Fig. 7.5).[193]

The duc d'Antin presented this exquisite piece of medallic art to the Russian tzar on behalf of the French king. Peter was impressed. He said he had never seen his own face so sharply modeled, implying that Russian minters had not or could not do the same, nor the ones in Holland from whom he had previously ordered historical medals. The tzar inspected and admired the bronze *balanciers*, their beauty, their mass (said to weigh 26,000 pounds), and how they functioned. He even worked the *balancier* himself with the *ricochons*. The party then proceeded to the so-called *cabinet des médailles*, the display room with all the *poinçons* and *carrés*, which the tzar inspected carefully, admiring all the work and skill involved. He was surprised to find a tray of 40 medals in silver and 40 in bronze of the same type as he had just received in gold. These were distrib-

193. For his success in this effort, Jean Duvivier was rewarded with an apartment in the Galeries du Louvre.

uted to the tzar's attendants and everyone in the party. The tzar then visited de Launay's private apartments—de Launay had yet to move across the river, and the tzar admired the paintings and antiques that he saw there. With that, the visit was over and everyone moved on.

Peter was so impressed by what he saw at the Monnaie des Médailles that eight days later, June 20, on the very eve of his departure from Paris, at two o'clock in the afternoon he sent a request to visit the Monnaie des Médailles again, and an hour later he showed up unaccompanied. He wanted to see the machines again, and he examined them more closely and at greater leisure than the first time. De Launay presented him with a gold medal of the king, a present de Launay reports the tzar clutched to his chest, expressing his eternal devotion to the young Louis XV.[194]

The visit to the Monnaie des Médailles in 1722 of the Spanish Infanta, the princess betrothed to Louis XV, had the same ceremonial character and symbolic overtones, but the Spanish princess being just four years old must have lent a surreal air to the affair.[195] She was accompanied by her governess, the duchesse de Ventadour, the princesse de Soubise, and a suite of other noble ladies. De Launay once again showed off the *balanciers* and struck copies of the medals that had been presented to the Infanta on her recent arrival in Paris. She saw the gallery of *poinçons* and *carrés*, and set out in a separate room were the many pieces of a rich and magnificent *toilette* that the king had made for this princess under de Launay's direction. On her departure from the Monnaie des Médailles de Launay presented the Infanta with the medals that had just been struck in the *balancier*, and these she distributed to the ladies in her suite.

Needless to say, lesser mortals, visitors, and tourists did not receive the same elaborate reception in visiting the Monnaie des Médailles as did the king, Peter the Great, or the Infanta, but the tour was a standard one for persons of appropriate standing. Voltaire was apparently impressed when he visited the Monnaie des Médailles.

This ends the story of the Monnaie des Médailles, the fount from which all jetons flowed. The Monnaie des Médailles was a remarkable institution with a long history, one absolutely central to the story of jetons and to what can be said of France in the *Âge classique* using jetons as the entrée into the subject.

194. In 1760 Jean Duvivier's son, Benjamin Duvivier created another medal, "Entrevue du roi Louis XV et du Czar Pierre le Grand," in 1760 as part of an historical series of the reign of Louis XV.

195. CAEF, Ms. F° 80, vol. 2, pp. 96–97. The marriage never took place, and the Infanta was sent back to Spain in 1725.

FIGURE CON.1: Bust of Louis XVI

CONCLUSIONS

Three cheers for the French Revolution! And please note that this is an historiographical hurrah, not a partisan one. Historians can only be grateful for 1789 and the radical break with the Old Regime brought about by the French Revolution and the Napoleonic era that followed. The year 1789 marks a rarity in the historical record, a clear boundary in the historical and historiographical landscape.[1] The Fourteenth of July, 1789, Bastille Day, signaled the end of one era and the start of another in France and the world. Changes didn't happen overnight, and there were continuities, to be sure, but the break was unequivocal. The old French monarchy (pace the Bourbon restoration, 1814–1830) and the distinctive social, political and economic organization of Old Regime society and government were things of the past. The quintessential intellectual, artistic, and affective culture of eighteenth-century France and the Enlightenment and the sensibilities and ways of seeing the world so typical of France in the Old Regime likewise didn't make it much past 1789.

Not surprisingly, the French Revolution also marks an abrupt and decisive turning point in the history of jetons and everything about jetons that had held true for at least the previous two hundred years. All just collapsed: Old Regime society and culture along with their jetons. What came afterward for jetons and for France post-1789 was historically and numismatically very different in all respects.

By 1789 there was less ardor surrounding jetons, but jetons were still entrenched in society in the ways explored in this study...in the guilds, in the academies, in the city councils, on counting tables, and so on. There is some evidence of public distaste over the expense of jetons prior to 1789, but the Monnaie des Médailles remained busy in 1789 stamping out jetons and the occasional historical medal, and that work continued for a while as the revolution gathered steam.[2] However, with such emblems of the monarchy that they bore, jetons came to be highly suspect. They became "tainted with royalism," and many jetons from earlier eras were melted down at this time.[3] A decree by the republican Convention of August 8, 1793, closed the academies and ended jeton practices there. Another sign of the times: a Convention decree

1. McClellan and Regourd (2011), p. 477.
2. Bonnet (1899) in Depeyrot (2007), pp. 84–85.
3. Quoting Barnard (1916), p. 89.

of September 15, 1793, invited members of the guild of haulers of charcoal to hand in their jetons and medals.[4]

The author has uncovered a tiny event that occurred in October of 1789 that sheds a gleaming light on the circumstances of that moment in the French Revolution and the place of jetons in the society undergoing such an upheaval. In particular, the event reveals what Louis XVI himself thought both of his jetons and his revolutionary circumstances. The episode discloses itself on the merest slip of paper in the archives.[5] It is a receipt for the jetons handed over on behalf of the king to an unnamed person. The receipt is dated October 30, 1789. Not three weeks earlier on the 5th and 6th of October the women of Paris had marched to Versailles and brought Louis XVI and Marie Antoinette back to Paris; the king, the queen, and a reduced royal court were newly and more or less forcibly installed in the Tuilleries palace. Events hadn't exactly spiraled out of control, but a violent political revolution was unfolding, and the 35-year-old Louis was right in the middle of it under extremely strained circumstances. It was at this very juncture—and the *conjuncture* is what is telling for the present account—Louis remembered the jetons he had in his personal possession. He may well have brought them with him from Versailles. He was already unsure enough of his situation that he thought about his jetons and sought to consign them to others for safekeeping. In October 1789 they were important enough to him that he would think of them and take action to safeguard them. If in this study we have tried to use jetons to evoke a larger history of Old Regime French society and culture, what could be more evocative than the vignette of Louis XVI acting in this way under these circumstances at this particular moment? The receipt reads as follows:

> The inventory of purses and jetons that were given to me on the part of the king by Monsieur Thierry [head of the Maison du Roi], 30 October 1789: Thirteen purses each containing two rolls of octagonal jetons. Note: I have not unrolled the rolls. One purse containing 100 jetons of the same shape for the Ville de Paris for 1784. Five purses each containing 100 round jetons of the Secrétaires du Roi.
>
> ----------------------------
>
> Empty purses: 11 velvet purses embroidered with silver thread; 1 purse embroidered with gold and silver; 1 purse of gold fabric; 8 purses of calfskin.[6]

The very specificity of the scene makes it fit for a movie, Louis handing his package of jetons to Monsieur Thierry there in the Tuilleries in October, 1789. There's something lonely and ominous about the gesture. Louis' stash of silver and presumably gold jetons was substantial indeed, and there were none of royal purses prettier, but the whole of what he handed over seems haphazard, and the list has a sad quality about it. A purse containing the 1784 jeton of the Ville de Paris? Eight empty calfskin purses? In October, 1789 both Louis and the Revolution had further to go, but the Old Regime and Old Regime jetons did not.[7]

4. J. Florange (1904), p. 65.
5. AN, K 506 #20 (8).
6. "Etat des Bourses et Jetons qui m'ont été données en garde de la part du Roi, par M. Thierry Le 30 8.bre 1789. 13 Bourses contenant chacune 2 Rouleaux de Jetons de forme octogone. N.B. Je n'ai point défait les Rouleau. 1 Bourse contenant 100. Jetons de même forme de la Ville de Paris, année 1784. 5 Bourses contenant chacune 100 Jetons des Secrétaires du Roi, de forme ronde. ------ Bourses vides. 11. Bourses de Velours. 1. Bourse ricotée or et argent. 1. Bourse d'étoffe d'or. 8 Bourses de peau blanche," AN, K 506 #20 (8).
7. Jetons and medals show up in other estates, including those of the great and mighty; Colbert's estate, for example, had 29 gold jetons and 23 velvet purses each with 100 silver jetons; Lerouge and Lerouge (1977), p. 41 n.; see also Wellington (2015a), p. 30.

Two sorts of conclusions follow from here. The first is temporal and concerns the subsequent history of jetons and medals in the nineteenth century and on to today. We can't have come this far without a nod to that story. The second is thematic and analytic and looks to confirm generalizations about the society and culture of Old Regime France that gave us such tiny and scattered relics in jetons.

The Nineteenth Century to Today

Only a comparative handful of jetons were struck through the period of Napoleon, not to ignore imperial propaganda otherwise.[8] One fact too easily overlooked is that the introduction of the decimal system in the French Revolution further hastened the end of the counting table as an instrument of calculation, and hence that function for jetons as counters—its original one in the Middle Ages, does not continue in the nineteenth century.[9] Jetons were not part of the Institut de France and its constituent academies. The government and the bureaucracy in the same way no longer dealt in jetons. There were no more guilds and no more jetons for them. Provincial estates had long since been dissolved into *départements* and their jetons consigned to history. Nobles and aristocrats no longer had personalized jetons. There was a decided revival of jetons and jeton culture after 1815, however.[10] Local learned societies came to use them again as *jetons de présence*.[11] The notaries and medical organizations had their jetons, too, which served the same purpose.[12] Although a monarchical or imperial bust might appear on one of these, they did not do so exclusively, and we are no longer, or at least not to the same extent, dealing with a cult of the monarchy, which was dead anyway soon after 1789. Marriage medals seem to be more common, but that savings banks and nineteenth-century insurance companies became prominent with their jetons is the real novelty of the day.[13] The medallic arts flourished in the nineteenth century and on to today, with the successor to the Monnaie des Médailles continuing to produce art and commemo-

FIGURE CON.2: Jeton of La Nationale life insurance company, 1830.

8. Feuardent passim. The Castorland jetons struck in 1796 for a proposed French colony in North America are to be noted here; see F.1585–1586; Guth (2007a), (2007b).
9. Barnard (1916), pp. 85, 89.
10. Gadoury (1986), p. 3; Müller (1984), pp. 811–12.
11. Banchereau (1909) provides the example of the *Société d'agriculture, sciences, belles-lettres et arts d'Orléans.*
12. Gillet (1897).
13. Gailhouste (2002).

rative coins and medals on the quai de Conti.[14] One way or another, as much as some functions for jetons did continue and new ones developed, iconographically, nineteenth-century French jetons and medals bear no resemblance to their eighteenth-century predecessors (compare Fig. Con.2). They do not look the same and are easily distinguishable as from two different eras and two different worlds.

The concept of what is a "jeton" expanded in new ways in the nineteenth century and in ways that continue to live on today in the French language and in French society and culture. The above-mentioned savings banks and insurance companies of the nineteenth century, soon joined by the modern corporation, transformed the word "jeton" to mean a straightforward cash payment to officers and directors of corporations as remuneration for attending corporate board meetings.[15] These were not token payments or simple *jetons de présence*, as of yore, distributed after, say, a meeting of the Académie Française or the king's council, and no real jetons were ever involved with this sort of "jeton." No, the "jeton" that boards of trustees voted for corporate officers entailed sizable sums; they were sometimes voted in secret as a form of compensation that had dicey tax consequences for those who received them.[16] The story of jetons here links with business history and the history of capitalism much more than with the antecedent, real jetons of the period prior to 1789.

Something similar happened regarding jetons for stockholders in corporations.[17] To effect shareholder loyalty and almost as a form of dividend, companies would pay a "jeton" to shareholders who attended and voted their shares at shareholder meetings. The Banque de France, for example, distributed actual jetons to those who attended shareholder meetings all the way through 1928, when real jetons turned into the corporate "jeton" and a cash payment.[18] Sometimes this "jeton" was paid out as a function of how many shares were owned, and individual shareholders could touch sizable sums.[19] In other situations shareholder jetons were per capita or up to a certain limit. The Société des Bains de Mer of Monaco, for example, in 1932/1933 gave out "jetons" worth two francs per share; the Paris streetcar company (Omnibus de Paris) paid 1 franc up to a limit of 100 francs.[20] One way or another substantial corporate budgets had to be set aside to cover the costs of this new kind of "jeton."

This form of "jeton" as cash payment extends in France and in the French language today to temporary or part-time workers in the theater (lighting, stagehands, and others) who pick up the "jeton" for a night's work. Liars, as in the Middle Ages, are still "*faux comme un jeton*" (false as a jeton). The word *jeton* has further entered the French language in a strange way with the expression, "*avoir les jetons*" (literally "to have the jetons"), meaning to have the jitters or be extremely nervous.[21]

To conclude chronologically, tokens galore proliferated as industrial civilization conquered the world in the nineteenth, twentieth, and now twenty-first century. Furthermore, poker chips and casino chips are jetons, and jetons are gaming pieces (*Spielmarken*).[22] Jetons used to show up

14. Rachline (1992), p. 66 calls the nineteenth century the "golden age" of medals and jetons.
15. Michel (1934) has made a study of corporate "jetons"; see here, pp. 14–18. On these points see also d'Affry de la Monnnoye (1878), p. xxii; Barnard (1916), p. 76.
16. Michel (1934), pp. 18ff., 35, 138.
17. Michel (1934) is again the source here, pp. 82–133.
18. Michel (1934), pp. 13–14; Blanchet (1890), vol. 2, p. 408; d'Affry de la Monnoye (1878), p. xxvii.
19. Michel (1934), p. 52.
20. Michel (1934), p. 93.
21. As in, "J'aiiii les jetons"; François Regourd, personal communication; the French expression "*un vieux jeton*" means an old chap.
22. Mackey (1994) is good on the development of casino chips.

FIGURE CON.3: A collage of modern tokens.

as subway and transit tokens, and let us not forget telephone tokens of an earlier, pre-smartphone period.[23] There are or have been tokens for toilets, and turnstile tokens, tokens for parking and for slot machines, ski tokens, commissary tokens, mail tokens, and more (Fig. Con.3).[24] Nineteenth- and twentieth-century French brothels used jetons to advertise themselves, and these are coveted by collectors. And, special purpose mintings continue to produce jetons for this or that occasion. Tokens have become a recognized branch of numismatics, but in all this richness a cogent sense of what were the jetons of Old Regime France increasingly fades away.

Old Regime France and its Jetons

The great French numismatist Adrien Blanchet summed up matters succinctly when he wrote, already in 1930, "The totality of the various series of jetons and *méreaux* brings with it a highly developed general view of the administrative, commercial and religious structure of the Old Regime."[25] As presented here, jetons expose the heart and soul of Old Regime France. Over decades and decades they accurately map the abstruse social order and complex institutional structure that constituted France up to the French Revolution. When looked at in their entirety the jetons we have encountered reflect contemporary society at every stage, and a dynamic and multifaceted civilization presents itself in the story of jetons from the high Middle Ages to the end of the eighteenth century.

That there was a distinct culture—albeit a multifaceted one—inherent in Old Regime France is not in any real dispute. Vast literatures support the point. Whether one thinks of early modern and Old Regime France from the point of view of the king or a peasant, Versailles or the village, or contemporary Paris, there is a "there" there. The spirit of the times certainly was not an ontologically transcendent *Zeitgeist*; it was dynamic, and it may have to be thought of as ineffable to an extent, but contemporaries recognized and we today recognize a certain unity to the French nation of that time, to the government of France, to the French language, to the history of France and to French customs and identity. The important regionalisms and particularisms affecting France were still French and not so much exceptions after all. France as it developed particularly under the Bourbon monarchs (1589–1792) was a very distinctive society with a significant role in Europe and on the world stage.

The present study about jetons doesn't change any of our previous understandings of Old Regime France or its earlier roots in the Middle Ages and early modern period, but it does extend them in revealing ways. Jetons do so directly and indirectly, and it is easy to summarize the main conclusions.

The first, again, is that with jetons we encounter the distinct culture and civilization of early modern and Old Regime France. Especially as particular customs and odd social practices came to surround them, jetons prove especially useful in helping us unpack the idiosyncratic nature of this culture. The many micro-examples we have encountered show the distinctiveness of Old Regime France, and the more they are unusual, unexpected, and odd, the more they reveal. From jetons on counting tables to jetons as perks of office and part of packages of *épices*, as New Year's presents handed to ladies at court along with *hypocras*, as bureaucratic emblems and tokens for attendance, for erudite literary games, as tokens displaying the insignia of nobles and aristocrats, as tossed at weddings and inaugurations, in their conceptualization as *devises* and manufacture as

23. See Topenot (1971) for examples of early twentieth-century "*jetons de transport*."

24. See, for example, jetons and sale catalogs issued by the Monnaie de Paris; ANS library pamphlet file.

25. "...l'ensemble des séries de jetons et méreaux apporte une vue générale, très développée, de l'organisation administrative, commerciale et religieuse de l'ancien régime," Blanchet (1930), p. iv.

jetons, as made objects or small hoards of silver, or in the singular case of Louis XVI worried about his fate and his jetons . . . examples like these abound and speak of a lost world.

In this connection one would be amiss not to come back to the counting table itself and underscore how jetons and the counting table continued to be used throughout the many centuries surveyed in this study. It would be too easy and condescending to overlook jetons and the unassuming counting table as a quiet, but enduring technology of arithmetic and accounting that merits its chapter in the history books.

On a more grandiose and more important level, the history of jetons reflects quite well the history and development of the modern nation-state of France. This much is plain in what we called the sixteenth-century transition and saw in the advent of jetons for the Conseil du Roi, the Chambre des Comptes, the Cour des Monnaies, and the other governing institutions that followed. The literal imprint of the monarchy under Louis XIV and his successors as witnessed on virtually every jeton proclaimed the nation-state of France. France was not alone among the emerging nation-states of Europe, but it was exemplary, and its jetons tell that tale. In so doing they reinforce Philip Grierson's notion that coinage is both connected to and documents the rise of the modern nation-state.[26]

But through to the end of the eighteenth century, France was not just any nation-state, and it was certainly not secular in any sense. It was Catholic, and the jetons and *méreaux* of religious institutions and organizations reveal that side of things. But so, too, did the symbolism (crosses and the like) found on the jetons of non-Church units, no less than the monarch as REX CHRISTIANISSIMUS, Most Christian King. Even more, jetons show the centrality of the monarchy in early modern and Old Regime France. The monarchy, like the state, has a history, but as it came to be fashioned under Louis XIV, we shouldn't forget that the French monarchy was in principle a divinely inspired institution with absolute power invested in the living person of the king. Jetons proclaimed the monarchy at every turn, and everything about jetons reeked of the monarchy. One of their main functions was to serve as monarchical propaganda and testimony of the submission of other authorities and institutions to the power and authority of the king. In this sense jetons give us a skewed view of the universe of the Old Regime. In particular, the various offices of the royal court are strongly represented, and they make plain that jeton culture was deeply rooted in the monarchy and the court. At the same time, jetons do present a virtual office directory, a "green book" of the government that nominally flowed from court and royal authority, and we need hardly call out the many offices and bureaus of royal government or the administrative jetons that were issued and paid for by the crown for 200 or so years. Out of like royal largesse and patronage, jetons allow us sharp insight into the world of the royal academies, a nontrivial adjunct to royal glory and the power of the French state.

The French king was an absolute monarch, but not-so-paradoxically the king was not everything, and there were other centers of power and authority in Old Regime France that did not exactly contend with the Crown, but which shared the political, social, and moral space with royal prerogative. Here, too, jetons provide a good guide to the lay of the land. The Church again shows up with its *méreaux* and jetons as an independent pole, as do, to varying degrees, provincial estates and towns with their privileges and their jetons, not to mention independent nobles and aristocrats of the Second Estate with their personalized jetons. The importance of Paris to the whole of the nation can be found not just in the series of jetons for the Ville de Paris, but the large number of other entities and individuals in Paris who had jetons struck for themselves and their offices.

26. Grierson (1951) and above, introduction.

As the case of Paris shows, and other cities and towns do as well, the culture of jetons may have been rooted and found its origins in the monarchy, but it became an urban phenomenon and the province of urban elites across the whole of France.

The guilds constituted an important slice of the urban milieu in Paris and elsewhere, and what we saw of the guilds and their jetons brings home the further point that for France in the Old Regime we are dealing with a preindustrial society and something more like a traditional, feudal economy. The steam engine had yet to make itself felt. In the eighteenth century and earlier the power of water and wind made things go, and horses and draft animals provided muscle to transport people and goods and do work, as did humans in the case of the *balancier*. As man-made objects and in their very manufacture at the Monnaie des Médailles jetons bespeak a level of technological development and economic organization characteristic of Old Regime France on the cusp of the Industrial Revolution and the birth of industrial civilization.

There were forces and factors in France at the time propelling modernization—mining, shipbuilding, textile manufacture, and trade in colonial products (notably sugar, coffee, cotton and indigo), but the backward-looking vision of the world wrapped up in the iconography of jetons and medals and the quasi veneration of classical antiquity we encountered in the Académie Royale des Inscriptions et Belles-Lettres and academicians composing their Latinist *devises* only reinforce the point that Old Regime France was still a traditional society—economically, politically, socially traditional and conservative. The boundary we find at the borderland of 1789 signals more than the arrival of political modernity. Industrialization is more truly a phenomenon of the nineteenth century for France. A whole new technological, economic, and social world awaited France, French history, and French numismatics in the nineteenth century that required a reorientation in the collective vision away from the past and on to an industrial future.

Jetons and the history of jetons show all this more or less directly in and of themselves and in how they were used in manifold situations of social practice. Beyond that, jetons show us the world of early modern and Old Regime France indirectly by their absence, where they do not appear, where we do not find them. Robert Proctor and Londa Schiebinger pioneered the concept of agnotology or the study of ignorance and how a lack of knowledge or mistaken knowledge can be a factor in history, and in a related way the concept proves useful here. The agnotology of jetons or, as it were, the shadows of jetons cast by their absence reveal the rest of contemporary French society and culture more dimly perhaps, but enough to see the outlines of areas beyond the ambit of jetons.[27]

Women, for example, have shown up to a degree in this account, particularly working women, laboring alone or in families, earning a living in the trades and licensed in the guilds, such as we have seen of purse makers and the *tailleresses* in the royal mints. A handful of elite women have made appearances, but the majority of French women of whatever station in the Old Regime are silent as far as jetons are concerned, as they were from so much else, although we know that many women, unrecorded by history, kept accounts with jetons on counting tables.

We previously mentioned the surprising general absence of universities and parlements in numismatic catalogs and the history of jetons. To this point we might suggest a possible geographical imbalance with jetons seemingly more deeply rooted in the north and east than in the south and west of France, and this circumstance, if true, may reflect other patterns in the political and social geography of contemporary France, including that not all areas of the kingdom used the counting table and some had shifted to modern arithmetical methods. But the utter absence of the

27. Proctor and Schiebinger, eds. (2008).

peasantry in this book and in the literature concerning jetons is what is most remarkable. If jetons formed part of the urban scene, and they did, they and the city thereby defined the rural outlands, and in the hexagon of France prior to the French Revolution the rural encompassed essentially all of the territory and eighty-five percent of the population. Jetons might be found in manor houses, and in odd instances peasants and plows show up in the iconography of jetons, but otherwise jetons did not penetrate *la France profonde*. The massive agricultural production that took place there and ultimately supported the thin film at the top does not appear in the story of jetons. This, in contradistinction to the much smaller level of craft and trade activity in the guilds. The vast rural landscape of Old Regime France, with its myriad of villages and its millions of peasants toiling in the fields and in the vineyards, stretched out across the large and fertile country beyond the walls of the city and beyond the world of jetons.

This book turns out to have had grand ambitions. It set out in the first place to see what the result might be when one combines history and numismatics in a case study of early modern and Old Regime France. Harking back to Philip Grierson's 1951 address, "Numismatics and History," one hopes that, as an exercise in how historians can enrich their work by incorporating numismatics, this one will not be thought a failure. The hope was also that numismatists might take note of this study as illuminating jetons and this one corner of numismatics, and about that we'll just have to see.

At the outset we invoked cultural history and the work of Clifford Geertz, Natalie Davis, and Robert Darnton as models for what this study might pretend to. The idea was, at least in part, to see jetons as completely woven into contemporary culture and to use jetons as tools to tease out unanticipated and revealing resonances with that culture. To see a world in a grain of sand, as Blake would have it. In writing these words the author is remembering the visit of Peter the Great to the Monnaie des Médailles in 1717 and that fellow from the Third Estate from Burgundy in 1737, kneeling before the king as part of the delegation from the Estates of Burgundy presenting its *cahiers de doléances* and its jetons. Through all the little vignettes involving jetons that we encountered, we can only hope to have succeeded in evoking at least something of those bygone days.

The largest ambition of the present work, it turns out, has been historiographical. In response to critiques of microhistories and the movement to write them, the effort here has aimed to combine microhistory and macrohistory over Fernand Braudel's *longue durée* in order to give a diachronic account of early modern and Old Regime France.[28] Ours is not just a *longue durée* history of jetons, mind you, but even more ambitiously of the *longue durée* of the cultures and societies of France that produced jetons and used them over the six hundred year period from 1200–1800. We have to employ the plural, cultures and societies (do we not?), in seeing the movement of jetons from the counting tables of the high Middle Ages through to the refinements at the top levels of society and the ruling order in France on the eve of the French Revolution. With these considerations in mind, the author's effort here has been as a pointillist painter, evoking via an illusion of dots a fleeting image and a passing feeling, in this case for the whole of French society and culture as they developed over time up until the French Revolution. The success or failure of that enterprise will have to be judged by others.

28. Above, introduction at n. 99; Cohen (2004).

In his book, *The Past is a Foreign Country*, David Lowenthal invites us to consider the residues of the past that have washed up on the shores of the present.[29] And jetons are certainly artifacts from the past. They are Blake's grains of sand. They have come down to us not quite willy-nilly, and they carry their own histories with them in the patinas, wear, and nicks they bear. Jetons bring with them, too, centuries' worth of collecting and numismatic study. But one way or another the jetons we have examined here all hark back to particular times and places—be that medieval, early modern, or Old Regime France, times and places far removed from our present, but ones we can glimpse again by dint of these tokens. What a glorious story jetons have to tell historians, numismatists, and anyone interested in France and its history.

29. Lowenthal (2015), original edition in 1985. Be it noted that historiography is another residue that we inherit from our forebears.

JETON: Experts des Bâtiments.

GLOSSARY[1]

"administrative jetons"—Jetons for various departments of the royal court, government offices, and the military ordered annually and paid for from royal funds for jetons.

ajusteurs—Mint workers who made the coin blanks and planchets used for striking as coins, medals or jetons.

balancier—A coin press.

Bâtiments du Roi—A bureaucratic division within the Maison du Roi that was responsible not only for the construction and maintenance of royal palaces and the king's buildings generally, but also for superintending royal manufactories, parks, and botanical gardens, and for providing patronage for the fine arts and the academies. The Bâtiments du Roi formally superintended the Monnaie des Médailles. See also **Maison du Roi** and **Monnaie des Médailles**.

billon—An alloy of copper and silver with less than 50% silver.

brassage, droit de brassage—Right of the director of a mint to set aside gold, silver, and copper from stock to pay for the costs of running the mint.

bureau de la ville—A town council.

Cabinet des Médailles—The official repository of coinage and the king's historical medals and jetons. The Cabinet des Médailles also housed the royal collection of ancient coins and medals. A subsidiary unit of the royal library, the Bibliothèque du Roi.

cahiers de doléances—A list of demurrals or grievances, most famously drawn up by the three estates in preparation for the convening of the Estates General in 1789, but these were also prepared by the estates of provinces for presentation to the king as part of biannual or triannual *voyages d'honneur*. See ***voyages d'honneur***.

carré—An intaglio die used to strike coins, medals, and jetons in the *balancier*. Could be sculpted *de novo* or by being struck with a *poinçon*. Also known as *coins*. See also ***balancier*** and ***poinçon***.

Chambre aux Deniers—Paying office for accounts of the royal household.

Chambre des Comptes—Royal accounting office. A royal department responsible for auditing royal and princely accounts. In Old Regime France several *chambres des comptes* were scattered throughout the kingdom; the Chambre des Comptes de Paris was the most important and the one referred to most often here.

confrérie—A brotherhood or religious confraternity of members of a guild or associated trade

1. On French weights and measures, see especially Zupko (1978); see also Newton (2014), pp. 277–78; McClellan (1992/2010), pp. xix–xx.

who worshiped together at a local parish church and who organized themselves for mutual aid and the provision of social welfare.

coupoir or ***découpoir***—A machine to cut out planchets for further handwork. See **planchet**.

Conseil du Roi—The king's council.

Cour des Monnaies—The sovereign royal monetary court.

denier—Unit of currency, equivalent to the pence in the Imperial system of pence, shillings and pounds. Twelve *deniers* made a *sol* and 240 *deniers* made a *livre*. The *denier* was also copper coinage denominated in 3- and 6-*denier* units valued at ¼- and ½-*sol* respectively; the 3-denier coin was also called a *liard*. See also ***sol*** and ***livre***.

devise—The ensemble of a jeton's design that principally included a symbolic image or emblem (*type*) and a legend and that carried an implicit message.

dessinateur—The in-house artist at the Académie des Inscriptions et Belles-Lettres who utilized the academy's suggestions for legends and designs for medals and jetons to create draft sketches for use by engravers at the Monnaie des Médailles.

échevin—An alderman or municipal councilor. Four *échevins* served along with the *prévôt des marchands* and 26 additional councilors on the city council of the city of Paris.

écu—A silver coin officially valued from 1726 onward at 6 *livres* tournois or 120 *sols*. Also as ½-, ¼-, and ⅛-*écu* denominations valued at 60, 30 and 15 *sols*. See also ***sol*** and ***livre***.

épices—Literally, "spices." A package of incidentals that came as a perk for individuals holding an office. These incidentals variously included paper, pens, penknives, inkstands, candles, matches, real spices, and jetons. Also as fees paid in court proceedings.

estates—Provincial governing units uniting representatives of the first, second, and third estates (the clergy, nobility, and commoners) that saw to certain tax liabilities and to the representation of the province before the seat of royal authority.

Étrennes, the Étrennes—The exchange of presents on the occasion of the new year. The occasion of such presents.

exergue—The small segmented space below the main design of jetons and medals, usually on the reverse and separated by a line from the main field above. An exergue may or may not contain an inscription or a date.

Extraordinaire des Guerres—The accounting department and disbursement ledger of the French army that handled the general staff of the army and other military units enlisted in the service of France but not in the personal service of the king. See also **Ordinaire des Guerres.**

flan—A prepared coin blank prior to striking; synonymous with planchet. The term is used more for ancient coins. See also **planchet**.

gros—A unit of weight equaling 1/8th of an *once*, 1/64th of a *marc* and 1/128th of a *livre*. See also ***marc*** and ***livre***.

histoire métallique—Metallic or medallic history or documenting and narrating history via a collection of medals.

Hôtel de Ville—Town hall; more in particular, the city hall of Paris.

Hôtel-Dieu—A hospital, notably the oldest hospital in Paris on the Île de la Cité.

Hôtel des Monnaies—Seat of the royal mint in Paris on the Quai de Conti.

huissier—An usher or bailiff.

jetonnier—Someone who attended meetings for the purpose of collecting *jetons de présence*.

jetonophilie—The hobby of jeton collecting; amateur enthusiasm for jetons.

jetons banaux—Common tokens, as in gaming tokens or counters. Plural of *jeton banal*.

jetons d'école—Tokens used for instructional purposes in schools teaching arithmetic and writing.

See also **Rechenmeister**.

jetons de compte—Accounting tokens for use on counting tables.

jetons de présence—Jetons given and received for attendance at meetings.

jetons d'étrennes—Jetons offered at the new year.

jurés—Sworn officers of towns or guilds.

livre—A unit of currency. The equivilant of a pound in the Imperial System of pounds, shillings and pence. The *livre tournois* (or Tours pound) became standard currency in the realm. Twenty *sols* made a *livre*, as did 240 *deniers*. Originally corresponding to the value of a pound of silver, which in the eighteenth century was worth approximately 100 *livres*. See also ***sol, denier***, and ***livre*** (unit of weight).

livre—A unit of weight. The principal unit of weight in Old Regime France. From the fourteenth century the *livre de Paris* weighed 489.5 g. It was subdivided into 2 *marcs*, 16 *onces* or 128 *gros*.

louis d'or—Standard French gold coin issued from 1640 in various denominations. From 1726 the value of the *louis d'or* was set at 24 *livres*. See ***livre*** (unit of currency).

magasinier—Storeroom attendant.

Maison du Roi—The administrative department superintending the royal court and the king's household. It included domestic, military and religious branches. It likewise subsumed the departments of the king's buildings (the Bâtiments du Roi) and the Menus-Plaisirs du Roi.

marc[2]—A unit of weight equaling half a *livre* or 244.75 g. A *marc* is further divided into 8 *onces*, 64 *gros*, 192 *deniers* or 4,608 *grains*. From "marck" or frontier, meaning an official weight within the borders of a territory. See also ***livre*** (unit of weight).

Marine Royale—The French Royal Navy.

Menus-Plaisirs du Roi—A department within the Maison du Roi and the royal court responsible for the lesser pleasures of the king. Its functions included organizing concerts, balls, theatrical performances, ballets, masquerades, religious ceremonies, and festivities of various sorts. The Menus-Plaisirs du Roi also superintended royal musicians.

Mercure galant—A monthly gazette published in Paris with a royal imprimatur from 1672, continuing from 1724 as the *Mercure de France* through to the French Revolution.

***Méreau, méreaux* (pl)**—Church or mass tokens.

Monnaie des Médailles—The official royal mint in the Galleries du Louvre in Paris chartered to strike medals and jetons.

Monnaie de Paris—The official royal mint chartered to strike coinage.

monnoyeurs—Mint workers who manually inserted planchets into the *balancier* and retrieved struck coinage, medals, or jetons. Before that, mint workers who hammered coinage. See ***balancier*** and ***ricochons***.

Ordinaire des Guerres—The accounting department and disbursement ledger of the French army that handled military units enlisted in the personal service of the king. See also **Extraordinaire des Guerres.**

parlement—A sovereign law court that judged on appeal and in the first instance and that formally registered royal edicts. There were 13 regional parlements in France prior to 1789. The Parlement of Paris was the most important, essentially a supreme court. Parlements had no directly legislative function. Seats were venal and inheritable.

planchet—A prepared coin blank ready for striking. See also **flan**.

2. On the *marc*, see previous note and Dillemann (1997), p. 14; Droulers (1989), p. 187; Guiffry (1881–1901), vol. 2, col. 537 [1684]; CAEF, Ms 8° 3, Ordonances des Monnoyes depuis 1680 Iusqu'en 1699, pp. 93–98.

Parties Casuelles—The bureaucratic royal office superintending venal offices. The Parties Casuelles kept records and collected the requisite fees on the sale, transmission, and holding legal title to offices. It also managed the collection of the annual fee (*La Paulette*) required to continue to hold non-hereditary offices.

Petite Académie—Dating from 1663, the subgroup initially selected by Colbert from among members of the Académie Française, whose purpose was to create *devises* for medals and jetons. The backbone of what became the Académie Royale des Inscriptions et Belles-Lettres in 1701.

plumatifs—Raw minutes of meetings.

petits monuments—"Little monuments." Among numismatists, another name for jetons, especially those of the seventeenth and eighteenth centuries.

prévôté—A unit of justice or a specialized court exercising local jurisdiction, often invested with police authority.

prévôt des marchands—Effectively, the mayor of Paris, elected for two-year terms by a selection of representatives from various neighborhoods in Paris. Originally dean of the city guilds, the *prévôt des marchands* chaired the city council and along with four *échevins* and 24 city councilors governed the civil affairs of the city. See also *échevin*.

poinçons—A "positive" die in relief used to create *carrés*. See also **carré**.

Rechenmeister—In German lands, the calculator at the counting table. German teachers at lower-level municipal schools or private schools teaching arithmetic and writing.

Rechenpfennig, Rechenpfennige (pl)—A jeton manufactured in Germany; a Nuremberg jeton.

ricochons—Mint workers who toggled the arms of the *balancier* to strike coins, medals and jetons. See ***balancier***.

serrurier—Locksmith, blacksmith.

Six Corps—The "Six Bodies" or six trades. The exclusive set of the top six Parisian guilds: the weavers, spice merchants, retail merchants, tanners, milliners, and silversmiths. The *Six Corps* set themselves above all other guilds. See also ***prévôt des marchands***.

sol, sols—A unit of currency, equivalent of a shilling in the Imperial system of pence, shillings and pounds. Twelve *deniers* make a *sol*; 20 *sols* make a *livre*. The *sol* (a copper coin) was denominated in 1- and 2-*sol* units. Also known as *sou* and *sous*. See also ***livre*** (unit of currency) and ***denier***.

tailleresse, tailleresses (pl)—Female mint workers who did the handwork preparing planchets. See also **planchet**.

Trésor Royal—The administrative department of the royal treasury.

voyages d'honneur—Biannual or triennual trips undertaken by delegations from the estates of outlying provinces to present their *cahiers de doléances* to the king. See ***cahiers de doléances***.

Jeton of the Académie Royale de Peinture et de Sculpture, 1764.

LIST OF FIGURES[1]

TITLE PAGE

Obverse of the jeton of the Académie Royale des Sciences, Belles-Lettres et Arts of Lyon, dated 1749.

F.10747.[2] Silver. 30.6 mm. 9.69 g. 6:00.[3] Planchet is 1.78 mm. Reeded edge. Signed: D.V. (Duvivier).

Obverse: Arms of Lyon with triple fleur-de-lis and seated allegorical nude figures representing the confluence of the Rhône and Saône rivers at Lyon. Date of 1749 is found below the shield and between the legs of the two figures. See also Prieur/Sombart catalog, *Jetons XIX* (2004), no. 207.

FRONTISPIECE

Drawing: Pen-and-ink and brush with gray wash. Title page of ms. "Jettons de France du Cabinet de Monsieur de Cotte a la Monnoye des Medailles." CAEF (Ministère de l'économie, des finances, de l'action et des comptes publics, service des archives, MEF-MACP, SAEF), YF-82, "Collection Robert de Cotte (1656–1735)," 1727. Licensed reproduction. See also Valin (2016), vol. 1, pp. 17, 394.

PREFACE

Jeton of the Marine Royale dated 1712.

F.1288. Silver. 29.0 mm. 6.72 g. 6:00. Planchet is 1.37 mm. Smooth edge.

Obverse: Bust of Louis XIV facing right (bust no. 245),[4] signed with monogram: T.B. (Thomas Bernard). Legend: LUDOVICUS MAGNUS REX (Louis the Great King).

1. Unless otherwise indicated, the images are of items in the author's collection. Unless otherwise indicated, photographs are by the author and © by the author. Some images have been retouched electronically by the author. Jetons, medals, and church tokens are depicted on a scale representing 5 cm.
2. References are to Feuardent numbers per Feuardent catalog (1904–1915). Also indicated are the jeton's metallic composition, diameter, weight, planchet thickness, quality of the edge, die alignment, and engraver, if known.
3. Die alignment. 6:00 indicates a die alignment (coin rotation) where the reverse is rotated 180° from the obverse and the "flip" is along the 6:00–12:00 axis; 12:00 indicates a medallic rotation where the reverse is rotated 0° from the obverse and the "flip" is along the 9:00–3:00 axis.
4. Royal busts and variants are known by the numbers assigned to them by Felix Feuardent (vol. 1, 1904, plates). Feuardent enumerates and reproduces 501 royal busts in his plates.

Reverse: A crowned Neptune with his trident plying the seas in his shell-backed and pedaled water chariot drawn by hippocampi. Legend: BELLO PACIQUE (In war and in peace). In the exergue: MARINE - 1712.

INTRODUCTION: HISTORIANS AND NUMISMATISTS

Fig. In.1. Jeton of the guild of architects (undated).
F.4913v. Silver. 29.6 mm. 9.46 g. 6:00. Planchet is 1.66 mm. Reeded edge.
Obverse: Bust of Louis XVI facing right (bust no. 395), signed: DU VIV (Duvivier). Legend: LUD. XVI. REX CHRISTIANISS. (Louis XVI, Most Christian King).
Reverse: Helmeted Pallas standing and facing left holding a level and a lance with the instruments of architecture at her feet beside a course of stone. The inscription is CONSOCIARE AMAT (It loves to bring together). From Horace, *Odes,* Book II, Ode 3, lines 9–11 (Quo pinus ingens albaque populous/umbram hospitalem consociare amant/ramis? /Quid obliquo laborat lympha fugax trepidare rivo?)

Fig. In.2. Jeton of the carpenters' guild (*charpentiers*), 1773.
F.5046. Silver. 28.3 mm. 8.17 g. 6:00. Planchet is 1.88 mm. Reeded edge.
Obverse: Bust of Louis XVI (bust no. 379) unsigned. Legend: LUDOV.XVI.REX.CHRISTIANIS. (Louis XVI, Most Christian King).
Reverse: Wooden crane and frameworks for a bridge and building. The inscription is SERIES JUNCTURAQUE POLLENT (The arrangement and connection are powerful) from Horace *Ars Poetica* 242. In the exergue: ART DE CHARPENTERIE/1773 (The art of carpentry).

Fig. In.3. Jeton: Académie de Stanislas, 1750.
F.7582. Silver. 33.2 mm. 15.36 g. 12:00. Planchet is 2.44 mm. Smooth edge. Restrike marked "cuivre" (*sic*).
Obverse: Bust of Stanisław Leszczyński (1677-1766), King of Poland, with laurels and in armor, signed Borrel. The inscription is STAN.ROY.DE.POL.G.D.D.LITH.D.DE.LOR.ET.DE.BAR., (Stanislas King of Poland, Grand Duke of Lithuania, and Duke of Lorraine and Bar).
Reverse: A shield with the arms of Lorraine and Bar. Legend: ACADEMIE DE STANISLAS 1750 (The Academy of Stanislas).

Fig. In.4. Photograph: Maison Platt, 49 rue Richelieu, Paris. Photo courtesy of Maison Platt.

Fig. In.5. Prieur/Sombart 2002 catalog, *Jetons XVII: Spécial Bourgogne*, entry no. 127, p. 16: 1746 jeton of the Estates of Burgundy.

1. AN HISTORICAL OVERVIEW OF EARLY JETONS

Fig. 1.1. *Jeton d'école*/School token.
Brass. 28 mm. 3.32 g. 12:00. Thin planchet (0.80 mm).
Obverse: Rechenmeister at the counting table.
Reverse: Alphabet in five ranks.

Fig. 1.2. An *Écu*, 1790.
Silver. 41mm. 29.10 g. 6:00. Planchet is 2.54 mm. Reeded edge.
Obverse: Bust of Louis XVI facing left. Legend: LUD.XVI.D.G.FR.ET NAV.REX. (Louis XVI by the Grace of God King of France and Navarre). Signed: B.DU.VIV.F (Benjamin Duvivier, *fils.*). Engraver's mark of François Alluaud: sheath of wheat.
Reverse: Crowned shield of France with laurels. Legend: SIT NOMEN DOMINI BENEDICTUM (Blessed be the name of the Lord; Job 1:21). Dated 1790. Mint mark "I" (Limoges).

Fig. 1.3. Medal: Treaty of Ryswick, 1697.
Original in BnF, DMMA; licensed reproduction.
Silver, 65mm. 12:00.
Obverse: Bust of Louis XIV facing right, signed "R" (Roettiers) Legend: LUDOVICUS.MAGNUS.REX. CHRISTIANISSIMUS (Great and Most Christian King Louis).
Reverse: France with shield and cloak with fleur-de-lis kneeling before Louis XIV who presents France with an olive branch, signed "R" (Roettiers). Legend: PATER PATRIÆ (Father of the country). In the exergue: PAX.CVM.GERM.HISP./ANGL.ET.BAT/1697 (Peace with Germany, Spain, England and Batavia, 1697).

Fig. 1.4. Méreau worth twelve *deniers* from Église de Saint-Omer, 1526.
F.7017. Copper. 23 mm. 3.86 g. 6:00. Thin planchet (1.2 mm). Smooth edge.
Obvserse: Arms of the church and convent of Saint Bertin of Saint-Omer with bishop's crosier. Legend: MO.ECC.SANTI.AUDOMARI (Ecclesiastical currency of Saint-Omer).
Reverse: The number XII (reverse lettering for the "X") and date 1526 in the center. Legend: PRESENTIBVS.DABITVR (To be given to the holder).

Fig. 1.5. Medieval jeton.
Brass. 23 mm. 2.03 g. Thin planchet (0.72 mm).
Obverse: quatrefoil with indecipherable legend.
Reverse: paschal lamb and indecipherable legend.

Fig. 1.6. Woodcut. Gregor Reisch's *Margarita philosophica*, (Freiburg: Joannē Schottũ Argeñ, 1503), frontispiece to Book IV. BnF, Site François-Mitterrand, Tolbiac; licensed reproduction.

Fig. 1.7. Illustration of casting counter and use of jetons for accounting purposes.

Fig. 1.8. Jeton of the Conseil du Roi, 1585.
F.55. Copper. 28.24 mm. 4.93 g. Planchet is 1.18 mm. Smooth edge. 6:00.
Obverse: Crowned shield of France with three lilies surrounded with the double chain of the royal orders. Legend: NIL.NISI.CONSILIO (Nothing save by counsel).
Reverse: Feuardent description: "An eagle on its airy at the top of a rocky outcropping, whence are falling eaglets overpowered by the sun." Legend: DEGENERES.SOL.ARGVIT (The sun reveals their baseness, an adaptation of Virgil, *Aeniad*, IV 13 [Degeneres animos timor arguit]) referring evidently to pretenders overpowered by the sun. In the exergue: 1585.

Fig. 1.9. Jeton of the royal counting house, the Chambre des Comptes, 1603.
F.1788. Copper. 5.75 g. 27.3 mm. Planchet is 1.27 mm. Smooth edge. 6:00.

Obverse: Crowned shields of France and Navarre with double chain of the royal orders. Legend: CAMERAE COMPVTOR REGIORVM (The Royal Counting House).
Reverse: A naked couple facing each other on an island strewn with lilies and with seven ships about. Legend: STAT.PROLE.HAC.ALTERA.DELOS (Another Delos is established by this birth). An allusion to birth of first child of Henri IV and Marie de Médicis, Élisabeth de France (1602–1644) on November 22, 1602. Delos was the mythical birthplace of Apollo and Artemis, offspring of Zeus and Leto, aka Henri IV and Marie de Médicis, here referencing Êlisabeth and the dauphin in birth order. In the exergue in three lines: SVB DVCENDIS/RATIONIBVS/1603 (To reckon accounts, 1603).

Fig. 1.10. Jeton of the Cour des Monnaies (*Curia Monetaria Franciæ*).
F.2155. Copper. 27.91 mm. 3.39 g. Thin planchet (0.84 mm). Smooth edge. 6:00.
Obverse. Crowned shield of France with the chain of St-Michel. Legend: CVRIA.MONETAR.FRANCIÆ (The Monetary Court of France).
Reverse: Boatman with oar on bark with lilies and overlapping waves. Legend: HOC.OPVS.HIC.LABOR (This is the task, this indeed is a toil.). The motto *hoc opus hic labor* is a quote from Virgil, *Aeneid* VI, 129, where Aeneas asks the Cumean Sybil how to enter Hades, and she replies that entering Hades is easy; returning is the hard part. In the exergue, the date 1577.

Fig. 1.11. Dutch jeton.
F.13776. Copper. 28.84 mm. 5.73 g. Planchet is 1.02 mm. Smooth edge. 6:00.
Obverse: David with sling and Goliath with a lance. Legend: FACIS.TV.SOLVS.DECVS.ET.MAGNA (You alone, God, do great things).
Reverse: A lion representing Flanders attacking a boar representing Spain. Legend: FIDE.DNO.ET.IPSE.EFFICIET (By faith in God even he shall bring it to pass). In the exergue, the date 1578.

Fig. 1.12. Jeton: A Nuremberg Rechenpfennig.
F.12968v. Brass. 25.45 mm. 1.84 g. Thin planchet (0.50 mm). 12:00.
Obverse: Bust of Louis XIV (bust no. 140 var.) with laurels facing right. Legend: LUD:XIIII.DG:FR:ET NAV:REX. (Louis XIV by the Grace of God King of France and Navarre).
Reverse: Crowned arms of France. Legend: CONRAD LAVFFERS RECHENPFENING (Token by Conrad Lauffer).

2. OLD AND NEW IN THE OLD REGIME

Fig. 2.1. Jeton of Louis XIV.
F.12893. Silver. 26.75 mm. 4.85 g. Planchet is 0.96 mm. Reeded edge. 6:00.
Obverse: Bust of Louis XIV (bust no. 241 var.), signed "T.B." (Thomas Bernard). Legend: LVDOVICUS.MAGNVS.REX (Louis the Great King).
Reverse: The figure of Aurora, the goddess of dawn, in an antique, two-wheeled horse-drawn racing chariot (*biga*) above clouds with a flaming torch. Legend: LATE.CVNCTA.PROFUNDIT (She pours forth all things copiously). No date.

Fig. 2.2. Jeton of the Estates of Brittany, 1770.
F.8779. Silver. 28.60 mm. 6.96 g. Planchet is 1.28 mm. Reeded edge. 6:00.

Obverse: Unsigned bust of Louis XV with laurels facing right (bust no. 352 var.). Legend: LUDOV. XV.REX CHRISTIANISS. (Louis XV, Most Christian King).
Reverse: Crowned oval shield with the arms of France and Brittany in quarters on an ermine cloak. Legend: JETON DES ETATS DE BRETAGNE 1770 (Jeton of the Estates of Brittany, 1770).

Fig. 2.3. Jetons for 1682 as illustrated in the *Mercure galant*, 1682(1), following p. 54. BnF, Site François-Mitterrand, Tolbiac; licensed reproduction.

Fig. 2.4. Engravng: Coronation of Louis XVI.
"La Cérémonie des Offrandes," engraving by Sieur Patas in Pichon (1775), following p. 136. Note sex segregation. BnF, Site François-Mitterrand, Tolbiac; licensed reproduction.

Fig. 2.5. Jeton/medalet for the coronation of Louis XIV, 1654.
F.7885. Silver restrike with smooth edge marked "argent." 30.10 mm. 9.06 g. Planchet is 2.25 mm. 12:00.
Obverse: Crowned bust of Louis XIV facing right in high relief (bust no. 142). Unsigned. Legend: LVD.XIIIII.D.G.FR.ET.NAV.REX.CHRISTIANISSIMVS (Louis XIV, By the Grace of God Most Christian King of France and Navarre).
Reverse in high relief: The Holy Spirit and Holy Ampule descending on the city of Reims. Legend: SACRAT.AC.SALUT.RHEMIS.IVNII.VII.1654 (Consecration and Sanctification at Reims, June 7, 1654).

3. GOVERNING BODIES

Fig. 3.1. Jeton of the Secrétaires du Roi, 1705.
F.321. Silver. 30.27 mm. 7.68 g. Reeded edge. Planchet is 1.19 mm. 6:00.
Obverse: Bust of Louis XIV facing right (bust no. 245) signed with monogram, "TB" (Thomas Bernard). Legend: LUDOVICUS MAGNUS REX. (Louis the Great King).
Reverse: King bee (*sic*) and swarm of bees rising to the sun. Legend: DVCEM.REGEM.QUE.SEQV-VNTVR. (They follow the royal leader). In the exergue in three lines: SECRETAIRE./DU.ROI./.1705. (The King's Secretaries, 1705).

Fig. 3.2. Jeton of the Bâtiments du Roi, 1756.
F.3182. Silver. 29.63 mm. 8.19 g. Reeded edge. Planchet is 1.63 mm. 6:00.
Obverse: Unsigned bust of Louis XVI facing right (bust no. 379). Legend: LUDOV.XVI.REX CHRISTIANIS. (Louis XVI, Most Christian King).
Reverse: Remodeling of the colonnade of the Louvre palace with crane and construction materials. Legend: MOX HOSPITE DIGNA (Soon worthy of its host). In the exergue in two lines: BATIMENS DU ROY/1756. (The King's Buildings, 1756). N.B. The reverse of this jeton was originally struck in 1756 with the bust of the reigning king, Louis XV; see F. 3161 and ff. Louis XVI became king only in 1774. The jeton in figure 3.2 is a much later striking ca. 1780 with Louis XVI on the obverse.

Fig. 3.3. Jeton of the Trésor Royal, 1742.
F.2057. Copper. 28.82 mm. 6.86 g. Planchet is 1.44 mm. Smooth edge. 6:00.
Obverse: Bust of Louis XV facing right (bust no. 319), signed "*fm*." (François Marteau). Legend: LUD.XV.REX.CHRISTIANISS. (Louis XV, Most Christian King).

Reverse: Noah's ark amid the waves with an open portal and a dove perched above holding an olive branch. Legend: IMMERSABILIS UNDIS (Unsinkable amidst the waves); proverbial from Horace, *Epistles* 1, 2, 22 (adversis rerum immersabilis undis). In the exergue: TRESOR ROYAL/1742 (The Royal Treasury, 1742).

Fig. 3.4. Jeton of the Colonies Françaises de l'Amérique, 1751.
F.1558v. Silver. 29.03 mm. 5.99 g. Planchet is 1.18 mm. Reeded edge. 6:00.
Obverse. Laureled bust of Louis XV in armor and facing right (bust no. 338A), signed "*fm.*" (François Marteau). Legend: LUD.XV.REX CHRISTIANISS. (Louis XV, Most Christian King).
Reverse: An Indian with bow and quiver of arrows, facing left and looking back right to a stand of lilies; an alligator shows its head from the river near his feet, a palm tree in the distance. Legend: SUB OMNI SIDERE CRESCENT (They grow beneath every star). In exergue in two lines: COL. FR. DE/LAM . 1751. (The French Colonies in America, 1751). A rare and otherwise undocumented jeton. Compare Feuardent no. 1558–1584; Betts (1894), p. 172, #385. The provenance provided to the author suggests that it was part of a shipment to Canada that was perhaps stolen.

Fig. 3.5. Jeton: General Assembly of French clergy, 1775.
F.3980. Octagonal in silver. Shortest diameter is 32.00 mm. 13.54 g. Planchet is 1.84 mm. Smooth edge. 12:00.
Obverse: Bust of young Louis XVI facing right (bust no. 375) signed DU VIV. (Jean Duvivier). Legend: LUD.XVI.REX CHRISTIANISS. (Louis XVI, Most Christian King).
Reverse: Inscription in five lines with cross above: CONVENTUS/CLERI GALLICANI/HABITUS/LUT. PARISIOR./MDCCLXXV. (The French Clergy Assembled at Lutetia of the Parisians, 1775).

Fig. 3.6. Jeton of the La Burgade family, 1715.
F.9237. Octagonal in copper. Shortest diameter is 29.76 mm. 7.46 g. Planchet is 1.27 mm. Smooth edge. 6 :00
Obverse: Arms and family crest of the de La Burgade family and date, 1715.
Reverse: Crowned monogram/cipher. Signed C.MAVELOT (C. Mavelot). Legend: EX FIDE PRETIVM (From Faith, Reward).

Fig. 3.7. Jeton of the Estates of the Languedoc, 1787.
F.11006. Silver. 30.42 mm. 10.31 g. Planchet is 1.94 mm. Reeded edge. 6:00.
Obverse: Bust of Louis XVI facing right (bust no. 387), signed GATTEAUX. (Nicolas Gatteaux). Legend: LUD.XVI.REX CHRISTIANISS. (Louis XVI, Most Christian King).
Reverse: Crowned Occitan Cross—aka the Cross of Languedoc or the Cross of Toulouse. Legend: COM.OCCIT.1787 (The Estates of Languedoc, 1787).

Fig. 3.8. Jeton of the Estates of Burgundy, 1785.
F.9867. Silver. 30.64 mm. 10.75 g. Planchet is 1.81 mm. Reeded edge. 6:00.
Obverse: Bust of Louis XVI (bust no. 395 var.), signed DUVIV. (Benjamin Duvivier). Legend: LUDOV.XVI.REX CHRISTIANISS. (Louis XVI, Most Christian King).
Reverse: Crowned coat of arms of Burgundy. Legend: COMITIA BURGUNDIÆ 1785 (The Estates of Burgundy, 1785).

Fig. 3.9. Jeton of the Ville de Paris.
F.3460. Silver. 28.04 mm. 5.88 g. Planchet is 1.14 mm. Reeded edge. 6:00.
Obverse: Bust of Louis XV with laurels facing right (bust no. 324), signed *R.Filius* (Roettiers *fils*). Legend: LUD.XV.REX.CHRISTIANISS. (Louis XV, Most Christian King).
Reverse: View of the City of Paris. Legend: LA VILLE DE PARIS (the City of Paris).

Fig. 3.10. Jeton of the Valenciennes' town council, 1726.
F.7356. Silver. 31.83 mm. 12.29 g. Planchet is 1.94 mm. Reeded edge. 6:00.
Obverse: Three coats of arms held by two storks: that of the provincial Governor, the Intendant, and the City of Valenciennes. Legend: EX CONCORDIA ET CANDORE FELICITAS URBIS (The felicity of the city through concord and candor).
Reverse: The assembled town council. Legend: LABORIS ASSIDUI PRAEMIUM (Reward for their hard work). In the exergue: CONSILIUM/VALENCENENSE/1726. (The Valenciennes Town Council, 1726).

4. ASSIDUITY AND THE ACADEMIES

Fig. 4.1. Jeton of the Académie Française, 1717.
F.4375. Silver. 30.21 mm. 7.19 g. Planchet is 1.23 mm. Reeded edge. 6:00.
Obverse: Bust of young Louis XV facing right (bust no. 275) signed with monogram "*jcr*" (Joseph-Charles Roettiers). Legend: LUDOVICUM XV. D. G. FRAN. ET NAV. REX. (Louis XV by the Grace of God King of France and Navarre).
Reverse: Legend: PROTECTEUR DE L'ACADEMIE FRANÇOISE. (Protector of the Académie Française) and in three lines in center encircled by laurels A/L'IMMOR/TALITÉ (To Immortality). In the exergue, the date 1717.

Fig. 4.2. Jeton of the Académie Royale des Sciences.
F.4408a. Silver. 30.12 mm. 7.69 g. Planchet is 1.34 mm. Reeded edge. 12:00.
Obverse: Laureled bust of Louis XV facing right (bust no. 346), signed C.N.R.FILIUS (Charles-Norbert Roettiers). Undated. Legend: LUD.XV.REX CHRISTIANISS. (Louis XV, Most Christian King).
Reverse: Seated Minerva surrounded by the instruments of science and with the Observatoire Royal in the background. Legend: INVENIT ET PERFICIT (It discovers and perfects). In the exergue in two lines: REGIA SCIENTIAR./ACADEM. (The Royal Academy of Science).

Fig. 4.3. Jeton of the Société Royale de Médecine.
F.4695. Silver. 28.72 mm. 7.86 g. Planchet is 1.51 mm. Reeded edge. 6:00.
Obverse: Bust of Louis XVI facing right (bust no. 379), unsigned. Legend: LUDOV.XVI.REX CHRISTIANIS (Louis XVI, Most Christian King).
Reverse: Blank field with SOCIÉTÉ/ROYALE/DE/MÉDECINE (Royal Society of Medicine) in four lines.

Fig. 4.4. Jeton of the Académie Royale de Chirurgie, 1751.
F. 4734. Silver. 30.34 mm. 9.81 g. Planchet is 1.67 mm. Reeded edge. 6:00.
Obverse: Laureled bust of Louis XV facing right (bust no. 336 var.), signed D.V. (Jean Duvivier). Legend: LUD.XV.REX CHRISTIANISS. (Louis XV, Most Christian King).
Reverse: A genie with shield standing before Minerva who transmits knowledge to it; surgical

instruments below. Signed D.V. (Jean Duvivier). Legend: COLIT ET COLITUR (It honors and is honored), and in the exergue in two lines: ACAD.REG.CHIR./MDCCLI. (Royal Academy of Surgery, 1751).

Fig. 4.5. Jeton of the Paris Faculté de Médecine, 1666.
F.4513. 27.49 mm. Reeded edge. 6:00. Original in BnF, DMMA; licensed reproduction.
Obverse: Draped bust of François Le Vignon (†1676) facing right. Legend: LE.VIGNON.CON^{R}.DES^{T}. ET.DOIEN (Le Vignon, Counselor, Official [*deservant*?] and Dean). Le Vignon was personal physician of the Duchesse of Lorraine from 1656 and of the Swiss Guards of Louis XIV.
Reverse: Three storks, the emblem of the Faculté de Médecine, beneath the sun. Legend: URBI. ET.ORBI. (For the city and the world). N.B.: This is the same address issued by the pope for apostolic blessings, i.e., the Paris Faculty is to medicine as the Pope is to the Catholic Church. In the exergue in two lines: FACUL.MEDIC./PARIS.1666. (Paris Faculty of Medicine, 1666).

Fig. 4.6. Jeton of the Académie Royale des Sciences, Belles-Lettres et Arts of Lyon.
F.10749. Silver. 30.67 mm. 11.31 g. Planchet is 2.01 mm. Reeded edge. 6:00.
Obverse: Historic altar of Lyon. Legend: ATHENAEUM LUGDUNENSE RESTITUTUM (The Lyon academy restored). In the exergue: M.D.CC. (1700).
Reverse: In the center surrounded by a crown of laurels in six lines: ACAD./LUGDUN./SCIENT./LITTER./ET/ARTIUM (The Lyon Academy of Science, Belles-Lettres and Arts).

Fig. 4.7. Jeton of the Académie Royale des Sciences, Belles-Lettres et Arts of Rouen.
F.6121. Silver. 30.06 mm. 8.85 g. Planchet is 1.60 mm. Reeded edge. 6:00.
Obverse: Laureled bust of Louis XV facing right (bust no. 324), signed *r.filius* (Roettiers *fils*). Legend: LUD.XV.REX CHRISTIANISS. (Louis XV, Most Christian King).
Reverse: Minerva with shield lounging on the right amid objects of art and science, pointing with her left hand towards a temple with three doors on a hill behind. Legend: TRIA LIMINA PANDIT (It opens three doors).

Fig. 4.8. Jeton of the Académie Royale de Marine, 1778.
F.1554. Silver. 30.43 mm. 9.57 g. Planchet is 1.62 mm. Reeded edge. 6:00.
Obverse: Bust of Louis XVI facing left (bust no. 380), signed I.P.DROZ.F. (J. P. Droz).
Reverse: A ship-rigged royal frigate under sail to the right, signed DU.V. (Duvivier). Legend: PER HANC PROSUNT OMNIBUS ARTES (By such proceed all the arts). In the exergue in three lines: ACADEMIE ROYALE/DE MARINE/1778 (Royal Academy of Naval Sciences, 1778).

Fig. 4.9. Medal/jeton of the Société Royale d'Agriculture, 1785.
F.4905. Silver, octagonal. Shortest diameter is 34.74 mm. 18.61 g. Planchet is 2.76 mm. Smooth edge. 12:00.
Obverse: Louis XVI standing beside a column and a curtain on the right in royal garb lifted to expose legs, offering with his right hand a group of cows to an allegorical figure representing the city of Paris, who is bowing and leaning on her escutcheon; a herdsman directing another cow in the background on the left. Signed DUPRÉ F. (Augustin Dupré). Legend: MUNIFICENCE ROYALE (Royal munificence). In the exergue in three lines: VACHES DISTRIBUÉES DANS LA/GÉNÉRALITÉ DE PARIS/M D CC LXXXV (Cows distributed in the district of Paris, 1785).
Reverse: An unharnessed plow in an empty field. Legend: EX UTILITATE DECUS (Worthy by its use-

fulness). In the exergue in two lines: SOC.ROY.DAGRIC./DE PARIS (The Royal Society of Agriculture of Paris).

Fig. 4.10. Jeton of the Cercle des Philadelphes, Cap François, Saint Domingue (Haïti), 1788.
Not in Feuardent. Silver. 33 mm. Original in BnF, DMMA; photo per McClellan (1992), p. 182.
Obverse: Bust of Louis XVI facing right (bust no. 395 var.), signed SIMON G. DU. ROI (Simon, royal engraver). Legend in French: LOUIS XVI ROI DE FRANCE ET DE NAVARRE 1788 (Louis XVI, King of France and Navarre, 1788).
Reverse: Hive and swarm of bees below a meridian sun. Legend: EXERCET SUB SOLE LABOR (Work is done under the sun). In the exergue in two lines: CERCLE DES PHILADELPHES/ETABLI AU CAP 1784 (The Cercle des Philadelphes founded at Cap François, 1784).

5. GUILDS AND YET OTHERS

Fig. 5.1. Jeton of the corduroy makers' and corduroy merchants' guild (*Corroyeurs du Corduroi*), 1755.
F.5076. Silver. 29.01 mm. 6.95 g. Planchet is 1.37 mm. Reeded edge. 6:00.
Obverse: Unsigned bust of Louis XVI facing right (bust no. 379). Legend: LUDOV.XVI.REX CHRISTIANIS (Louis XVI, Most Christian King).
Reverse: Four gentlemen porters, masters of the corduroy guild, carrying the reliquary of Saint Merri (Medericus, †700), patron saint of the Right Bank in Paris and presumably the patron saint of the guild. Legend: M.T CORROYEURS PORTEURS DE LA CHASSE DE S.T MERRY (Master corduroy makers carriers of the reliquary of Saint Merri). In the exergue: tools of the corduroy trade and the date, 1755. N.B., the nominal date on the reverse of this jeton does not match the reign of Louis XVI on the obverse; see also above, figure 3.2.

Fig. 5.2. Jeton of the distillers' guild.
F.5078. Silver. 27.87 mm. 5.55 g. Planchet is 1.11 mm. Reeded edge. 6:00.
Obverse: Laureled bust of Louis XV facing right (bust no. 308A var.), signed DU VIV. (Duvivier). Legend: LUD.XV.REX.CHRISTIANISS. (Louis XV, Most Christian King).
Reverse: Legend: TOTUM IN SPIRITU CORPORE NIHIL (Everything is in the spirit, nothing in the body – note pun). Saint Louis (Louis IX) in royal cloak kneeling and praying to the Holy Spirit in the form of a radiant dove at 1:00; to the right is an alembic. In the exergue in three lines in French: COMMUNAUTÉ DES/DISTILLATEURS M.DS/ D'EAU DE VIE. (Guild of distillers and schnapps merchants).

Fig. 5.3. Jeton of the weavers of Elbeuf.
F.6400. Silver. 29.86 mm. 10.11 g. Planchet is 2.02 mm. Reeded edge. 6:00.
Obverse: Laureled bust of Louis XV (bust no. 341 var.), signed B. DUVIV. F (Benjamin Duvivier). Legend: LUDOVICUS XV ATRIUM PROTECTOR (Louis XV, Protector of the Arts).
Reverse: The cross of Lorraine entwined with a vine. Legend: TALI FULCIMINE CRESCET (With such support shall it grow), signaling the support of the house of Lorraine. In the exergue: MANUFACTURE D'ELBEUF (the Elbeuf manufactory).

Fig. 5.4. A royal jeton purse.
Lined red velvet purse, handstitched in fine petit point needlework. Embroidered with fleurs-de-lys and with a drawstring and wooden end ornaments covered with metallic thread. Embroidered crowned shield and fleurs-de-lys on the flat-bottom underside with embroidered medallion featuring a small collared dog on a blue background. Oval: 12 x 14cm. BnF, DMMA; licensed reproduction.

Fig. 5.5. Jeton of the Chambre de Commerce of Rouen, 1712.
F.6304. Silver. 31.00 mm. 10.48 g. Planchet is 1.69 mm. Reeded edge. 6:00.
Obverse: Laureled bust of Louis XIV facing right (bust no. 241 var.), signed H R (Jérôme Roussel).
Legend: LUDOVICUS MAGNUS COMMERCII PROTECTOR (Louis the Great, Protector of Commerce).
Reverse: Apollo on the docks at Rouen with bundles; city and cathedral of Rouen in background.
Legend: FIRMATA CONSILIO COMMERCIA (Commerce fortified with deliberation). In the exergue in French in three lines: CHAMBRE DE COMMERCE/DE ROUEN/1712 (The Rouen Chamber of Commerce, 1712).

Fig. 5.6. Token of the united merchants.
F.5437. Copper. 28.13 mm. 6.74 g. Planchet is 1.40 mm. Smooth edge. 6:00. Unsigned.
Obverse: Seven standing figures holding up the world. Legend: STAT MUTUIS VIRIBUS (It stands firm through mutual strength).
Reverse: Bee hive and swarm of bees. Legend: SOCIETATIS BENE UNITAE (Belonging to a well unified society).

Fig. 5.7. Jeton of the Société Philanthropique, 1781.
F.4478. Silver. 30.57 mm. 8.37 g. Planchet is 1.54 mm. Reeded edge. 6:00.
Obverse: A hand watering plants with a watering can beneath the Zodiac; a fortified emplacement is to the right. Unsigned. Legend: DONEC E CŒLO DESCENDAT (Until it comes down from Heaven). In the exergue: 1781.
Reverse: Surrounded by a crown of foliage, in five lines: MAISON/PHIL/ANTROPIQUE/DE/PARIS (The Philanthropic Society of Paris).

Fig. 5.8. Jeton of the *Arbusquiers* (marksmen) of Lyon, 1741.
F.10787. Silver. 30.25 mm. 9.71 g. Planchet is 1.69 mm. Reeded edge. 6:00.
Obverse: Arms of the city of Lyon with allegorical standing figures representing the Rhône and Saône Rivers.
Reverse: A lance and two muskets crosswise in center, target on left, city of Lyon in the background. Signed DUBOSC (Dubosc). Legend: EX JOCIS ET BELLO (Both in sport and in war.) In the exergue in three lines: ACAD LUGDUNENSIS/SCOPELATORIA/1741 (The Lyon Shooting Academy, 1741).

Fig. 5.9. Masonic jeton: *Pont Parfait de l'Or de Paris, 5760* (1759).
F.5479. Copper. 29.01 mm. 10.05 g. Planchet is 2.43 mm. Reeded edge. 6:00.
Obverse: On the right a man in antique garb leaning on a table with a sphere and an unfurled scroll. On the left, compasses, stars, a pyramid, and a tree. Legend: DU POINT PARFAIT A L'O. DE PARIS 5760 (At the crux of the Orient of Paris 5760 [1759]).
Reverse: Surrounded by a snake eating its tail is a table on which are a bee hive and swarm of bees

beneath the sun with a square, compass and other instruments. Legend: ÉTERNITÉ CONSTANCE (Eternity [and] constancy). Below a trowel and a hammer.

6. AESTHETICS AND THE SPECIAL ROLE OF THE ACADÉMIE ROYALE DES INSCRIPTIONS ET BELLES-LETTRES

Fig. 6.1. Jeton of the Académie Royale des Inscriptions et Belles-Lettres.
F.4398. Silver. Undated. 32.00 cm. 12.53 g. Planchet is 1.94 mm. Reeded edge. 6:00.
Obverse: Bust of Louis XVI facing left (bust no. 400), signed DU VIV. (Jean Duvivier). Legend: LUDOVICUS XVI.FR.ET NAV.REX (Louis XVI, King of France and Navarre).
Reverse: A female figure in antique garb leaning on a cippus with an unfurled scroll in her left hand and a crown of laurels in her outstretched right hand. An obelisk and an urn spilling coins on the right; on the left a broken architrave with the inscription L. VERO. CON Q. VIX. ANN. (To Lucius Verus, Con[sul III], who died in his [39th] year). [Lucius Verus ruled Rome as emperor jointly with Marcus Aurelius (161–169).] The cippus is signed DU VIV. (Jean Duvivier). Legend: VETAT MORI (She preserves from death) is from Horace *Odes* 4.9.27 (dignum laude virum Musa vetat mori –The Muse prevents the worthy man to die).

Fig. 6.2. Jeton of the Estates of Artois.
F.6892. Silver. 30.93 mm. 9.08 g. Planchet is 1.63 mm. Smooth edge. 6:00.
Obverse: Bust of Louis XV facing right (bust no. 309), signed with monogram "*jcr*" (Joseph-Charles Roettiers). Legend: LUD. XV.REX CHRISTIANISS. (Louis XV, Most Christian King).
Reverse: The crowned arms of the Artois supported by a lion and a greyhound. Legend: COMITIA ARTISIÆ (The Estates of Artois).

Fig. 6.3. Jeton of the Artillerie, 1731.
F.1070v. Silver. 29.15 mm. 5.99 g. Planchet is 1.05 mm. Smooth edge. 6:00.
Obverse: Laureled bust of Louis XV facing right (bust no. 308A), signed DU VIVIER (Duvivier). Legend: LUD.XV.REX CHRISTIANISS. (Louis XV, Most Christian King).
Reverse: Two shepherdesses on the left encounter cannons and mortars at a foot of a tree on right with barrel of mortar behind. Legend: ET MUTA MINANTUR (And silent they threaten). In the exergue in two lines: ARTILLERIE/1731.

Fig. 6.4. Jeton of the Extraordinaire des Guerres, 1755.
F.851. Silver. 29.32 mm. 7.27 g. Planchet is 1.32 mm. Reeded edge. 6:00.
Obverse: Laureled bust of Louis XV facing right (bust no. 338A), signed *f.m.* (François Marteau). Legend: LUD.XV.REX CHRISTIANISS. (Louis XV, Most Christian King).
Reverse: Two knights in full amour jousting in a tournament. Legend: ET LUDUS IN ARMIS (Also a game in arms). In the exergue in three lines in French: EXTRAORDINAIRE/DES GUERRES/1755 (The Extraordinaire des Guerres, 1755).

Fig. 6.5. Jeton of the Extraordinaire des Guerres, 1772.
F.886. Silver. 29.46 mm. Planchet is 1.48 mm. Reeded edge. 6:00.
Obverse: Laureled bust of Louis XV (bust no. 343), unsigned. Legend: LUDOVICUS XV.REX CHRISTIANISS. (Louis XV, Most Christian King).

Reverse: Contemporary soldier standing with musket, a cannon on the left and a plow on the right. Legend: AD UTRUMQUE PARATUS (PREPARED FOR BOTH), a sentence attributed to Virgil. In the exergue in three lines in French: EXTRAORDINAIRE/DES GUERRES/1772 (The Extraordinaire des Guerres, 1772).

7. THE MONNAIE DES MÉDAILLES

Fig. 7.1. Jeton struck by the Monnaie des Médailles for the Rouen Mint, 1787.
F.6281. Silver. 30.96 mm. 8.23 g. Planchet is 1.46 mm. Reeded edge. 6:00.
Obverse: Bust of Louis XVI facing right (bust no. 395), signed DU VIV. (Jean Duvivier). Legend: LUDOCIVUS XVI.REX CHRISTIANISS. (Louis XVI, Most Christian King).
Reverse: A coin press in a crowned cartouche with a balance on the right, tools on the left and baskets of minted coins on both sides; hammer and die at the bottom. Legend across the top of the jeton: HINC PONDUS ET PRETIUM (Hence comes weight and value). Legend at the bottom of the jeton: MONN. DE ROUEN 1787 (The Rouen Mint, 1787).

Fig. 7.2. Engraving: The Monnaie des Médailles in the Louvre.
P. Faure, "Plan de la paroisse royale de St. Germain L'Auxerrois en Octobre 1739" (and detail). BnF, Département des Estampes, Va 441 F[t] 6 (H186985). Mazerolle (1907a) reproduces a detail from this engraving at p. 31. Licensed reproduction.

Fig. 7.3. Engraving: The coin press (the *Balancier*).
From d'Alembert and Diderot, *Encyclopédie*, article Monnoyeur. See Gillispie, ed. (1959).

Fig. 7.4. A die.
19th century. 77 mm in diameter. BnF, DMMA; licensed reproduction.

Fig. 7.5. Medal commemorating the visit of Tzar Peter the Great to Paris, 1717.
Silver. 49mm. BnF, DMMA; licensed reproduction.
Obverse: Bust of Tzar Peter the Great, signed DU VIVIER F. (Jean Duvivier). Legend: PETRUS ALEXIEWITZ TZAR MAG. RUSS IMP. (Peter Alexiewitz, Great Tzar of Imperial Russia).
Reverse: Winged fame with two trumpets flying from the north and in the legend the words Virgil used to describe fame: VIRES ACQUIRIT EUNDO (It gains strength in going). In the exergue: LUTET PARISI, MDCCXVII. (Lutetia of the Parisians, 1717).

CONCLUSIONS

Fig. Con.1. Altered obverse of a jeton of the Académie Française.
F.4381. Silver. 32.40 mm. 13.53 g. Planchet is 2.04 mm. Reeded edge. 6:00.
Bust of Louis XVI facing left (bust no. 388A), signed DU VIV. (Jean Duvivier). Legend: LOUIS XIV ROI DE FR. ET DE NAVARRE (Louis XVI, King of France and Navarre). Undated [1786].

Fig. Con.2. Jeton of La Nationale life insurance company, 1830.
Octagonal silver jeton, restrike marked with "hand" and "argent" on one edge. 36.52 mm across

shortest diameter. 20.89 g. Planchet is 2.80 mm. Smooth edge. 12:00.
Obverse: Across a strongbox a parent and child exchange a purse for an unscrolled written policy with an antique figure with quill, signed DESNOYERS (Auguste Gaspard Louis Desnoyers). Legend: LA PREVOYANCE DOTE L'AVENIR (Foresight funds the future).
Reverse: Surrounded by a crown of laurels in six lines LA/NATIONALE/COMPAGNIE/D'ASSURANCES/ SUR LA VIE/HUMAINE (The Nationale Life Insurance Company).

Fig. Con.3. A collage of twentieth-century jetons. Collage and photograph by the author.

GLOSSARY

Jeton of the Experts des Bâtiments.
F.3196. Silver. 28.69 mm. 8.22 g. Planchet is 1.53 mm. Reeded edge. 6:00.
Obverse: Laureled bust of Louis XV facing right (bust no. 333), signed *r.filius* (Roettiers *fils*). Legend: LUD.XV.REX CHRISTIANISS. (Louis XV, Most Christian King).
Reverse: Two palaces under construction; the one on the right with a plumb line suspended from its cornice. Legend: RECTI IRREQUIETA CVRIDO (The unfulfilled desire for rectitude). In the exergue in two lines: EXPERTS DES/BASTIMENS (Building Specialists).

LIST OF FIGURES AND TABLES

Jeton of the Académie Royale de Peinture et de Sculpture, 1764.
F.4419. Copper. 30.18 mm. 9.41 g. Planchet is 1.79 mm. Smooth edge. 12:00.
Obverse: Bust of Louis XV facing right (bust no. 324A var.), signed R FILIUS (Roettiers *fils*). Legend: LUCOVICUS XV ARTIUM PARENS (Louis XV, Father of the Arts).
Reverse: Illuminated by the sun on the left allegorical figures of Painting and Sculpture are seated and shaking hands. Legend: AMICÆ QUAMUIS ÆMULÆ (Friends in earnest rivalry). In the exergue in three lines: ACAD. ROYALE DE/ PEINT. ET DESCULPT/1764. (The Royal Academy of Painting and Sculpture, 1764).

ABBREVIATIONS

Jeton of the Académie Royale d'Architecture [ca. 1780].
F.4426. 30.82 mm. 7.66 g. Planchet is 1.33 mm. Reeded edge. 6:00.
Obverse: Bust of Louis XVI facing left (bust no. 380), signed J.P.DROZ.F. (J. P. Droz). Legend: LUDOV: XVI . REX CHRISTIANISS: (Louis XVI, Most Christian King).
Reverse: Three lines in a blank field: ACADEMIE/ROYALE/D'ARCHITECTURE (Royal Academy of Architecture).

BIBLIOGRAPHY

Jeton of the Printers' and Booksellers' Guild, 1723.
F.5264. Silver. 30.28 mm. 9.20 g. Planchet is 1.57 mm. Reeded edge. 6:00.

Obverse: Crowned arms of the Booksellers' and Printers' Guild supported by two sphinx bookends. No legend. In the exergue: M.DCC.XXIII (1723).
Reverse: An open book amid clouds and beneath a meridian sun. Legend: EX UTROQUE LUX (Light from Both). In the exergue in three lines: BIBLIOPOLÆ/ET TYPOGRAPHI PARISE.^{ES} (Booksellers and Printers of Paris).

INDEX

Jeton of the guild of notaries, 1683.
F.3291. Silver. Possible restrike. 29.30 mm. 9.88 g. Planchet is 2.07 mm. Reeded edge. 6:00.
Obverse: Bust of Louis XIV facing right (bust no. 209), signed H.B. (unidentified by Feurardent, vol. 3, p. 264). Legend: LUDOVICUS. MAGNUS.REX. (Louis, the Great King).
Reverse: Gnomon on a stand. Legend: LEX EST QUODCUMO NOTAMUS (The law is what we write down). In the exergue in three lines: CON$^{.ERS.}$DV.ROY.ET./NOTAIRES/1683 (Royal Counselors and Notaries, 1683).

BACK COVER

Author at the Pont Neuf Métro station, Paris. Photograph © Jacqueline A. McClellan.

LIST OF TABLES

Jeton: Académie Royale d'Architecture.

ABBREVIATIONS

AI. Institut de France, Archives de l'Institut, Paris.

AIBL. Académie des Inscriptions et Belles-Lettres.

AIBL, H&M. *Histoire et Mémoires de l'Académie Royale des Inscriptions et Belles-Lettres.* 51 volumes, 1710–93. (Paris: 1710–1843.)

AN. Archives Nationales, France.

ANS. American Numismatic Society, New York.

BM-Rouen. Bibliothèque municipale, Rouen.

BnF. Bibliothèque nationale de France, Paris.

BnF, DMMA. BnF, Département des Monnaies, Médailles et Antiques.

CAEF. Centre des archives économiques et financières, Savigny-le-Temple, France.

CAEF, Ms. F° 80. Nicolas de Launay, "Journal historique de la Monnoye des Médailles, 1697–1725," 2 vols.; vol. 2 labeled, "Journal historique de la Monnoye des Médailles Regie par M. de Launay Directeur sous l'Inspection de Monseigneur le Duc d'Antin sur Intendant des Bâtimens du Roy, depuis le Regne de Sa Majesté. 1715."

CARAN. Centre d'Accueil et de Recherche des Archives Nationales, Paris, France.

F.1234. Feuardent number, per Feuardent (1904–1915).

PA. Archives of the Académie des Sciences, Institut de France, Paris, France.

PA-PV. PA-Procès-verbaux.

PA-PS. PA-Pochette de séance.

SHM-Vincennes. Service historique de la Marine, Vincennes, France.

JETON: Printers and booksellers' guild, 1723.

BIBLIOGRAPHY

Académie de Marine (France). 2002. *250e Anniversaire de l'Académie de Marine, 1752–2002*. Paris: Académie de Marine.

Académie Royale des Inscriptions et Belles-Lettres. 1710–1843. *Histoire de l'Académie royale des inscriptions et belles-lettres*, 51 vols. Paris: Imprimerie royale.

Amandry, Michel, ed. 2001. *Dictionnaire de numismatique*. Paris: Larousse.

Amardel, Gabriel. 1901. *Les Jetons des Etats du Languedoc du musée de Narbonne*. Narbonne: Caillard. [Extract of the *Bulletin de la Commission archéologique de Narbonne*.]

Amaton, Ferdinand. 1871. *Précis historique et chronologique sur l'établissement de la commune et des vicomtes mayeurs de Dijon, suivi de la description de ces jetons*. Dijon: Lamarche.

Antérion, Dominique. 2003. "Les jetons des XIVe–XXe siècles: question de formes, question 'd'usages,'" in Béatrice Coullaré (ed.), *L'essor de la médaille aux XIVe et XXe siècles* [*Moneta* 31 (2003)], pp. 174–85.

Anonymous. 1982. "Jetons et calcul," *La vie numismatique* 32 (1982), pp. 96–104. [French translation of chapter 2 ("Het rekenen met penningen,") of Provinciale Dienst voor het Kunstpatrimonium, *Munt in Limburg*, pp. 29–37.]

Babelon, Jean. 1958. "Adrien Blanchet, 1866–1957," *Revue numismatique* (1958), pp. 7–10.

__________. 1901–1932. *Traité des monnaies grecques et romaines*, 4 vols. Paris: Ernest Leroux. [Vol. 1 (1901): *Théorie et doctrine*.]

Babelon, Jean and Josèphe Jacquiot. 1951. *Histoire de Paris d'après les médailles de la Renaissance au XXe siècle*. Paris: Éditions de la Propagande Artistique.

Babut de Rosan, Auguste. 1909. *Quelques mots sur les jetons et la médaille des caisses d'escomptes sous l'Ancien Régime*. Mâcon: Protat Frères. [Extract of the *Procès-verbaux de la Société française de numismatique*.]

Balmuth, Miriam S., ed. 2001. *Hacksilber to Coinage: New Insights into the Monetary History of the Near East and Greece. A Collection of Eight Papers Presented at the 99th Annual Meeting of The Archaeological Institute of America* [*Numismatic Studies*, no. 24]. New York: The American Numismatic Society.

Banchereau, Jules. 1909. *Notes sur les jetons de présence de la Société d'agriculture, sciences, belles-lettres et arts d'Orléans*. Orléans: A. Gout.

Barbeau, Albert. 1900. *Les Six cents mariages célébrés aux frais de la ville de Paris en 1751*. Nogent-le-Rotrou: Daupely-Gousernau. [Extract of the *Bull. Sté. Hist. de Paris & Ile de France*.]

Barbiche, Bernard. 1999. *Les institutions de la monarchie française à l'époque moderne XVIe–XVIIIe siècle*. Paris: PUF.

Barnard, Francis Pierrepont. 1916. *The Casting-Counter and the Counting-Board, A Chapter in the History of Numismatics and Early Arithmetic*. Oxford: The Clarendon Press; reprint: Castle Cary, UK: Fox, 1981.

Barthélemy, Anatole de. [1875]. *Documents sur la fabrication des jetons aux XIVe, XVe et XVIe siècles*. S.l., S.n. [Extract from R.F.N; pp. 227–63.]

Barthélemy, Dominique. 2012. *La France des Capétiens, 978–1214*. Paris: Éditions du Seuil.

Bauer, Geo. J. 1894. "The Jetons of France: Their Importance in Numismatics: Historical and Otherwise," *The Numismatist* 7 (1894), pp. 136–38.

Belaubre, Jean. 1965. "Les teinturiers," *Bulletin du club français de la médaille* #9 (1965), pp. 24–25.

Belaubre, Jean and Bruno Collin. 1992. *Les monnaies de France: Histoire d'un peuple*. Paris: Perrin.

Belhoste, Bruno. 2011. *Paris savant. Parcours et rencontres au temps des Lumières*. Paris: Armand Colin.

Bély, Lucien, et al. 2003. *Dictionnaire de l'Ancien Régime, Royaume de France XVIe–XVIIIe siècle*. Paris: Quadridge/PUF.

Berry, George. 1977. "Jetons of Louis XIII & Cardinal Richelieu," *Coins & Medals* 14 (1977), pp. 26–28.

__________. 1968. "Mediaeval Jettons and the Exchequer," *Coins & Medals* 10 (1968), pp. 544–46.

Betts, Charles Wyllys. 1894. *American Colonial History Illustrated by Contemporary Medals*. New York: Scott Stamp and Coin Company.

Bizot, Pierre. 1687. *Histoire métallique de la République de Hollande*. Paris: Daniel Horthemels.

Blanchard, Raphaël. 1901. "Études de numismatique médicale française (méreaux, jetons, armoiries)," *Revue d'Europe* 6 (1901), pp. 81–101, 161–72, and 257–68.

Blanchet, Adrien [aka Jules-Adrien (1866–1957)]. 1954. "Inventaire de Joseph Roettiers en 1703," *Revue numismatique* (1954), pp. 173–78.

__________. 1950. "Jetons de l'ordinaire des guerres, de 1651 à 1653," *Revue numismatique* (1950), pp. 197–201.

__________. 1939. "Le collectionneur Duvau: Notes sur des amateurs du XVIIIe siècle," *Revue numismatique* (1939), pp. 189–206.

__________. 1932. "Une critique de divers jetons de 1715," *Revue numismatique* (1932), pp. 231–34.

__________. 1930. *Médailles, Jetons, Méreaux* is vol. 3 of Blanchet and Dieudonné, *Manuel de numismatique française*.

__________. 1930b. "Jeton de Marchand du XVe siècle," *Revue numismatique* (1930), pp. 255–58.

__________. 1924. "Les jetons banaux fabriqués par les monnayeurs de Nancy en 1636," *Revue numismatique* (1924), pp. 233–35.

__________. 1920a. "Graveurs de jetons et cachets (XVIIe et XVIIIe siècles)," *Mémoires et notes de numismatique*, 2e sér. (Paris: Auguste Picard), pp. 202–06.

__________. 1920b. "Jetons et méreaux," *Mémoires et notes de numismatique, 2e sér.* (Paris: Auguste Picard), pp. 207–19.

__________. 1920c. "Recherches sur l'atelier monétaire de Bayonne (1488–1837)," *Mémoires et notes de numismatique*, 2e sér. (Paris: Auguste Picard), pp. 131–201.

__________. 1917–1918. "Méreaux de la corporation des drapiers-chaussetiers de Paris," *Revue numismatique* (1917–1918), pp. 173–78.

__________. 1912a. "Communication d'Adrien Blanchet sur la fabrication des jetons au XVIIIe siècle," *Procès-verbaux de la Société française de numismatique*, séance du 6 juillet 1912, p. lxi; séance du 5 juin 1937, p. xxxiii.

__________. 1912b. "Jetons frappés à Rouen et à Paris, en 1744," *Revue numismatique* (1912), pp. 281–82.

__________. 1909a. "Documents numismatiques relatifs à la Compagnie des Indes de Law," in *Mémoires et notes de numismatique* (Paris: E. Leroux), pp. 396–405.

__________. 1909b. "Jetons de corporations parisiennes au XVIIIe siècle," in *Mémoires et notes de numismatique* (Paris: E. Leroux), pp. 218–22.

__________. 1909c. "Les Pelhauquins ou jetons de jeu du XIVe siècle," in *Mémoires et notes de numismatique* (Paris: Ernest Leroux), pp. 298–300.

__________. 1892. "Médailles et jetons du sacre des rois de France," in *Études de numismatique*, vol. 1 (Paris: Rollin and Feuardent), pp. 191–220. [Extract of the *Bulletin de numismatique et d'archéologie* 6 (1890), pp. 149–70.]

__________. 1890. *Nouveau manuel de numismatique du moyen âge et moderne*, 2 vols. + 'atlas'. Paris: Roret.

__________. 1888. "Jean Warin," *Annuaire de numismatique* 12 (1888), pp. 84–90.

Blanchet, Adrien and Adolphe Dieudonné. 1912–1936. *Manuel de numismatique française*, 4 vols. Paris: A. Picard et fils. [Vol. 1: *Monnaies frappées en Gaule depuis les origines jusqu'à Hugues Capet* by Adrien Blanchet (1912); vol. 2: *Monnaies royales françaises depuis Hugues Capet jusqu'à la Révolution* by A. Dieudonné (1916); vol. 3: *Médailles, jetons, méreaux* by A. Blanchet (1930); vol. 4: *Monnaies féodales françaises* by A. Dieudonné (1936).]

Bluche, François. 1966. *Les Magistrats de la Cour des monnaies de Paris au XVIIIe siècle, 1715–1790*. Paris: Les Belles lettres.

Berchem, Denis van. 1936. "Tessères ou calculi?: Essai d'interprétation des jetons romains en plomb," *Revue numismatique* (1936), pp. 297–305.

Bolton, Henry Carrington. 1890. *Contributions of Alchemy to Numismatics*. New York: Privately printed.

Bonnet, Émile. 1899. "Les jetons des États Généraux de Languedoc," *Bulletin archéologique du Comité des travaux historiques et scientifiques* (1899), pp. 241–329. [Reprinted in Depeyrot (2007).]

Bordeaux, Paul. 1905. *Médaille et jeton frappés à l'occasion de la réunion de Lille à la France en 1713*. Paris: Rollin & Feuardent. [Extract of the *Revue numismatique française* (1905).]

Bouclier, Jacques. 1924. *La Cour des Monnaies de Paris à la fin de l'Ancien Régime*. Paris: Rousseau.

Boudeau, Émile. 1908. *Jetons des compagnies de notaires de France*. Paris: Boudeau.

Boutin, Serge. 1966. "Le collectionneur de jetons," *Cahiers numismatiques* 8 (1966), pp. 240–41.

Brichaut, Auguste. 1884. *Numismatique maçonnique*. Brussels: F. Gobbaerts. [Extract from the *Revue belge de numismatique* 1884.]

Burke, Peter. 1992. *The Fabrication of Louis XIV*. New Haven and London: Yale University Press.

Chapitre de l'Eglise Métropolitaine de Reims. 1654. *Le Sacre et couronnement de Louys XIV, Roy de France et de Navarre, Dans l'Eglise de Reims, le septième Iuin 1654*. Reims: Veuve François Bernard.

Chalon, Renier. 1867. "Jetons de mariage," *Revue belge de numismatique* 23 (1867), pp. 333–46; 25 (1869), pp. 189–98.

Charton, Fabrice. 2011. *<u>Vetat Mori</u>. Une institution au service du Prince, de la Petite Académie à l'Académie royale des Inscriptions et Belles-Lettres (1663–1742)*. Ph.D. dissertation, EHESS, Paris.

Charvet, Léon [E.-L.-G.]. 1907–1909. *Médailles et jetons de la ville de Lyon*. Chalon-sur-Saône: Émile Bertrand.

__________. 1878. *La Médaille et les jetons de la Chambre de Commerce de Lyon*. Lyon: Dessollins.

Chassant, Alphonse and Henri Tausin. 1878–1895. *Dictionnaire des Devises Historiques et Héraldiques*, 5 vols. Paris. J.-B. Dumoulin.

Chautard, Jules. [1900]. *Note relative aux graveurs des jetons des galères de Louis-Joseph duc de Vendome*. Vendôme: Empaytaz, s.d. [1900].

__________. 1896. "Quelques sceaux, jetons et armoires concernant les corporations de médecins, chirurgiens, barbiers aux XVIIe et XVIIIe siècles," *Revue belge de numismatique* 52 (1896), pp. 78–91.

__________. 1890. "Étude sur les jetons au point de vue de la reproduction du type du revers," *Revue belge de numismatique* 46 (1890), pp. 469–503.

__________. [1883]. *Jetons d'Alexandre, de Louis et de François, princes de Vendôme*. S.l., S.n. [Extract of the *Bulletin de la Société archéologique, scientifique et littéraire du Vendômois*.]

__________. [1882]. *Jetons de César duc de Vendôme*. S.l., S.n.

__________. 1881. *Notice sur les jetons des galères frappées au nom et aux armes de Louis Joseph duc de Vendôme*. Vendôme: Lermercier & fils. [Extract of the *Bulletin de la Société archéologique, scientifique et littéraire du Vendômois* 20 (1881), pp. 329–47.

Chauvin, J. M. 1857. *Louis XIII à Rouen. Jeton historique en argent de Louis XIII frappé aux armes de Rouen & trouvé dans la Seine*. Rouen: Brière, [1857]. [Extract of the *Journal de Rouen*, 3 August 1857.]

Christ, Karl. 1958. "Adrien Blanchet," *Gnomon* 30 (1958), pp. 478–79.

Clain-Stefanelli, Elvira E. 1965. "Numismatics—An ancient science: a survey of its history," *Contributions from the Museum of History and Technology*, Paper 32. Washington, DC: Smithsonian Institution.

Cohen, Thomas V. 2014. *Love and Death in Renaissance Italy*. Chicago and London: The University of Chicago Press.

Coornaert, Émile. 1941. *Les corporations en France avant 1789*. Paris: Gallimard.

Corlieu, Auguste. 1887. *Les jetons des doyens de l'ancienne faculté de médecine de Paris d'après la médaillier de la faculté*. Paris: J.-B. Baillière et fils.

Corre, Pierre. 1986. *Corpus des jetons armoriés de personnages français, Deuxième édition revue et complétée*. Paris: Le Léopard d'Or.

d'Affry de la Monnoye, Joseph Lambert Alfred. 1878. *Les jetons de l'échevinage parisien: documents pour servir à l'histoire métallique du Bureau de la ville et de diverses institutions parisiennes recueillis par feu A. d'Affry de la Monnoye*. Paris: Imprimerie nationale.

Daniel, Joseph. 1980. *Les jetons des Etats de Bretagne*. Nantes: B. Gallier.

Darnis, Jean-Marie. 1996–2005. *Catalogue des fonds d'archives de la Monnaie de Paris: 1418–1996*, 2 vols. (unitary pagination). Paris: Monnaie de Paris, 1996 and 1999; vol. 3: *Catalogue de la Bibliothèque historique de la Monnaie de Paris*, 2005.

__________. 1988. *La Monnaie de Paris. Sa création et son histoire du Consulat et de l'Empire à la Restauration 1795–1826)*. Levallois, FR: Centre d'Études Napoléoniennes.

__________. 1982. "Le graveur mécanicien Jean-Pierre Droz Ier adaptateur des viroles pleines et brisées aux mécanismes du balancier monétaire," *Proceedings of the 9th International Congress of Numismatics* (1982). Luxembourg: Louvain-la-Neuve.

Darnton, Robert. 1984. *The Great Cat Massacre and Other Episodes in French Cultural History*. New York: Basic Books.

__________. 1968. *Mesmerism and the End of the Enlightenment in France.* Cambridge, MA: Harvard University Press.

Dauban, Charles-Aimé. 1859. *Histoire des Maires de la ville de Tours par les jetons.* Paris: Paul Dupont.

__________. 1858. *Les jetons des maires de Nantes.* Nantes: Guéraud.

Davis, Natalie Zemon. 1983. *The Return of Martin Guerre.* Cambridge, MA: Harvard University Press.

de la Marche, Olivier. 1825. *Les Mémoires de Messire Olivier de la Marche,* 2 vols. Paris: Foucault. [Original edition 1562.]

de Turckheim-Pey, Sylvie. 2004. *Médailles du Grand Siècle: Histoire métallique de Louis XIV.* Paris: Bibliothèque nationale de France.

__________. 1997. *Le jeton au Moyen Âge, vers 1250–1498.* Paris: Bibliothèque nationale de France.

de Witte, Alphonse Félix François Charles. 1899. *Le jeton d'étrennes pour l'année 1773 aux Pays-Bas autrichiens.* Chalon-sur-Saône: Bertrand. [Offprint from the *Gazette numismatique française,* 1899.]

__________. 1893. *État actuel de la science numismatique.* Bruxelles: J. Goemaere.

Delcamp, Marion. 2015a. "Les tailleresses de la Monnaie de Paris au travail (1640–1789)," *Revue numismatique* (2015), pp. 491–515.

__________. 2015b. "Les jetons de cuivre armories de la Compagnie des monnayeurs de Paris," *Bulletin de la Société française de Numismatique* 70 (2015), pp. 219–26.

Delcamp, Marion and Jérôme Jambu. 2018. "Café sucré chez les monnayeurs de Bordeaux à la fin du XVIIIe siècle," *Bulletin de la Société française de Numismatique* 73 (2018), pp. 66–72.

Delgrange, Dominique. 1987. *Jetons de Flandre, Artois, Hainaut, Cambrai: Les pays formant le Nord-Pas-de-Calais, France, de la fin du règne de Louis XIV & époque de Louis XV & Louis XVI, (1712 à 1789).* Lille: D. Delgrange.

Depeyrot, Georges. 2007. *Les jetons des États du Languedoc (17e et 18e siècles).* Wetteren, BE: Moneta.

__________. 1982. "Numismatique du canal du Midi," in Jean-Denis Bergasse, ed., *Pierre-Paul Riquet et le canal du Midi dans les arts et la littérature* (Cessenon, FR: J.-D. Bergasse), pp. 79–84.

__________. 1980. *Les jetons des États du Languedoc (XVIIe et XVIIIe siècles).* Cahors: Privately printed.

Desavis, Madeleine. 1933. *Description des jetons décanaux de la collection du professeur Gilbert avec commentaires historiques et biographiques des doyens (1638–1793).* Paris: Jouve et Cie. [Thèse pour le doctorat en médecine.]

Devlin, Keith. 2011. *The Man of Numbers: Fibonacci's Arithmetic Revolution.* New York: Walker & Co.

Dillemann, Georges. 1977. *Jetons et médailles pharmaceutiques (Corporations, Sociétés, Écoles).* Paris: Éditions de la Porte Verte.

__________. 1976. "Le jeton de présence de la Société de pharmacie de Paris," *Annales pharmaceutiques françaises* 34 (1976), pp. 17–18.

Dillemann, Georges and Paul Roux-Fouillet. 1976. "Les Jetons de la Société pharmaceutique de Paris," *Revue d'histoire de la pharmacie* 33 (1976), pp. 186–90.

Drewing, Clyde J. 1970. "Jetons relating to the Dutch Revolt," *TAMS Journal* 10 (1970), pp. 82–94.

Droulers, Frédéric. 1991. *A Short History of the European* Écu *and of Previous Monetary Unions from the Middle Ages to the Present Day.* Lagny-sur-Marne, FR: Éd du Donjon.

__________. 1989. *Encyclopédie pratique de numismatique et de finances royales, 1610–1792.* Rouen: Editions du Cabinet Numismatique Albuquerque.

Druet, J. 1892. *Le jeton de l'ancienne manufacture d'Elbeuf.* Elbeuf: Saint-Denis.

Du Cange, Charles Du Fresne. 1840–1850. *Glossarium ad scriptores mediæ et infimæ latinitatis*, 7 vols. Paris: Firmin Didot, 1840–1850. Originally published in 1688.

Dubois de Saint-Gelais, Louis-François. 1893. *Pierre le Grand à la manufacture des Gobelins, 12 mai – 15 juin 1717.* Paris: Manufacture des Gobelins, 1893. [Extract of *Histoire journalière de Paris en 1716 et 1717.*]

__________. 1885. *Histoire journalière de Paris (1716–1717).* Paris: Société des bibliophiles françois.

__________. 1717. *Histoire journalière de Paris (1716–1717)*, 2 vols. Paris: Estienne Ganeau.

Duclos, Charles. *Mémoires secrets sur les règnes de Louis XIV et de Louis XV*, 2nd ed. 1791. Paris: Buisson.

Dugniolle, Docteur Jean-François. 1880. *Le jeton historique des dix-sept provinces des Pay-Bays: XVIIe et XVIIIe siècles*, 4 vols. Bruxelles.

Elliott-Smith. D. 1952. "Counters or Jetons," *The Numismatist* 65 (1952), pp. 1174–75.

Erlanger, Herbert J. [HJE] (1959). Annotation in *Numismatic Literature* 46 (1959), p. 378.

Estienne, Henri. 1645. *L'Art de faire des devises.* Paris: Jean Paslé.

Even, Ch. 1877. *Le jetonophile, précédé de notions élémentaires d'héraldique.* Saint Brieux.

Évrard de Fayolle, Alexis de. 1902. "Médailles et jetons municipaux de Bordeaux," *Gazette numismatique française* 6 (1902), pp. 69–89, 125–88, 295–306, 393–410.

__________. 1901. *Monographie des jetons médicaux bordelais.* Chalon-sur-Saône: Émile Bertrand. [Extract of the *Gazette numismatique française* 5 (1901), pp. 369–94.]

__________. 1900. *Histoire numismatique de la Chambre de Commerce de Bordeaux (1705–1898).* Bordeaux: Gounouilhou.

[Faludi, Géza.] 1929. *Medicine in numis: Sammlung Dr. Med Géza Faludi.* Budapest: Globus Druckerei.

Fenaille, Maurice. 1923. *Notice sur la vie et les travaux de M. Jules Guiffrey.* Paris: Firmin-Didot.

Feuardent, Félix-Bienaimé. 1904–1915. *Jetons et méreaux depuis Louis IX jusqu'à la fin du Consulat de Bonaparte*, 3 vols. Paris and London: Rollin & Feuardent. [Vol. 1 (1904); vol. 2 (1907); vol. 3 (1915).]

Feuardent, Félix-Bienaimé. 1995. *Jetons et méreaux depuis Louis IX jusqu'à la fin du Consulat de Bonaparte*, 4 vols., Gérard Barré and Daniel Renaud, eds. Paris: Maison Platt.

Fidière, Octave. 1885. *Les femmes artistes à l'Académie royale de peinture et de sculpture.* Paris: Charavet frères.

Firmin, Gwenola, Francine-Dominique Liechtenhan, and Thierry Sarmant. 2017. *Peter the Great: A Tzar in France* [Exhibition Catalogue]. Versailles, FR: Château de Versailles.

Florange, Charles. 1933. *Les jetons des doyens de l'ancienne faculté de médecine de Paris.* Paris: Jules Florange.

Florange, Charles and Stéphane Strowski. 1927. *Les Assemblées du Clergé de France avant 1789 et leurs jetons commémoratifs.* Lille: Douriez-Bataille; Paris: Jules Florange.

Florange, Jules. 1936. *L'Amour et le mariage dans la numismatique.* Moulins: Crépin-Leblond.

__________. 1904. *Essai sur les jetons et médailles de mines françaises.* Paris: Chez l'auteur. [Extract of the *Annales des Mines*, February 1904.]

__________. 1902, 1907. *Armorial du jetonophile: Guide de l'Amateur des Jetons Armoriés*, 2 vols. Paris: J. Florange; 2nd ed., 1921.

__________. 1899. *Souvenirs numismatiques du tir français avant 1789.* Paris: J. Florange.

Fontenay, Joseph-Étienne de. 1854. *Manuel de l'amateur de jetons.* Paris: Dumoulin.

__________. 1850. *Nouvelle étude des jetons.* Autun: Michel Dejussieu.

__________. 1847. *Fragments d'histoire métallique*. Autun: F. Dejussieu.

Forgeais, Arthur. 1874. *Numismatique des corporations parisiennes*. Paris: Auguste Aubry.

Fournié, Docteur Henri. 1907. *Les jetons des doyens de l'ancienne faculté de médecine de Paris*. Chalon-sur-Saône: E. Bertrand.

Freeman, Sarah Elizabeth. 1964. *Medals Relating to Medicine and Allied Sciences in the Numismatic Collection of The Johns Hopkins University, A Catalogue*. Baltimore: The Evergreen House Foundation.

__________. 1946. "The Jetons of the Deans of the Old Faculty of Medicine in Paris," *Bulletin of the History of Medicine* 19 (1946), pp. 48–95.

Fisch, A. [1905?] *Traité pratique des impressions photo-mécaniques*. Paris: Charles Mendel, s.d.

Froeschlé-Chopard, Marie-Hélène. 1990. "Les confréries du Saint-Sacrement aux XVIIe et XVIIIe siècles," in Maria Terese Maiullari, ed. *Storiografia francese ed italiana a confronto sul fenomeno associativo durante XVIII e XIX secolo* (Turin: Fondazione Luigi Einaudi), pp. 41–66.

Frossard, Ed. 1899. *Franco-American Jetons, Fully Described and Illustrated*. New York: Privately printed.

Gadoury, Victor. 1986. *Jetons – 1986 (Liste à Prix Fixes, No: 57)*. Monaco: Privately printed.

Gailhouste, Raymond. 2002. *Numismatique de l'Assurance française: 1670–2002*. Castres: Raymond Gailhouste.

Gallinato, Bernard. 1992. *Les corporations à Bordeaux à la fin de l'Ancien Régime: Vie et mort d'un mode d'organisation du travail*. Bordeaux: Presses Universitaires de Bordeaux.

Galster, Georg. 1936. *Danske og norske Medailler og Jetons ca. 1533–ca. 1788*. Copenhaven: Selskabet til Udgivelse af danske Mindesmaerker.

Ganière, Paul. 1964. *L'Académie de Médecine: Ses origines et son histoire*. Paris: Librairie Maloine.

Gay, Peter. 1966–1969. *The Enlightenment: An Interpretation. Vol. 1: The Rise of Modern Paganism* (1966); *vol. 2: The Science of Freedom* (1969). New York: Knopf.

Geertz, Clifford. 1973. *The Interpretation of Cultures: Selected Essays*. New York: Basic Books.

Gelfand, Toby. 1980. *Professionalizing Modern Medicine: Paris Surgeons and Medical Science and Institutions in the 18th Century*. [Contributions in Medical History, Number 6.] Westport, CT: Greenwood Press.

Gilleman, Charles. 1934. "La genèse des jetons d'inauguration sous la domination espagnole," *Revue belge de numismatique* 86 (1934), pp. 77–81.

Gillet, Lucien. 1906. *Nomenclature des médailles concernant l'histoire de Paris ayant figuré aux divers salons depuis 1699*. Chalons-sur-Saône: Émile Bertrand.

Gillet, R. 1897. *Jetons de présence des compagnies de notaires de France*. Joinville: Rosenstiel.

Gillispie, Charles C. 1980. *Science and Polity in France at the End of the Old Regime*. Princeton: Princeton University Press.

Gillispie, Charles C., ed. 1959. *A Diderot Pictorial Encyclopedia of Trades and Industry*, 2 vols. New York: Dover Publications, Inc., 1959.

Goldfiem, Jean de. 1976. "Le cheval et l'âne à la Monnaie de Paris," Le Club Français de la Médaille, *Bulletin* 50 (1976), pp. 69–96.

Gombert, Jean-Michel. 1974. "L'Hercule à travers les jetons, les monnerons et médailles," *Archéonumis* 12 (1974), pp. 44–46.

Grierson, Philip. 1991. *The Coins of Medieval Europe*. London: Seaby.

__________. 1951. *Numismatics and History*. London: Published for the Historical Association by G. Philip. [Historical Association, Pamphlets, General Series, G19.]

Grierson, Philip and Mark A. S. Blackburn. 2006. *Medieval European Coinage. 1, The Early Middle Ages*. Cambridge and New York: Cambridge University Press.

Gros de Boze, Claude. 1747. *Histoire de l'Académie Royale des Inscriptions et Belles Lettres depuis son établissement jusqu'à présent.* 3 vols. Paris, 1747.

Guichon de Grandpont, Alfred. 1862. *Notice sur des jetons français à emblèmes maritimes.* Brest: Lefournier.

__________. 1854. *Notice sur les jetons de la marine et des galères.* Paris: Paul Dupont. [Extract of the *Nouvelles Annales de la Marine et des Colonies,* Numéro d'avril 1854.]

Guiffrey, Jules [Jules-Joseph]. 1885. "Mémoire concernant les médailles," *Revue numismatique* (1885), pp. 208–09.

__________. 1884–1888. "La Monnaie des Médailles: Histoire métallique de Louis XIV et de Louis XV," *Revue numismatique* (1884), pp. 465–89; (1885), pp. 82–115, 187–209, 432–60; (1886), pp. 86–100; (1887), pp. 281–320; (1888), pp. 306–34.

__________. 1881–1901. *Comptes des bâtiments du Roi sous le règne de Louis XIV,* 5 vols. Paris: Imprimerie nationale.

__________. 1877–1880. *Les comptes des bâtiments du Roi (1528–1571),* 2 vols. Paris: J. Baur.

Guth, Ron. 2007a. "The 1796 Castorland jetons," *MCA Advisory* 10, #8 (2007), pp. 15–21.

__________. 2007b. "The 1796 Castorland jetons. Part II," *MCA Advisory* 10, #9 (2007), pp. 7–12.

Hahn, Roger. 1971. *The Anatomy of a Scientific Institution: The Paris Academy of Sciences, 1666–1803.* Berkeley, Los Angeles, London: University of California Press.

Haskins, Charles Homer. 1924. "The Abacus and the Exchequer," *Studies in the History of Medieval Science* (Cambridge: Harvard University Press), chapter XV pp. 327–35.

Harmand, Alexandre. 1834. *Recherches sur les monnaies, médailles et jetons dont la ville de Saint-Omer a été l'objet.* S.l.

Hoc, Marcel. 1950. "Les *jetons d'étrennes* de Marie-Élisabeth gouvernante générale des Pays-Bas (1725–1741)," *Revue belge de numismatique* 96 (1950), pp. 57–88.

__________. 1947. "Les *jetons d'étrennes* de Charles de Lorraine," *Revue belge de numismatique* 93 (1947), pp. 77–108.

Jacquiot, Josèphe. 1989a. "Nicolas de Launay et les balanciers de la Monnoye de Paris," *The Medal* 15 (1989), pp. 7–10.

__________. 1989b. "Virgile dans les devises de médailles et de jetons de la Renaissance au XVIIIe siècle," *Revue belge de numismatique* 135 (1989), pp. 97–109.

__________. 1980. "Le latin des légendes de médailles et jetons au XVIIe siècle," *Acta conventus neo-latini Turouensis* (Paris: Vrin), vol. 2, pp. 857–69.

__________. 1976. "L'allégorie aux revers de médailles et de jetons du XVe au XVIIe siècle," *Cahiers de l'Association internationale des études françaises,* 28 (1976), pp. 51–63.

__________. 1971. "Hercule au revers de médailles et de jetons de la Renaissance au début du XIXe siècle," *Revue belge de numismatique* 117 (1971), pp. 243–53.

__________. 1968–1970. *Médailles et jetons de Louis XIV après le manuscript de Londres Add. 31.908,* 4 vols. Paris: Imprimerie Nationale, C. Klincksieck.

__________. 1962. "Devises pour les médailles et jetons de Louis XIV composées par Boileau, historiographe du roi, membre de la petite Académie," *Comptes-rendus des séances de l'Académie des Inscriptions et Belles-Lettres,* April–December, 1962.

__________. 1961. "Devises pour les médailles et jetons de Louis XIV composées par Jean Racine, historiographe du roi, membre de la petite Académie," *Actes du 1er Congrès international racinien,* Uzès.

Jacquiot, Josèphe et al. 1970. [Monnaie de Paris.] *La médaille au temps de Louis XIV.* Paris: Hôtel de la Monnaie. [Catalog of an exhibition.]

Jambu, Jérôme. 2015. "La Société française de Numismatique: Des collectionneurs antiquaires aux chercheurs érudits," Académie des Inscriptions & Belles-Lettres, *Comptes Rendus* 2015 (II), pp. 849–66.

__________. 2014. "Le jeton des étrennes de Sa Majesté pour l'année 1671," *Bulletin de la Société française de Numismatique* 69 (2014), pp. 16–19.

__________. 2013a. "Le jeton des officiers la Monnaie de Lille," *Bulletin de la Société française de Numismatique* 68 (2013), pp. 182–89.

__________. 2013b. *Tant d'or que d'argent: la monnaie en Basse-Normandie à l'époque moderne, XVIe–XVIIIe siècle*. Rennes: Presses Universitaires de Rennes.

__________. 2012. "La circulation de la fausse monnaie métallique dans le royaume de France, du milieu du XVIIe à la fin du XVIIIe siècle (diffusion, perception, utilisation, élimination," in Olivier Caporossi and Bernard Traimond, eds., *La fabrique du faux monétaire du Moyen Age à nos jours* (Toulouse: Méridiennes), pp. 83–102.

__________. 2010. "Une histoire de la Monnaie de Caen à l'époque moderne (1693–1772), *Revue numismatique* (2010), pp. 509–36.

Jambu, Jérôme and Matthieu de Oliveira. 2014. *La monnaie en Flandre et dans le Nord. XVIe–XIXe siècle*. Lille: Université Lille 3.

Jaumonet, Victor. 1985–1986. "Les jetons - Bibliographie," *Vie numismatique* 35 (1985), pp. 270–74; 36 (1986), pp. 9–13.

_______________. 1985. "Du jeton de compte au jeton de présence," *Vie numismatique* 35 (1985), pp. 18–20.

Joly, Raymond. 1970. "Les techniques de la médaille au temps de Louis XIV," in Jacquiot et al., 1970, pp. 371–99.

Jones, Mark. 1983. "The Medal as an instrument of propaganda in late 17th and early 18th century Europe," *Numismatic Chronicle* 143 (1983), pp. 202–13 and plates 36–41.

__________. 1979a. *The Art of the Medal*. London: British Museum Publications.

__________. 1979b. *Medals of the Sun King*. London: British Museum Publications.

Josset, P. 1978. "Problèmes posés par l'émission des jetons des administrations royales en 1737: présentation d'un jeton inédit des Bâtiments," *Bulletin de la Société française de Numismatique* 33 (1978), pp. 337–40.

Jouvet, Ludovic. 2016. Review of Wellington, *Antiquarianism and the Visual Histories of Louis XIV* in *Revue de l'Art* 193 (2016), p. 74.

Julien, Pierre. 1973. "La Confrérie des Saints Côme et Damien Luzarches," *Revue d'histoire de la pharmacie* 61 (1973), pp. 505–18.

Justin, Émile. 1935. *Les Sociétés royales d'agriculture au XVIIIe siècle*. Saint-Lô.

Kluyskens, Hippolyte. 1884. *Les Jetons et médailles de la Faculté de médecine de Paris*. Brussels: S.n.. [Extract from the *Revue belge de numismatique*, 1884.]

La Tour, Henri de. 1899–1910. *Catalogue de la Collection Rouyer*, 2 vols. Vol 1: *Jetons & méreaux du Moyen Age*. Paris: Ernest Leroux, 1899; vol 2: *Jetons & méreaux de la Renaissance et des temps modernes*. Paris: Ernest Leroux, 1910.

__________. 1897. *Catalogue des jetons de la Bibliothèque Nationale: Rois et reines de France*. Paris: C. Rollin & Feuardent.

Labrot, Jacques. 1989. *Une histoire économique et populaire du moyen âge: Les jetons et les méreaux*. Paris: Éditions Errance.

Lacronique, Docteur R. 1902a. "Appendice à l'étude historique sur les jetons de l'Académie royale de chirurgie," *Gazette numismatique française* 6 (1902), pp. 411–17.

__________. 1902b. *Étude historique sur les médailles et jetons de l'Académie Royale de Chirurgie (1731–1793)*. Chalon-sur-Saône: E. Bertrand.

Lafaurie, Jean. 1976. "Jean Tricou, 1890–1977," *Revue numismatique* 18 (1976), pp. 7–10.

Lapassade, Robert. 1970. "L'administration de la Monnaie des Médailles," in Jacquiot (1970), pp. 351–69.

Le Fur, Didier. 2010. *Le royaume de France en 1500*. Paris: rmn.

Le Moyne, Pierre. 1666. *De l'Art des devises*. Paris: S. Cramoisy.

Lerouge, Marie-Élisabeth and Joseph Lerouge. 1979. *Jetons et méreaux*. Saint Omer: Société académique des antiquaires de la Morinie.

__________. 1977. *Jetons et documents numismatiques du notariat français*. Paris: Institut International d'Histoire du Notariat.

Lespinasse, René de. 1897. *Jetons et armoiries des métiers de Paris*. Nevers: G. Vallière, 1897.

__________. 1886–1897. *Les métiers et corporations de la ville de Paris*, 3 vols. Paris: Imprimerie nationale.

Lespinasse, René de and François Bonnardot, eds. 1879. *Les métiers et corporations de la Ville de Paris: XVIIIe siècle, Le livre des métiers d'Étienne Boileau*. Paris: Imprimerie Nationale.

Lesueur, Boris. 2014. *Les Troupes coloniales d'Ancien Régime*. Paris: SPM.

Levinson, Robert A. 1998. "Rechenmeister Jetons of Nuremberg," *Num.* 111 (1998), pp. 47–53.

Lhôte, Jean-Marie. 1994. *Dictionnaire des jeux de société*, Paris: Flammarion.

Lippens, Jan. 1969. "Les jetons du Bureau des finances," Cercle d'études numismatique, *Bulletin* 6 (1969), pp. 33–36.

Loach, Judi. 1995. "Reverses of Consular Jetons in Seventeenth-Century Lyons: Locally Specific Propaganda in Absolutist France," *The Medal* 27 (1995), pp. 24–56.

Loir, Joseph Napoléon. 1859. *Recherches sur des monnaies, méreaux, sceaux, jetons historiques de la ville de Mantes*. Paris: Rollin.

Longman, W. 1916. *Tokens of the Eighteenth Century Connected with Booksellers & Bookmakers*. London: Longmans, Green and Co.

Longpérier, Adrien de. 1863. "Jetons composés par Sully," *Revue numismatique*, (1863) pp. 425–50.

Lousse, Émile. 1943. *La société d'ancien régime: Organisation et représentation corporatives*. Louvain: Éditions Universitas.

Lowenthal, David. 2015. *The Past is a Foreign Country – Revisited*. New York: Cambridge University Press. Original edition 1985.

Mackay, James. 1994. "Jetons and Counters," *Coin News* 31 (April, 1994), pp. 52–53.

Mahudel, Nicolas. 1729. "De l'origine & de l'usage des Jettons," *Histoire et mémoires de l'Académie royale d'inscriptions et belles-lettres*, 5 (1729), pp. 259–65. [N.B. This is actually a report of Mahudel's paper of 1724 in the *Histoire* section.]

Mainguy, Anne. 1999. *Du corps du roi aux corps de l'état: Les jetons des grandes administrations (1695–1758)*. Thèse du doctorat sous la direction de Jacques Revel, École des Hautes Études en Sciences Sociales, Paris.

Marion, Marcel. 1923. *Dictionnaire des institutions de la France aux XVIIe et XVIIIe siècles*. Paris: A. & J. Picard.

Martin Saint-Léon, Étienne. 1941. *Corporations de métiers depuis leurs origines jusqu'à leur suppression en 1791, avec un appendice de bibliographie critique par Émile Coornaert*, 4th edition. Paris: Presses Universitaires de France.

Mattingly, Harold, et al., eds. 1923–1981. *The Roman Imperial Coinage*, 9 vols. London: Spink.

Mazard, Jean. 1992. "The psychoanalysis of the numismatist," Robert Turfboer, trans., *The Celator* 6, #7 (1992), pp. 38–39.

Mazerolle, Fernand. 1932. *Jean Varin: Conducteur de la monnaie du moulin, Tailleur général des monnaies, Controleur général des poinçons et effigies. Sa Vie - Sa Famille - Son Oeuvre (1596–1672)*, 2 vols. Paris: Ét. Bourgey, Jean Schemit.

__________. 1907a. *L'Hôtel des Monnaies: Les batiments, le musée, les ateliers*. Paris: Renouard.

__________. 1907b. *Les jetons des avocats au conseil*. Paris: H. Bouilant.

__________. 1903–1907. "Inventaire des poinçons et des coins de la Monnaie des Médailles (1697–1698)," *Gazette numismatique*, 1903, pp. 259–90, 367–404; 1904, pp. 45–54, 119–206, 225–408; 1905, pp. 47–92, 187–302, 383–416; 1906, pp. 193–204; 1907, pp. 29–80. [BNF, DMMA, 30001 MAZ, regroups these separate publications.]

__________. 1903. "Les débuts de la monnaie du Moulin," *Gazette numismatique française* 7 (1903), pp. 113–35.

__________. 1902–1904. *Les médailleurs français du XVe siècle au milieu du XVIIe*, 3 vols. Paris: Imprimerie Nationale.

__________. 1897. *Visites de Pierre le Grand et de Nicolas II à la Monnaie des médailles*. Paris: Aux Bureaux de la *Gazette des Beaux-Arts*.

__________. 1888. "Jetons de la Maison du Roi," *Annuaire de la Société française de numismatique et d'archéologie* 12 (1888), pp. 350–73 and 450–88.

McClellan, James E. III. Forthcoming. "Science and the City." Proceedings of the International Symposium on Civilization, Cities, and Architecture, Istanbul, Turkey, 12–13 April 2016.

__________. 2005. "Accident, Luck and Serendipity in Historical Research," *Proceedings of the American Philosophical Society* 149 (2005), pp. 1–21.

__________. 2003. "Specialist Control: The Publications Committee of the Académie Royale des Sciences (Paris), 1700–1793," *Transactions of the American Philosophical Society*, vol. 93, pt. 3.

__________. 2000. "L'Historiographie d'une académie coloniale: Le Cercle des Philadelphes (1784–1793)," *Annales historiques de la Révolution française 2000* (#2), pp. 77–88.

__________. 1999. "Les réseaux savants: Les colonies des Lumières," *Les Cahiers de Science & Vie* (April, 1999), pp. 22–28.

__________. 1993. "Europe des Académies: Forces centripètes, forces centrifuges," *Dix-Huitième Siècle* 23 (1993), pp. 153–65.

__________. 1992/2010. *Colonialism and Science: Saint Domingue in the Old Regime*. Baltimore and London: The Johns Hopkins University Press, 1992; reprint ed. with a new forward by Vertus Saint-Louis, Chicago: University of Chicago Press, 2010.

__________. 1985. *Science Reorganized: Scientific Societies in the Eighteenth Century*. New York: Columbia University Press.

McClellan, James E. III and François Regourd. *The Colonial Machine: French Science and Overseas Expansion in the Old Regime*. Turnhout, BE: Brepols, 2011.

Meadows, Andrew and Kirsty Shipton. 2001. *Money and Its Uses in the Ancient Greek World*. Oxford: Oxford University Press.

Ménestrier, Claude-François. 1686a. *Histoire du roy Louis le Grand par les medailles, emblêmes, devises, jettons, inscriptions, armoiries, et autres monumens publics*. Paris: I. B. Nolin.

__________. 1686b. *La Science et l'art des devises, dressez sur de nouvelles règles, avec six cens devises sur les principaux évenemens de la vie du Roy et quatre cens devises sacrées dont tous les mots sont tirés de l'éciture sainte*. Paris: Robert J. B. de la Caille.

__________. *1682–1683. La philosophie des images, composée d'un ample recueil de devises, et du jugement de tous les ouvrages qui ont été faits sur cette matière*, 2 vols. Paris: R.-J.-B. de la Caille. Vol. 2 separately titled: *Devises des princes, cavaliers, dames, sçavans et autres personages illustres de l'Europe ou la philosophie des images*.

Ménétra, Jacques-Louis. 1982. *Journal de ma vie: Jacques-Louis Ménétra, compagnon vitrier au 18e siècle*, presented by Daniel Roche. Paris: Montalba.

Menin, Nicolas. 1724. *Traité historique et chronologique du sacre et couronnement des rois...avec la relation du sacre de Louis XV*. Amsterdam: Jean van Septeren.

Michel, Etienne-Désiré. 1934. *Les jetons de présence*. [Université d'Aix-Marseille. Faculté de droit d'Aix, Thèse pour le doctorat.] Marseille: GED.

Minard, Philippe. 1998. *Le fortune du colbertisme: État et industrie dans la France des Lumières*. Paris: Fayard.

Mitchiner, Michael. 1988–1998. *Jetons, Medalets & Tokens*, 3 vols. 1988–1998. [Volume one: *The Medieval Period and Nuremberg* (1988); volume two: *The Low Countries and France* (1991); volume three: *British Isles Circa 1558 to 1850* (1998).] London: Hawkins Publications.

Morin-Pons, Henry. 1900. *Numismatique de l'Académie des sciences, belles-lettres et arts de Lyon*. Lyon: Rey.

Mörkholm, Otto. "A history of the study of Greek Numismatics," *Scandinavian Numismatic Journal* 1979–1980, pp. 5–14.

Moulin, Christian. 1995. *Recueil des Arts & Métiers dépendant des Corporations Rouennaises sous l'Ancien Régime*. Rouen: Cercle Généalogique Rouen Seine-Maritime.

Mousnier, Roland E. 1990–1992. *Les institutions de la France sous la monarchie absolue, 1598–1789*, 2 vols., 2nd ed. Paris: Presses Universitaires de France. Original edition, 1974.

Mowat, Robert. 1908. "Emploi frauduleux des jetons," *Revue numismatique* (1908), pp. 403–05.

Mukerji, Chandra. 2009. *Impossible Engineering: Technology and Territoriality on the Canal du Midi*. Princeton: Princeton University Press._

Müller, Gerhard F. 1984. "Die Jetons und ihre Bedeutung im Wandel der Zeit," *Münzen Revue* 16 (1984), pp. 808–12.

Newsome, Daniel. 2011. *Quadrivial Pursuit: An Interdisciplinary History of the Mathematical Arts in the Late Middle Ages and Renaissance*. Ph.D. dissertation; City University of New York.

Newton, William Ritchey. 2014. *Vivre à Versailles: Derrière la façade, la vie quotidienne au château*. Paris: Seuil, Champs Histoire.

Nivaille, Jean. 1987. "Les jetons des secrétaires du roi et le type de l'essaim d'abeilles," *Revue belge de numismatique et de sillographie* 133 (1987), pp. 155–78 and plates XXIII–XXV.

Nocq, Henry. 1911. *Les Duvivier: Jean Duvivier (1687–1761), Benjamin Duvivier (1703–1819)*. Paris: Société de propagation des livres d'art.

__________. 1910. *Mélanges: Une histoire métallique de la Faculté de médecine de Paris publiée en 1790*. [Extract of the *Gazette numismatique française*, 14 (1910), pp. 166–69.]

__________. 1908. "Mélanges: Compte des Monnaies en 1720, Idée du Cabinet du Roi, Médailles du sacre de Louis XV," *Gazette numismatique française* 12 (1908), pp. 423–35.

Pages, Alfred. 1931. *Numismatique des colonies françaises*. Paris: Imp. Girault.

Paris, Daniel, Jr. 1887. "Dutch jetons of the sixteenth century," *American Numismatic Society Proceedings* (1887), pp. 41–47.

Parsons, Jotham. 2014. *Making Money in Sixteenth-Century France: Currency, Culture and the State*. Ithaca: Cornell University Press.

Pastoureau, Michel. 1984. *Jetons, Méreaux et Médailles*. Turnhout, BE: Brepols.

Patin, Charles. 1695. *Histoire des médailles, ou Introduction à la connoissance de cette science*. Paris: Vve Mabre Cramoisi.

Pédery-Hunt, Dora de. 1979. "Nos médailles et jetons: beaux livres d'histoire," *Médailles* (1979), pp. 41–43.

Pichon, Thomas-Jean. 1775. *Sacre et couronnement de Louis XVI à Rheims, le 11 juin 1775*. Paris: Chez Vente et chez Patas.

Piossens, Chevalier de. 1729. *Mémoire de la Régence de S. A. R. Mgr le duc d'Orléans durant la minorité de Louis XV*, 3 vols. La Haye: J. Van Duren.

Poncet, Ernest. 1883. *Recherches sur les jetons consulaires de la ville de Lyon*. Lyon: A. Brun.

Pradel, Pierre. 1936. *Catalogue des jetons de la bibliothèque nationale. Jetons des princes et princesses de la Maison de France*. Paris: Éditions de la Bibliothèque Nationale.

Préau, Charles. 1890a. "Étude sur la Cour des monnaies de France," *Revue belge de numismatique* 46 (1890), pp. 97–127.

__________. 1890b. *Jetons inédits de quelques administrations françaises du* XIIIe *au* XVIe *siècle*. Paris: Troublé.

__________. 1887. *Jeton inédit de la corporation des maçons-tailleurs de pierre, plastriers-mortelliers au* XIVe *siècle*. Paris: Blanpain.

Prieur, Michel and Stéphan Sombart. 2004. *Jetons XIX: Spécial Lyon et le Lyonnais, Quart Sud-Est de la France*. Paris: Comptoir Général Financier.

__________. 2002. *Jetons XVII: Spécial Bourgogne*. Paris: Comptoir Général Financier.

Proctor, Robert N. and Londa Schiebinger, eds. (2008). *Agnotology: The Making and Unmaking of Ignorance*. Stanford, CA: Stanford University Press.

Prou, Maurice. 1888. Review of F. Mazerolle "Jetons rares ou inédits," *Revue numismatique* 6 (1888), pp. 346–48.

Provinciale Dienst voor het Kunstpatrimonium. 1981. "Het rekenen met penningen," in *Kunst & Oudheden in Limburg 26: Munt in Limburg* [Sint-Truiden, Netherlands: Provinciaal Museum voor Religieuze Kunst], pp. 29–37.

Rachline, Michel. 1992. *La Monnaie de Paris*. Paris: Albin Michel/Monnaie de Paris.

Raimbault, Maurice. 1903. *Les Médailles et les Jetons des États de Provence*. Chalon-sur-Saône: Émile Bertrand.

Regourd, François. 1998. "La Société royale d'agriculture de Paris face à l'espace colonial (1761–1793)," *Bulletin du Centre d'Histoire des Espaces Atlantiques* 8 (1998), pp. 155–94.

Renauldin, Léopold Joseph. *Études historiques et critiques sur les médecins numismatistes*. Paris: J. B. Ballière, 1851.

Restif de la Bretonne, Nicolas-Edme. 1977. *Le Paysan perverti, ou les dangers de la ville*. Lausanne: Éditions de l'Age d'Homme. Original edition 1775.

Robert, Pierre-Charles. 1890. *Monnaies, jetons et médailles des évêques de Metz*. Mâcon: Protat Frères. [Extract of the *Annuaire de la Société française de Numismatique*.]

__________. 1886. *Monnaies et jetons des évêques de Verdun*. Mâcon: Protat Frères. [Extract of the *Annuaire de la Société française de Numismatique* 1886.]

__________. 1860. *Monnaies de Mâcon*. Paris: Thunot. [Extract of the *Revue numismatique* 5 (1860).]

__________. 1853. *Recherches sur les monnaies et les jetons des maîtres-échevins et description des jetons divers*. Metz: Nouvian.

Roche, Daniel. 1978. *Le Siècle des Lumières en province: Académies et académiciens provinciaux, 1680–1789*, 2 vols. Paris and La Haye: EHESS and Mouton.

Roelandt, Claude, Stéphan Sombart, and Michel Prieur, with the participation of Alain Schärlig. 2004. *Jetons & méreaux du moyen âge*. Paris: Editions Les Chevau-légers. [Includes reprint of Rouyer and Hucher of 1858.]

Roman, Joseph. 1911. *Les Jetons du Dauphiné*. Paris: E. Leroux.

__________. [1880.] *Méreaux et jetons ecclésiastiques du Dauphiné*. Grenoble: S.n. [Extract from the *Bulletin de l'Académie Delphinale* (1880), pp. 376–408.]

Rondot, Natalis. 1904. *Les médailleurs et les graveurs de monnaies, jetons et médailles de France*. Paris: A. Leroux.

__________. 1885. "Les médailleurs et les graveurs de monnaies français," *Revue numismatique* (1885), pp. 210–13.

Roovers, Olga N. (1957). "Gildepenningen," *De Geuzenpenning Munt- en Penningkundig Nieuws* 7, #3 (July, 1957), pp. 25–27.

Rossignol, Claude. 1851. *Des libertés de la Bourgogne d'après les jetons de ses États*. Autun: Dejussieu et L. Villedet.

Rostron, P. R. 1993. "French Jetons," *Canadian Numismatic Journal* 38 (1993), pp. 188–89.

________. 1977. "Jetons reflect royal passion for gambling," *Coin World International* 18 (1977), pp. 65–66.

Rothney, John, ed. 1969. *The Brittany Affair and the Crisis of the Ancien Regime*. New York: Oxford University Press.

Rouyer, Jules. 1879. *Les Jetons municipaux de Paris du XVe au XVIIIe siècle, Examen de l'ouvrage d'Alfred d'Affry de la Monnoye intitulé Les Jetons de l'Échevinage parisien*. Paris: Arnous de Rivière, 1879. [Extract from *Mélanges de Numismatique* 3 (1878), pp. 284–306.]

__________. 1893. "L'Œuvre du médailleur Nicolas Briot en ce qui concerne les jetons," *Revue belge de numismatique* 49 (1893), pp. 175–214, 488–516.

__________. 1892. *Rectification biographique relative au médecin publiciste Théophraste Renaudot. A propos de la description d'un jeton frappé pour un fils portant le même prénom*. Paris: Rollin et Feuardent.

Rouyer, Jules and Eugène Hucher. 1982. *Histoire du jeton au moyen-âge*, with preface by Michel Pastoureau. Paris: Le Léopard d'Or.

__________. 1978. *Histoire du jeton au moyen-âge*. Facsimile reprint: Paris: P. de La Perrière, 1978.

__________. 1858. *Histoire du jeton au moyen-âge*. Paris: Rollin.

Roy-Chevrier, Joseph. 1919. *Chalon métallique. Description, étude et reproduction de jetons, médailles, plombs et méreaux de Saône-et-Loire, antérieurs à 1789 et relatifs au territoire des anciens Bailliages et Diocèses de ce département*. [*Mémoires de la Société d'histoire et d'archéologie de Chalon-sur-Saône*, 2e série, tome VIII.] Chalon-sur-Saône: Émile Bertrand.

Rulau, Russell. 1964. "Jetons of the Great Dutch Revolt," *Calcoin News* 18, #3 (1964), pp. 79–82.

Safier, Neil. 2008. *Measuring the New World: Enlightenment Science and South America*. Chicago and London: The University of Chicago Press.

Sargent, Thomas J. and François R. Velde. 2002. *The Big Problem of Small Change*. Princeton: Princeton University Press.

Sarmant, Thierry. 2003. *La République des médailles: Numismates et collections numismatiques à Paris du Grand Siècle au Siècle des Lumières*. Paris: Honoré Champion.

__________. 1994. *Le Cabinet des Médailles de la Bibliothèque nationale, 1661–1848*. Paris: École des Chartes.

Sarmant, Thierry and François Ploton-Nicollet. 2010–2014. *Jetons des institutions centrales de l'Ancien Régime, Catalogue*, 2 vols. Paris: Bibliothèque nationale de France._

Sarmant, Thierry et al. 2007. *Le palais, le registre et la balance: Médailles et jetons autour de la Cour des comptes (exposition au Palais Cambon 25 avril – 31 mai 2007)*. Paris: La Cour des Comptes.

Sauval, Henri. 1724. *Histoire et recherches des antiquités de la ville de Paris*. Paris: chés Charles Moette, Jacques Chardon.

Schärlig, Alain. 2003. *Compter avec des jetons: Tables à calculer et tables de compte du Moyen Age à la Révolution*. Lausanne: Presses polytechniques et universitaires romandes.

Schindler, Leo. 1952. "Das Vorbild der ersten italienischen und französischen Jetons," *Mitteilungen der Österreichischen Numismatischen Gesellschaft* 7 (1952), pp. 145–49.

Schodt, Alphonse de. 1873. *Le Jeton considéré comme instrument de calcul*. Brussels: Gobbaerts.

Schulman, Jacob. 1912. *Pax in Nummis. Médailles, Jetons et Monnaies ayant rapport aux divers Traités de paix*. Amsterdam: J. Schulman. [Deluxe and annotated sale catalog of the "Collection Le Maistre" that took place in Amsterdam in October, 1913.]

Schultz, Emily A. and Robert H. Lavenda. 2013. *Cultural Anthropology: A Perspective on the Human Condition, Ninth Edition*. New York: Oxford University Press.

Serrure, Raymond. 1897. "Le vol au jeton," *Bulletin de numismatique* 4 (1897), pp. 55–56.

Shattuck, Gary. 2015. "7 Gold Medals of America's Revolutionary Congress," *Journal of the American Revolution*, online version, April 7, 2015.

Smith, David Eugene. 1921. *Computing Jetons*. Numismatic Notes and Monographs 9. New York: American Numismatic Society.

__________. 1920. "The Casting-Counter and the Counting Board," *Bulletin of the American Mathematical Society* 27 #3 (1920), pp. 129–31. [Review of Barnard (1916).]

Sombart, Stéphan. 1993. *Médailles et Jetons du sacre des Rois de France à Reims*. Reims: Privately printed, 1993.

Spooner, Frank C. 1954. *L'économie mondiale et les frappes monétaires en France 1493–1680*. Paris: A. Colin.

Stahl, Alan Michael. 2000. *Zecca: The Mint of Venice in the Middle Ages*. Baltimore and London: Johns Hopkins University Press and the American Numismatic Society.

Storer, Horatio R. 1887. "The Medals, Jetons, and Tokens Illustrative of Sanitation," *The Sanitarian* 215 (1887), pp. 316–39.

Swann, Julian. 2003. *Provincial Power and Absolute Monarchy: The Estates General of Burgundy, 1661–1790*. Cambridge: Cambridge University Press.

Swetz, Frank J. 1987. *Capitalism and Arithmetic: The New Math of the 15th Century*. LaSalle, IL: Open Court.

Tarrade, Jean. 1972. *Le Commerce colonial de la France à la fin de l'Ancien Régime: L'évolution du régime de "l'exclusif" de 1763 à 1789*, 2 vols. Paris: Presses Universitaires de France.

Thompson, David. 1990. "Makers of English jetons gave up in face of output from Continental artisans," *World Coin News* 17 (1990), pp. 18, 20.

Tilly, Charles. 1992. *Coercion, Capital, and European States, 990–1992*. Cambridge, MA: Blackwell Publishing.

Topenot, P. 1971. "Jetons de la C° O.T.L. à Lyon," *Bulletin de la Société française de Numismatique* 28 (1971), p. 140.

Tricou, Jean-Louis. [Pseudonym of Epipoy Gardenote.] 1973. "Les jetons de Pierre Grassin, Directeur général des Monnaies," in *Bulletin de la Société française de Numismatique* 28 (1973), p. 380.

__________. 1972. "Le commerce de jetons à Lyon au XVIIIe siècle," *Bulletin de la Société française de Numismatique* 27 (1972), p. 197.

__________. 1967. "À propos d'une étude récente. L'Académie des Inscriptions et les médailles et jetons lyonnais du XVIIIe siècle," *Bulletin de la Société française de Numismatique* 22 (1967), pp. 179–80.

__________. 1957. *Numismatique des corporations, des métiers et du commerce lyonnais de l'ancien régime*. Paris: É. Bourgey.

__________. 1955. *Les Jetons Consulaires de Lyon*. Paris: É. Bourgey, 1955.

__________. 1954. "Jetons offerts par la ville de Lyon à diverses compagnies, XVIIe–XVIIIe siècles," *Revue numismatique* 16 (1954), pp. 151–71.

Vallier, G. 1879. *Essai sur les jetons de la chambre des comptes du Dauphiné*. Brussels: Gobbaerts. [Extract of the *Revue belge de numismatique* 1880.]

van Beek, Bert. 1986. "Jetons: Their Use and History," in Saul B. Needleman, ed., *Perspectives in Numismatics: Studies Presented to the Chicago Coin Club* (Chicago: Ares Publishers), pp. 195–220.

Van Hende, Édouard. 1858. *Numismatique lilloise, ou Description des monnaies, médailles, jetons, méreaux, etc., de Lille*. Lille: A Quarré.

van Loon, Gerard. 1732–1737. *Histoire métallique des dix-sept provinces des Pays-Bas, depuis l'abdication de Charles-Quint jusqu'à la paix de Bade en 1716*, A.-F. Prévost and J. Van Effen, trans., 5 vols. The Hague: P. Gosse, J. Neaulme, and P. de Hondt.

__________. 1723–1731. *Beschrijving der Nederlandsche Historiepenningen of beknopt verhaal van 't gene sedert de overdracht der heerschappye van keyzer Karel den Vyfden op koning Philips zynen zoon, tot het sluyten van den Uytrechtschen vreede, in de zeventien Nederlandsche gewesten is voorgevallen*, 4 vols. 's-Gravenhage: C. Van Lom, I. Vaillant, P. Gosse, R. Alberts, and P. de Hondt.

van Mieris, Frans. 1733–1735. *Histori der Nederlandsche vorsten*, 3 vols. 's Graavenhaage: Pieter de Hondt.

van Wie, Paul D. 1999. *Image, History and Politics: The Coinage of Modern Europe*. Lanham, MD: University Press of America.

Valin, Sabrina. 2016. *Les jetons royaux frappés sous les règnes de Louis XIII et de Louis XIV (1610–1661): inventaire et analyse*. Ph.D. dissertation, Université de Paris Ouest Nanterre - La Défense.

Veillon, Marie. 2008. *Histoire de la numismatique ou la science des médailles*. Paris: Éditions Errance.

__________. 2005. "L'Iconographie metallique d'Hercule dans la France du seizième siècle," *The Medal* 46 (2005), pp. 31–39.

__________. 1994. "L'art de la devise sous le règne de Louis XIV: l'exemple des devises de jetons," *The Medal* 25 (Autumn 1994), pp. 20–49.

Vian, Pierre-Carlo. 1949. "Les jetons des Etats du Languedoc," in Bibliothèque Nationale, Cabinet des Médailles, *Exposition-Concours de Numismatique* (Paris: Imprimerie Nationale), pp. 36–39.

Villela-Petit, Inès. 2015a. "Documents du Minutier central des notaires de Paris pour servir à l'histoire des Roettiers," *Revue numismatique* 172 (2015), pp. 193–228.

__________. 2015b. "L'Histoire du règne par la médaille," in Gérard Sabatier and Béatrix Saule, eds., *Le Roi est mort - Louis XIV - 1715* (Versailles: Éditions Tallandier/Château de Versailles), pp. 150–151.

__________. 2012a. "Le Cabinet des médailles à l'époque des Babelon: des collections historiques à la promotion de la médaille contemporaine," in Edouard Papet and Catherine Chevillot, eds., *Au creux de la main: la médaille aux XIXe et XXe siècles* (Paris: Skira Flammarion/Musée d'Orsay), pp. 21–29.

__________. 2012b. "Médaille de Louis XII," in Béatrice de Chancel-Bardelot, et al., eds. *Tours 1500, capitale des arts* (Paris: Somogy éditions d'art, 2012), pp. 69–71.

Waller, Richard. 1988. "The Académie des inscriptions in 1705: A year in the life of a learned institution," *Enlightenment Essays in Memory of Robert Shackleton* (Oxford: The Voltaire Foundation), pp. 299–318.

Wallon, Henri. 1897. *Jetons et médailles de la Chambre de commerce de Rouen*. Rouen: Cagniard.

Weil, Alain. 1988. "Les jetons de 1785 pour la Compagnie des Indes," *Revue numismatique* 30 (1988), pp. 248–53.

Wellington, Robert. 2018. "Notes from the field: A Louis XV jeton for Martinique, not Canada." Blog post dated February 8, 2018: https://robertwellington.com/blog/; accessed September 15, 2018.

__________. 2016. "Médailles en mouvement: La réception des médailles de Louis XIV à la croisée des cultures," in Yvan Loskoutoff, ed., *Les médailles de Louis XIV et leur livre* (Mont-Saint-Aignan: Presses Universitaires de Rouen et du Havre), pp. 290–305.

__________. 2015a. *Antiquarianism and the Visual Histories of Louis XIV: Artefacts for a Future Past*. Farnham, UK; Burlington, VT: Ashgate.

__________. 2015b. "Louis XIV's medal cabinet at Versailles," *The Medal* 67 (2015), pp. 12–25.

Welter, Gerhard. 1976. *Cleaning and Preservation of Coins and Medals*. New York: Sanford J. Durst.

Zay, Ernest. 1892. *Histoire monétaire des colonies françaises, d'après les documents officiels*. Paris: J. Montarier.

__________. 1889. "Canada: Médailles d'honneur pour les Indiens," *Annuaire de la Société française de Numismatique* 13 (1889), pp. 294–301.

Zupko, Ronald Edward. 1978. *French Weights and Measures Before the Revolution: A Dictionary of Provincial and Local Units*. Bloomington, IN: Indiana University Press.

Zysberg, André. 1987. *Les galériens: Vies et destins de 60 000 forçats sur les galères de France, 1680–1748*. Paris: Éditions du Seuil.

Jeton: Notaries, 1720.

INDEX[1]

A

1. Page numbers in **bold** reference a figure or a table.

B

C

D

E

F

I

J

K

L

M

N

O

P

Q

R

S

T

U

V

W